CHEMICAL
GRAPH
THEORY

SECOND EDITION

CHEMICAL GRAPH THEORY

SECOND EDITION

Nenad Trinajstić, Ph.D.
Professor of Chemistry
The Rugjer Bošković Institute
Zagreb
The Republic of Croatia

CRC Press
Taylor & Francis Group
Boca Raton London New York

CRC Press is an imprint of the
Taylor & Francis Group, an **informa** business

CRC Press
Taylor & Francis Group
6000 Broken Sound Parkway NW, Suite 300
Boca Raton, FL 33487-2742

© 1992 by Taylor & Francis Group, LLC
CRC Press is an imprint of Taylor & Francis Group, an Informa business

First issued in paperback 2019

No claim to original U.S. Government works

ISBN-13: 978-0-367-45039-7 (pbk)
ISBN-13: 978-0-8493-4256-1 (hbk)

Visit the Taylor & Francis Web site at
http://www.taylorandfrancis.com

and the CRC Press Web site at
http://www.crcpress.com

Library of Congress Card Number 91-31571

Library of Congress Cataloging-in-Publication Data

Trinajstić, Nenad, 1936–
    Chemical graph theory/Nenad Trinajstić.—2nd rev. ed.
        p.   cm.
    Includes bibliographical references and index.
        ISBN 0-8493-4256-2
        1. Graph theory.   2. Chemistry—Mathematics.   I. Title.
    QD39.3.G73T75   1992
    540′.1′5115—dc20   91-31571
                                                                          CIP

Mathematical Chemistry Series

Editors

Douglas J. Klein
Texas A&M at Galveston

Milan Randić
Drake University

Associate Editors

Nenad Trinajstić
Rugjer Bošković Institute, Zagreb

Paul G. Mezey
University of Saskatchewan

Advisory Editors

Alexandru T. Balaban
Polytechnic Institute, Bucharest

Gregg S. Ezra
Cornell University

Haruo Hosoya
Ochanomizu University

Josef Paldus
University of Waterloo

The objective of this series is to provide a range of high-quality introductory, intermediate, and advanced research-oriented monographs on all aspects of mathematical chemistry. Topics of interest for this series include mathematical and computational modeling of molecular structures; mathematical theory of chemical transformations; novel approaches to structure-property relationships, fractal characterization of supramolecular microstructures; renormalization-group technology for chemistry; computational algorithms for analysis of chemical data; chemical applications of group theory; topological and differential geometric characterization of molecules; and chemical enumerations and combinatorial methods in chemistry. The series is naturally open to the variety of traditional topics of quantum chemical methods, statistical mechanical theory, and classical mechanical models in chemistry. Applications in closely allied fields such as crystallography, molecular biology, and medicinal chemistry are also welcome. Texts for self-study or for classroom use are of special interest.

This book is dedicated to the memory of those brave people who died defending freedom and democracy in the Republic of Croatia.

PREFACE TO THE SECOND EDITION

And God said, Let there be light:
and there was light.

Genesis 1:3

Since the appearance of the first edition of this book, the discipline of chemical graph theory has continued an explosive growth. A large number of papers reporting new developments and applications have appeared by authors from all over the world. There have been a number of international symposia on chemical graph theory and mathematical chemistry, e.g., 1983 — Athens (Georgia, U.S.); 1985 — Dubrovnik (The Republic of Croatia); 1987 — Athens (Georgia, U.S.); 1988 — Los Angeles (California, U.S.); 1989 — Galveston (Texas, U.S.); 1991 — Bled (The Republic of Slovenia). In this period a major event was the appearance in 1987 of the *Journal of Mathematical Chemistry* (published by J. C. Baltzer AG, Scientific Publishing Company, Basel, Switzerland), initially under the Editorship of Dennis H. Rouvray. This journal soon became the most important international medium for publishing reports in mathematical chemistry and likewise in chemical graph theory. Several symposia books have also been published on the subject[1-4] and special issues and volumes of various journals have appeared with new developments and applications of graph theory and topology in chemistry.[5-8] In this way chemical graph theory has become an important part of mathematical and theoretical chemistry.

This rapidly increasing interest in mathematical chemistry in general and chemical graph theory in particular, together with the modest popularity of the first edition of this book among chemical graph-theorists, mathematical chemists, computational chemists, and theoretical chemists, has motivated me to prepare a second edition. The second edition contains corrections, revised and up-dated material, and new developments in those areas of chemical graph theory that were discussed in the first edition. In addition, the part of the book dealing with applications is also enlarged to embrace exciting new developments in this field. In the revision some chapters have also been shortened and two chapters in particular (i.e., Isospectral Molecules and Subspectral Molecules) have been completely removed from the second edition because of their somewhat specialized content. Therefore, the second edition is rewritten in its entirety and most of the drawings have been prepared anew. The list of references in each chapter is considerably expanded, in order to accommodate most of the relevant advances since 1983 regarding the content of the chapter in question.

I have again to thank many friends and colleagues for valuable advice and assistance. I am grateful, especially for their suggestions and criticisms (particularly the latter!) to Mr. D. Babić (Zagreb), Professor A. T. Balaban

(Bucharest), Professor K. Balasubramanian (Tempe), Professor D. Bonchev (Burgas), Professor S. D. Bosanac (Zagreb), Dr. S. Carter (Reading), Professor S. J. Cyvin (Trondheim), Professor J. R. Dias (Kansas City), Professor A. Graovac (Zagreb), Dr. B. Jerman-Blažič (Ljubljana), Professor L. Klasinc (Zagreb), Dr. E. C. Kirby (Pitlochry), Professor D. J. Klein (Galveston), Professor J. V. Knop (Düsseldorf), Professor Z. B. Maksić (Zagreb), Professor P. G. Mezey (Saskatoon), Dr. D. Plavšić (Zagreb), the late Professor O. E. Polansky (Mülheim/Ruhr), Professor V. Prelog (Zürich), Professor M. Randić (Ames, Des Moines), Professor D. H. Rouvray (Athens, Georgia), Professor A. Sabljić (Zagreb), and Professor T. Živković (Zagreb). I also wish to thank Dr. D. Amić, Dr. M. Barysz, Dr. B. Bogdanov, Mr. D. Horvat, Mr. M. Ivanušević, Dr. A. Jurić, Mr. Z. Mihalić, Dr. S. Nikolić, and Ms. B. Panzova, my graduate students since 1983, for their zeal, high quality work, and many discussions on the role of graph theory and topology in chemistry.

Mr. Darko Babić (Zagreb), Professor Alexandru T. Balaban (Bucharest), Professor K. Balasubramanian (Tempe), Professor Sven J. Cyvin (Trondheim), Professor Ante Graovac (Zagreb), Dr. E. C. Kirby (Pitlochry), and Professor Douglas J. Klein (Galveston) read several chapters of the book and made a number of valuable comments. I am thankful to them for their kind help. The entire book was read in manuscript form by Dr. Sonja Nikolić (Zagreb) who has made many useful comments leading to the improvement of the book. This is a fitting occasion to thank her for very critical, but amiable discussions, comments, and help.

Finally, I should also like to thank my publishers, CRC Press, for transforming the manuscript into this beautiful book.

REFERENCES

1. **King, R. B., Ed.**, *Chemical Applications of Topology and Graph Theory*, Elsevier, Amsterdam, 1983.
2. **Trinajstić, N., Ed.**, *Mathematics and Computational Concepts in Chemistry*, Horwood, Chichester, 1986.
3. **King, R. B. and Rouvray, D. H., Eds.**, *Graph Theory and Topology in Chemistry*, Elsevier, Amsterdam, 1987.
4. **Rouvray, D. H., Ed.**, *Computational Chemical Graph Theory*, Nova Science Publishers, Commack, NY, 1990.
5. **Trinajstić, N. and Orville-Thomas, W. J., Eds.**, special issue (Vol. 185, 1989), *J. Mol. Struct. (Theochem.)* entitled *Topics in Mathematical Chemistry*.
6. **Klein, D. J. and Randić, M., Eds.**, Volume No. 4 (1990) of *J. Math. Chem.* entitled *Methods of Mathematical Chemistry*.
7. **Náray-Szabó, G. and Weinstein, H., Eds.** Volume No. 1 (Issue 2, 1990) of *Reports in Molecular Theory*.
8. **Bonchev, D. and Rovray, D. H., Eds.**, Volume No. 1 (1991) of *Math. Chem.* entitled *Chemical Graph Theory: Introduction and Fundamentals*.

THE AUTHOR

Nenad Trinajstić is a Research Professor at the Rugjer Bošković Institute in Zagreb, the Republic of Croatia and a Professor at the Department of Chemistry, Faculty of Natural Sciences and Mathematics, the University of Zagreb. He received the B. Techn. (1960), M.Sc. (1966), and Ph.D. (1967) degrees from the University of Zagreb. During the period of 1964 to 1966, he was a predoctoral fellow with Professor John N. Murrell at the University of Sheffield and the University of Sussex, Brighton, England, and has collaborated with him in research on MO interpretation of electronic spectra of conjugated molecules and on the development of criteria for producing localized orbitals. Professor Trinajstić's postdoctoral years (1968 to 1970) were spent with Professor Michael J. S. Dewar at the University of Texas at Austin, Texas. His main research interest during this period was the development of a convenient semi-empirical MO theory for studying the ground states of large molecules. After returning to Zagreb from the U.S., he initiated research on chemical applications of graph theory and has contributed greatly to the revival of the uses of graph theory in chemistry.

In the last two decades, he has spent some time at various universities such as the University of Trieste, Trieste, Italy (collaborating with Professor Vinicio Galasso); the University of Utah, Salt Lake City, Utah (collaborating with Professor Frank E. Harris); the University of Oxford, England (collaborating with Dr. Roger B. Mallion); the University of South Carolina in Columbia, (collaborating with Professor Benjamin M. Gimarc); the Heinrich Heine University, Düsseldorf, Germany (collaborating with Professor Jan V. Knop); the Iowa State University, Ames; and Drake University in Des Moines, (collaborating with Professor Milan Randić); Higher Institute of Chemical Technology, Burgas, Bulgaria (collaborating with Professor Danail Bonchev); the University of Sussex, Brighton, England (collaborating with Professor John N. Murrell); the University of Reading, England (collaborating with Dr. Stuart Carter); and Texas A&M University at Galveston, Texas (collaborating with Professor Douglas J. Klein). He has taught a number of B.Sc., M.Sc., and Ph.D. students, and has published over 400 research papers and several books in the fields of theoretical organic chemistry, quantum chemistry, mathematical chemistry, computational chemistry, history of chemistry, and philosophy of science. His present main research interest is the chemical applications of graph theory and topology. In 1972, he received the City of Zagreb Award for research. In 1982 he received the Rugjer Bošković Award (Croatian National Award for Science) for his work in chemical graph theory. In 1986, he received the MASUA (Mid-America State Universities Association) Distinguished Foreign Scholar Award and in 1989, was named the University of Zagreb Distinguished Scientist.

Professor Trinajstić is a member of the Editorial Boards of several journals (e.g., *Croatica Chemica Acta, Journal of Molecular Structure — Theochem, Symmetry, Computers & Chemistry)* and is the coeditor-in-chief of the *Journal of Mathematical Chemistry.* He is married (Judita), has two children (Regina and Dean), and two grandsons (Sebastijan and Mateo).

TABLE OF CONTENTS

CHEMICAL GRAPH THEORY

SECOND EDITION

Chapter 1

INTRODUCTION

Graph theory is a branch of (discrete) mathematics that deals with the way objects are connected.[1] Thus, the connectivity in a system is a fundamental quality of graph theory. The principal concept in graph theory is that of a *graph*. For a mathematician, a graph is the application of a set on itself, i.e., a collection of elements of the set and the binary relations between these elements.[2] Graphs are one-dimensional objects,[3] but they can be embedded or realized in spaces of higher dimensions.

Graph theory is related to topology (in fact graph theory is one-dimensional topology[3]), matrix theory, group theory, set theory, probability, combinatorics, and numerical analysis. It has been used in such diverse fields as economics[4] and theoretical physics,[5] psychology[6] and nuclear physics,[7] biomathematics[8] and linguistics,[9] sociology[10] and zoology,[11] technology[12] and anthropology,[13] computer science[14] and geography,[15] biology[16] and engineering,[2] etc.

Chemical graph theory is a branch of mathematical chemistry that is concerned with analyses of all consequences of a connectivity in a chemical graph. *Chemical graph* serves as a convenient model for any real or abstract chemical system (molecule or reaction scheme in a chemical transformation).[17,18] In other words, chemical graph theory is concerned with all aspects of the application of graph theory to chemistry. By the use of the term *chemical* it is emphasized that one is allowed in chemical graph theory, unlike in graph theory, to rely on the intuitive understanding of many concepts and theorems rather than on formal mathematical proofs.

The recent years have witnessed a remarkable growth of chemical graph theory.[19-43] There are many reasons for the increasing popularity of graph theory in chemistry.[17-21,25-28,30,34,37,38,41,42,44-46] First, there is hardly any concept in the natural sciences which is closer to the notion of graph than the structural (constitutional) formula of a chemical compound,[47] because a graph is, simply said, a mathematical model[48] which may be used directly to represent a molecule when the only property considered is the internal connectivity, i.e., whether or not a chemical bond joins two atoms in a molecule. Since almost all discussions in chemistry are carried out by means of graphic display of compounds and reactions, chemists manipulate graphs on a daily basis, albeit many a chemist is not aware of this fact. Thus, it appears that the natural language of chemistry by which chemists communicate is provided by (chemical) graph theory. Second, graph theory provides simple rules by which chemists may obtain many qualitative predictions about the structure and reactivity of various compounds. All these predictions can be reached using nothing more than pencil and paper. Furthermore, the obtained results have in many cases a general validity and may be formulated as theorems and/or rules which can then be applied to a variety of similar problems without any

TABLE 1
Correspondence Between the Graph-Theoretical and Chemical Terms

Graph-theoretical term	Chemical term
Chemical (molecular) graph	Structural formula
Vertex (point)	Atom
Weighted vertex	Atom of a specified element
Edge (line)	Chemical bond
Weighted edge	Chemical bond between the specified elements
Valency (degree) of a vertex	Valency of an atom
Tree graph	Acyclic structure
Chain	Linear alkane or polyene
Cycle	Cycloalkane or annulene
Bipartite (bichromatic) graph	Alternant chemical structure
Nonbipartite graph	Nonalternant chemical structure
1-Factor (Kekulé graph)	Kekulé structure
Adjacency matrix **A**	Hückel (topological) matrix
Characteristic polynomial	Secular polynomial
Eigenvalue of **A**	Eigenvalue of Hückel matrix
Eigenvector of **A**	Hückel (topological) molecular orbital
Positive eigenvalue	Bonding energy level
Zero eigenvalue	Nonbonding energy level
Negative eigenvalue	Antibonding energy level
Graph-spectral theory	Hückel theory

further numerical or conceptual work. Third, graph theory may be used as a foundation for the representation, classification, and categorization of a very large number of chemical systems.[40,49-51] Fourth, graphs appear as useful devices for computer-assisted synthesis design.[52,53] Fifth, the development of simple valence-bond resonance-theoretic schemes is mostly influenced, knowingly or unknowingly, by graph-theoretical work on the enumeration of valence structures.[54,55] One more reason must be mentioned here: that is the explosive use of graph-theoretical invariants in the structure-property-activity relationships.[23,31,56,57] Perhaps this is an area of chemistry in which graph theory could be the most useful theoretical tool. Chemists also know and use a number of graph-theoretical theorems without being aware of this fact in many cases.[58] A classic example is provided by the concept of alternant hydrocarbons, used in chemistry[59,60] since 1940, which is for graph-theoretists the two-color problem.[1,61] Therefore, chemists can easily grasp the basis of graph theory. However, the language of graph theory is different from that of chemistry. Since there is no unique graph-theoretical terminology,[1,2,62,63] a glossary is offered in Table 1 which contains the terminology of graph theory which is proposed for standard use in chemistry and the corresponding chemical terms.

In relating graph-theoretical terms and chemical terms, a certain caution is needed, because one set of terms is taken from the abstract theory and the other from the concrete models used in chemistry. Thus, in one case, for example, trees may be used to represent acyclic structures and in another, certain reaction schemes.

REFERENCES

1. **Wilson, R. J.**, *Introduction to Graph Theory*, Oliver & Boyd, Edinburgh, 1972.
2. **Johnson, D. E. and Johnson, J. R.**, *Graph Theory with Engineering Applications*, Ronald Press, New York, 1972.
3. **Biggs, N. L.**, *Algebraic Graph Theory*, University Press, Cambridge, 1974.
4. **Avondo-Bodino, G.**, *Economic Applications of the Theory of Graphs*, Gordon & Breach, New York, 1962.
5. **Harary, F., Ed.**, *Graph Theory and Theoretical Physics*, Academic Press, New York, 1967; **Capra, F.**, *Am. J. Phys.*, 47, 11, 1979.
6. **Cartwright, D. and Harary, F.**, *Psychol. Rev.*, 63, 277, 1963.
7. **Mattuck, R. D.**, *A Guide to Feynman Diagrams in the Many-Body Problem*, McGraw-Hill, New York, 1967.
8. **Lane, R.**, *Elemente der Graphentheorie und ihre Anwendung in den biologischen Wissenschaften*, Akademischer Verlag, Leipzig, 1970.
9. **Čulik, K.**, *Application of Graph Theory to Mathematical Logic and Linguistics*, Czechoslovak Academy of Sciences, Prague, 1964.
10. **Flament, C.**, *Applications of Graph Theory to Group Structure*, Prentice-Hall, Englewood Cliffs, NJ, 1963.
11. **Lissowski, A.**, *Acta Protozool.*, 11, 131, 1971.
12. **Korach, M. and Haskó, L.**, *Acta Chem. Acad. Sci. Hung.*, 72, 77, 1972.
13. **Hage, P. and Harary, F.**, *Structural Models in Anthropology*, Cambridge University Press, Cambridge, 1983; *J. Graph Theory*, 10, 353, 1986.
14. **Even, S.**, *Graph Algorithms*, Pitman, London, 1979.
15. **Cliff, A., Haggett, P., and Ord, K.**, in *Applications of Graph Theory*, Wilson, R. J. and Beineke, L. W., Eds., Academic Press, London, 1979, 293.
16. **Roberts, F., Ed.**, *Applications of Combinatorics and Graph Theory to the Biological and Social Sciences*, Springer-Verlag, New York, 1989.
17. **Turro, N. J.**, *Angew. Chem. Int. Ed. Engl.*, 25, 882, 1986.
18. **Trinajstić, N.**, in *MATH/CHEM/COMP 1987*, Lacher, R. C., Ed., Elsevier, Amsterdam, 1988, 83.
19. **Rouvray, D. H.**, *Roy. Inst. Chem. Rev.*, 4, 173, 1971.
20. **Gutman, I. and Trinajstić, N.**, *Topics Curr. Chem.*, 42, 49, 1973.
21. **Balaban, A. T., Ed.**, *Chemical Applications of Graph Theory*, Academic Press, London, 1976.
22. **Wilson, R. J.**, in *Colloquia Mathematica Societatis János Bolyai*, Vol. 18, *Combinatorics*, Keszthely, Hungary, 1976, 1147.
23. **Kier, L. B. and Hall, L. H.**, *Molecular Connectivity in Chemistry and Drug Research*, Academic Press, New York, 1976.
24. **Graovac, A., Gutman, I., and Trinajstić, N.**, *Topological Approach to the Chemistry of Conjugated Molecules*, Springer-Verlag, Berlin, 1977.
25. **Slanina, Z.**, *Chem. Listy*, 72, 1, 1978.
26. **Rouvray, D. H. and Balaban, A. T.**, in *Applications of Graph Theory*, Wilson, R. J. and Beineke, L., Eds., Academic Press, London, 1979, 177.
27. **King, R. B., Ed.**, *Chemical Applications of Topology and Graph Theory*, Elsevier, Amsterdam, 1983.
28. **Balasubramanian, K.**, *Chem. Rev.*, 85, 599, 1985.

29. **Trinajstić, N., Ed.,** *Mathematics and Computational Concepts in Chemistry,* Horwood, Chichester, 1986.
30. **Gutman, I. and Polansky, O. E.,** *Mathematical Concepts in Organic Chemistry,* Springer-Verlag, Berlin, 1986.
31. **Kier, L. B. and Hall, L. H.,** *Molecular Connectivity in Structure-Activity Analysis,* Research Studies Press, Letchworth, Hertfordshire, 1986.
32. **Slanina, Z.,** *Contemporary Theory of Chemical Isomerism,* Academia, Prague, 1986.
33. **Dias, J. R.,** *Handbook of Polycyclic Hydrocarbons. Part A. Benzenoid Hydrocarbons,* Elsevier, Amsterdam, 1987.
34. **King, R. B. and Rouvray, D. H., Eds.,** *Graph Theory and Topology in Chemistry,* Elsevier, Amsterdam, 1987.
35. **Dias, J. R.,** *Handbook of Polycyclic Hydrocarbons. Part B. Polycyclic Isomers and Heteroatom Analogs of Benzenoid Hydrocarbons,* Elsevier, Amsterdam, 1988.
36. **Lacher, R. C., Ed.,** *MATH/CHEM/COMP 1987,* Elsevier, Amsterdam, 1988.
37. **Hansen, P. J. and Jurs, P. C.,** *J. Chem. Educ.,* 65, 575, 1988.
38. **Gutman, I. and Cyvin, S. J.,** *Introduction to the Theory of Benzenoid Hydrocarbons,* Springer-Verlag, Berlin, 1989.
39. **Graovac, A., Ed.,** *MATH/CHEM/COMP 1988,* Elsevier, Amsterdam, 1989.
40. **Rouvray, D. H., Ed.,** *Computational Chemical Graph Theory,* Nova Science Publishers, Commack, NY, 1990.
41. **Polansky, O. E.,** in *Theoretical Models of Chemical Bonding,* Maksić, Z. B., Ed., Springer-Verlag, Berlin, 1990, 29.
42. **Bonchev, D. and Rouvray, D. H., Eds.,** *Chemical Graph Theory: Introduction and Fundamentals,* Abacus Press/Gordon & Breach Science Publishers, New York, 1991.
43. **Trinajstić, N.,** *Rep. Mol. Theory,* 1, 185, 1990.
44. **Balaban, A. T.,** *J. Chem. Inf. Comput. Sci.,* 25, 334, 1985.
45. **Balaban, A. T.,** *J. Mol. Struct. (Theochem),* 120, 117, 1985.
46. **Trinajstić, N., Klein, D. J., and Randić, M.,** *Int. J. Quantum. Chem.: Quantum Chem. Symp.,* 20, 699, 1986.
47. **Prelog, V.,** Nobel Lecture, December 12, 1975; reprinted in *Science,* 193, 17, 1976.
48. **Chartrand, G.,** *Graphs as Mathematical Models,* Prindle, Weber & Schmidt, Boston, 1977.
49. **Lynch, M. J., Harrison, J. M., Town, V. G., and Ash, J. E.,** *Computer Handling of Chemical Structure Information,* MacDonald, London, 1971.
50. **Carthart, R. E., Smith, D. H., Brown, H., and Sridharan, N. S.,** *J. Chem. Inf. Comput. Sci.,* 15, 124, 1975.
51. **Trinajstić, N., Nikolić, S., Knop, J. V., Müller, W. R., and Szymanski, K.,** *Computational Chemical Graph Theory: Characterization, Enumeration and Generation of Chemical Structures by Computer Methods,* Simon & Schuster, Chichester, 1991.
52. **Corey, E. J.,** *Q. Rev. Chem. Soc.,* 25, 455, 1971.
53. **Hendrickson, J. B., Grier, D. L., and Toczko, A. G.,** *J. Am. Chem. Soc.,* 107, 5288, 1985.
54. **Ruščić, B., Křivka, P., and Trinajstić, N.,** *Theor. Chim. Acta,* 69, 107, 1986.
55. **Cyvin, S. J. and Gutman, I.,** *Kekulé Structures in Benzenoid Hydrocarbons,* Springer-Verlag, Berlin, 1988.
56. **Rouvray, D. H.,** *Sci. Am.,* 254, 40, 1986.
57. **Trinajstić, N., Nikolić, S., and Carter, S.,** *Kem. Ind. (Zagreb),* 38, 469, 1989.
58. **Mallion, R. B.,** *Chem. Br.,* 9, 242, 1973.
59. **Coulson, C. A. and Rushbrooke, G. S.,** *Proc. Cambridge Philos. Soc.,* 36, 193, 1940.
60. **Mallion, R. B. and Rouvray, D. H.,** *J. Math. Chem.,* 5, 1, 1990; *J. Math. Chem.,* 8, 399, 1991.
61. **Trinajstić, N.,** in *Semiempirical Methods of Electronic Structure Calculation. Part A. Techniques,* Segal, G. A., Ed., Plenum Press, New York, 1977, 1.
62. **Essam, J. W. and Fisher, M. E.,** *Rev. Mod. Phys.,* 42, 272, 1970.
63. **Harary, F.,** *Graph Theory,* Addison-Wesley, Reading, MA, 1971, 2nd printing.

Chapter 2

ELEMENTS OF GRAPH THEORY

This chapter contains the basic definitions, theorems, and concepts of graph theory. Since this book is designed for the chemical community at large, mathematical rigor is omitted whenever possible. The precise details of graph theory can be found in a number of mathematical books dealing with the subject.[1-9]

I. THE DEFINITION OF A GRAPH

The central concept in graph theory is that of *graph*. A convenient way to introduce this concept is by giving the definition of a *simple graph*.[6]

A simple graph G is defined as an ordered pair [V(G),E(G)], where V = V(G) is a nonempty set of elements called *vertices* (or points) of G and E = E(G) is a set of unordered pairs of distinct elements of V(G) called *edges* (or lines). Sets V(G) and E(G) are called the *vertex-set* and the *edge-set* of G, respectively. In most cases of chemical interest the vertex-set V(G) and the edge-set E(G) are finite. The number of elements N in V(G) is called the *order* of G and the number of elements M in E(G) is the *size* of G. Throughout this book the symbols N and M will be used for the number of vertices and the number of edges, respectively.

A very attractive feature of graph theory is that a graph can be visualized by means of a diagram when the vertices are drawn as small circles or dots and the edges as lines or curves joining the appropriate circles. Since a diagram of a graph completely describes the graph, it is customary and convenient to refer to the diagram of the graph as the graph itself. Mainly because of their diagrammatic representation, graphs have appeal as structural models[8-11] in science in general and in chemistry in particular. In this respect it is worth mentioning the following historical detail: the term graph itself was proposed by Sylvester[12] in 1878 on the basis of the structural formulae (which are really 2-dimensional models of molecules)[10,13] used by the chemists of his day. A nice account of the contribution of James Joseph Sylvester (born: London, September 3, 1814 — died: London, March 15, 1887) to chemical graph theory (and mathematical chemistry) is given by Rouvray.[14]

An example of a simple graph is given in Figure 1. It is a diagram of the labeled simple graph G whose vertex-set V(G) is the set $\{v_1, v_2, v_3, v_4\}$ (or in the simplified notation $\{1,2,3,4\}$) and whose edge-set E(G) consists of pairs $\{v_1, v_2\}$, $\{v_2, v_3\}$, $\{v_2, v_4\}$, and $\{v_3, v_4\}$ (or in the simplified notation $\{1,2\}$, $\{2,3\}$, $\{2,4\}$, and $\{3,4\}$). In this book both notations, more explicit and simplified, will be used depending on a particular need. The edge $\{v_i, v_j\}$ is said to join the vertices v_i and v_j. Elements in the edge-set may also be denoted by $\{e_1, e_2, \ldots, e_M\}$. Note that since E(G) is a set rather than a family, there can

FIGURE 1.   A diagram of a simple graph G.

never be more than a single line joining a given pair of vertices in a simple graph G. The term *family* in this context means a collection of elements, some of which may occur several times;[6] for example, {1,2,3,4} is a set, but {1,1,2,2,2,3} is a family. A graph G is *labeled* when its vertices are distinguished from one another by labels which may be letters or numbers.

Many results which can be proved for simple graphs may be extended without difficulty to more general graphs in which two vertices may have more than one edge joining them. In addition, it is often convenient to remove the restriction that any edge must join two distinct vertices, and to allow the existence of loops, i.e., edges joining vertices to themselves. A *general graph* (in which multiple edges and loops are allowed) is defined as follows:[6] A general graph G is an ordered pair [V(G), E(G)], where V(G) is a nonempty set of elements called vertices and E(G) is a family of unordered pairs of (not necessarily distinct) elements of V(G), called edges. Note that the use of the term family permits the existence of multiple edges. Now V(G) is called the vertex-set of G and E(G) the edge-family of G. General graphs may be split into two classes: (1) multigraphs[4] (in which multiple edges are allowed) and (2) loop-multigraphs[9] or pseudographs[4] (in which both multiple edges and loops are allowed). In Figure 2 are given two examples of labeled general graphs (i.e., a multigraph G_1 and a loop-multigraph G_2). The vertex-set $V(G_1)$ and the edge-family $E(G_1)$ are given, respectively, as $V(G_1) = \{1,2,3,4\}$ and $E(G_1) = \{(1,2),(1,2),(2,3),(2,3),(2,4),(3,4)\}$. Similarly, the vertex-set $V(G_2)$ and the edge-family $E(G_2)$ are given, respectively, as $V(G_2) = \{1,2,3,4\}$ and $E(G_2) = \{(1,1),(1,2),(2,3),(2,4),(2,4),(3,4),(4,4),(4,4)\}$.

The first multigraph in the history of graph theory[15] was the graph (shown in Figure 3) of the Königsberg bridge problem set by Euler[16] in 1736. (The story of the Königsberg bridge problem will be given in Section VIII of this chapter.)

A special class of graphs is a class of directed graphs or *digraphs*. A digraph D is defined[6] as an ordered pair [V(D), A(D)], where V(D) is a nonempty set of elements called vertices and A(D) is a family of (not necessarily distinct) ordered pairs of elements of V(D) called *arcs* or *directed edges*. V(D) is called the vertex-set and A(D) the *arc-family* of D. An arc

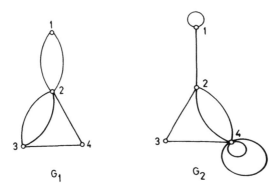

FIGURE 2.   Diagrams of a multigraph G_1 and a loop-multigraph G_2.

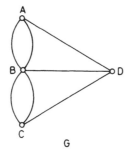

FIGURE 3.   Euler's graph G of the Königsberg bridge problem.

whose first element is v_i and whose second element is v_j is called an arc from v_i to v_j and is written as $\{v_i,v_j\}$ or $\{i,j\}$. Note that two arcs of the form $\{i,j\}$ and $\{j,i\}$ are different. In Figure 4 is given as an illustrative example a labeled digraph D with the vertex-set $V(D) = \{1,2,3,4\}$ and the arc-family $A(D) = \{(1,2),(2,1),(2,3),(2,4),(4,3),(4,4)\}$.

In this book, if not otherwise stated, a graph will be understood as a finite undirected graph without multiple edges and loops.

At this point it is convenient to introduce the concepts of adjacency and incidence. Two vertices v_i and v_j of a graph G are said to be *adjacent* if there is an edge joining them; the vertices v_i and v_j are then said to be *incident* to such an edge. Similarly, two distinct edges e_i and e_j of G are *adjacent* if they have at least one vertex in common. In the graph G, given in Figure 5, for example, the vertices v_1 and v_3 are adjacent, but v_2 and v_4 are not. Likewise edges e_1 and e_4 are adjacent, but e_1 and e_5 are not.

Some other more general definitions of a graph may be found in mathematical literature.[1,17]

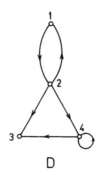

FIGURE 4.   A diagram of a labeled directed graph D.

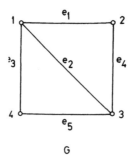

FIGURE 5.   A labeled graph G to illustrate the concepts of adjacency and incidence.

II. ISOMORPHIC GRAPHS AND GRAPH AUTOMORPHISM

Two graphs $G' = [V(G'), E(G')]$ and $G'' = [V(G''), E(G'')]$ are said to be *isomorphic* (written as $G' \cong G''$ or sometimes $G' = G''$) if there exists a one-to-one mapping f,

$$fv' = v''; \quad v' \in V(G'), \quad v'' \in V(G'') \tag{1}$$

such that $(f v_i', f v_j') \in E(G'')$ if, and only if, $(v_i', v_j') \in E(G')$. The procedure to recognize isomorphic graphs is simple for small graphs (like G_1 and G_2 in Figure 6), but it is very difficult to recognize the complex graphs (like G_3 and G_4 in Figure 6) as isomorphic graphs.[18]

Both pairs of graphs: G_1 and G_2, and G_3 and G_4, are isomorphic because mapping f defined as

$$fv_i' = v_i'' \text{ for each } i; \quad v_i' \in V(G'), \quad v_i'' \in V(G'') \tag{2}$$

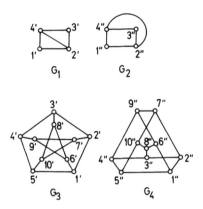

FIGURE 6. Two pairs of labeled isomorphic graphs.

preserves adjacency. It follows from the definition that the isomorphic graphs are identical, but differently drawn, graphs.

Generally, the problem of recognizing isomorphic graphs is one of the grand unsolved problems of graph theory. Construction of all N! possible mappings from one graph to another, although obviously impractical, remains as the only secure check for isomorphism of graphs.

An *invariant* of a graph G is a quantity associated with G which has the same value for any graph isomorphic to G. Consequently, graph invariants are quantities independent of the labeling of the vertices of a graph. Hence, the number of vertices and the number of edges are invariants. A complete set of invariants determines a graph up to isomorphism. For example, the numbers N and M constitute such a set for all simple graphs with less than four vertices.[4]

An isomorphic mapping of the vertices of a graph G onto themselves (which also preserves the adjacency relationship) is called an *automorphism* of a graph G.[4,6] Evidently, each graph possesses a trivial automorphism which is called the *identity automorphism*. For some graphs, it is the only automorphism; these are called *identity graphs*. The set of all automorphisms of a graph G forms a *group* which is called the *automorphism group* of G.

III. WALKS, TRAILS, PATHS, DISTANCES, AND VALENCIES IN GRAPHS

A *walk* of a graph G is an alternating sequence of vertices and edges v_0, $e_1, v_1, \ldots, v_{\ell-1} e_\ell, v_\ell$, beginning and ending with vertices, in which each edge is incident with two vertices immediately preceding and following it.[4] This walk joins the vertices v_0 and v_ℓ, and may also be denoted by $v_0 v_1$ $v_2 \ldots v_\ell$ (the edges being evident by context). The length ℓ of a walk is the number of occurrences of edges in it; v_0 is called the *initial vertex* of the walk, while v_ℓ is called the *terminal vertex* of the walk. A *closed walk* is a

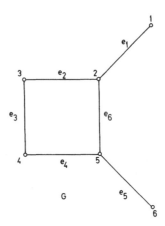

FIGURE 7.   A labeled graph G to illustrate walks, trails, and paths.

v_i–v_i walk, i.e., a walk which starts and ends at the same vertex v_i. Otherwise, a walk is said to be an *open walk*. A walk is a *trail* if all the edges are distinct and a *path* if all vertices (and thus necessarily all edges) are distinct.

In a labeled graph G (see Figure 7) a sequence of vertices v_1 v_2 v_3 v_2 v_5 is a walk which may also be denoted as a walk 12325. One of the paths of length four in G is, for example, the sequence v_1 v_2 v_3 v_4 v_5, which may be denoted as 12345. Walks 12521 and 121 are closed, while walks 345, 123256, and 12543 are open. Walks 345 and 12543 are paths of length 2 and length 4, respectively. A walk 234521 is the trail of length 5.

The length of the shortest path between vertices v_i and v_j in a graph G is called the *distance* between these two vertices and is denoted as $d(v_i, v_j)$ or $d(i, j)$. The distance d is a non-negative quantity and has only integral values. It has the following properties:

$$d(i,j) = 0 \quad \text{if, and only if, } i = j \tag{3}$$

$$d(i,j) = d(j,i) \tag{4}$$

$$d(i,j) + d(j,k) \geq d(i,k) \tag{5}$$

$$d(i,j) = 1 \quad \text{if, and only if, } (i,j) \in E(G) \tag{6}$$

If every pair of vertices in G is joined by a path, G is a *connected graph*. If there is no path between two vertices in G, i.e., $d(i, j) = \infty$, G is a *disconnected graph* and these two vertices belong to different *components* of a graph. The number of components of a graph G is denoted by $k = k(G)$. Examples of graphs with one, two, and three components are given in Figure 8.

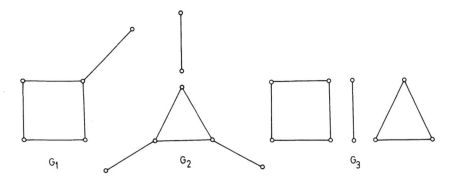

FIGURE 8.   Graphs G_1, G_2, and G_3 with one [$k(G_1)$ = 1], two [$k(G_2)$ = 2], and three [$k(G_3)$ = 3] components, respectively.

Once the concept of the distance in a graph is introduced, it is easy to define the graph-theoretical *valency* or the *degree* of a vertex. All vertices among which the distance is unity (i.e., adjacent vertices) are the first neighbors. The second, third, and higher neighbors are defined in analogous ways. The number of the first neighbors of a vertex i is the valency or degree of a vertex i. It is denoted as $D_1(i) = D(i)$. $D(i)$ equals the number of edges incident with i. Any vertex of valency zero is called an *isolated vertex*. A vertex of valency one is a *terminal vertex* or *endpoint*.

The total sum of valencies of all vertices in a graph G equals twice the number of edges, because in the summation each edge is counted twice,

$$\sum_{i=1}^{N} D(i) = 2M \qquad (7)$$

This result was known to Euler[16] and is referred to as the *handshaking lemma*, because it means that if several persons shake hands, the total number of hands shaken must be *even*. This is so because two hands are involved in each handshake. An immediate corollary of the handshaking lemma is that in any graph the number of vertices of odd degree must be even. An example to illustrate the handshaking lemma is given in Figure 9.

In addition, if the number of vertices of valencies, 1,2,3, . . . are denoted by F, S, T, . . . , the following identity holds,

$$F + 2S + 3T + ... = 2M \qquad (8)$$

Note,

$$F + S + T + ... = N \qquad (9)$$

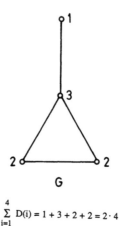

$$\sum_{i=1}^{4} D(i) = 1 + 3 + 2 + 2 = 2 \cdot 4$$

FIGURE 9.   An example to illustrate the handshaking lemma. A digit at each vertex of a graph G is its valency.

IV. SUBGRAPHS

A *subgraph* G′ of a graph G is a graph having all its vertices and edges contained in G.[4] More formally, if V(G′) is a subset of V(G) and E(G′) a subset of E(G),

$$V(G') \subseteq V(G) \tag{10}$$

$$E(G') \subseteq E(G) \tag{11}$$

then the graph G′ = [V(G′), E(G′)] is a subgraph of G. A graph can be its own subgraph. The removal of only one edge e from G results in a subgraph G − e which consists of all vertices and M − 1 edges of G. The removal of a vertex v from G results in a subgraph G − v consisting of N − 1 vertices and M − D(v) edges of G. A subgraph G − (e) is obtained by deletion of, respectively, an edge e and its two incident vertices u and v and its adjacent edges from G. The subgraph G − (e) consists of N − 2 vertices and M − [D′(u) + D′(v)] + 1 edges of G. A *spanning subgraph* is a subgraph containing all the vertices of G. An example of a spanning subgraph is G − e. Examples of subgraphs are given in Figure 10.

If a complete set of subgraphs G − v_i, i = 1,2, . . . ,N for a graph G (with N ≥3) is formed and if the same set of subgraphs is obtained for a graph G′, the *Ulam conjecture*[19] states that G and G′ are isomorphic graphs. This is proved for trees[20] and for some special classes of graphs,[21] but for arbitrary graphs it remains unsolved.

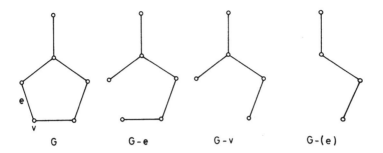

FIGURE 10.   A set of subgraphs belonging to a graph G.

V. REGULAR GRAPHS

A graph in which every vertex has the same valency is called a *regular graph*. If every vertex has degree (valency) D, the graph is called *regular graph of degree D*.

If all vertices in a connected graph are of degree 2, the graph is called a *ring* or a *cycle* or a *circuit*. Cycles are denoted by C_N ($N \geq 3$). A cycle is *even* if N = even and *odd* otherwise.

For a regular graph of degree D,

$$M = (1/2)ND \tag{12}$$

The meaning of relation (12) is that a regular graph exists only if N and/or D are even.

If $D = 1$, there exists only one connected (regular) graph. This is a graph consisting of two vertices joined by a single edge, called a *complete graph of degree one*, and denoted by K_2. A regular graph G with N vertices and $D = N - 1$ is called a *complete graph* and is denoted by K_N. In such a graph every pair of its vertices is adjacent and the number of its edges is given by,

$$M = \binom{N}{2} = \frac{N(N - 1)}{2} \tag{13}$$

In Figure 11 are given examples of regular graphs and complete graphs.

VI. TREES

A graph is *acyclic* if it has no cycles. A *tree* is a connected acyclic graph. If a tree has N vertices, then it contains $N - 1$ edges. Obviously, by deleting any of its edges, a tree can be partitioned into two parts. A *rooted tree* is a tree in which one vertex has been distinguished in some way from others.[4] This vertex is usually called the *root-vertex* or simply, the *root*.[22] Examples of a tree T and a rooted tree T* are given in Figure 12.

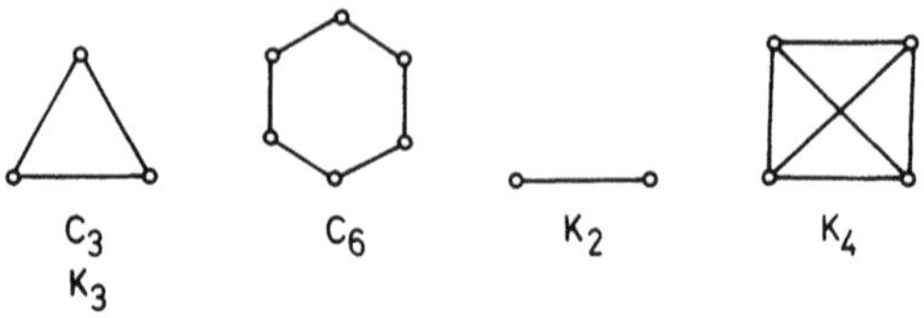

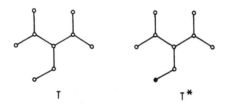

C_3
K_3 C_6 K_2 K_4

FIGURE 11.   Examples of regular graphs and complete graphs.

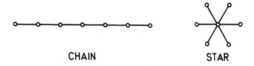

T   T*

FIGURE 12.   Examples of a tree T and a rooted tree T*. The root-vertex in T* is denoted by black dot.

CHAIN   STAR

FIGURE 13.   Examples of a chain and a star with seven vertices.

A tree necessarily possesses vertices of degree one. These are end-vertices and are called *terminals*. The tree with the minimum number of terminals (two of them) is the *chain*, while the tree with the maximum number of terminals (N − 1 of them) is the *star*. Examples of a chain and a star are given in Figure 13.

One of the important structural features of trees is *branching*. However, branching is an intuitive concept not uniquely defined,[23] though it can be identified through the appearance of vertices with valencies three or higher as the sites of ramification. These vertices are sometimes called the *branching vertices*. A tree T in Figure 12 and a star in Figure 13 are examples of the branched trees. The central vertex in a star is an example of the maximum branching vertex. All trees but chains are branched trees.

VII. PLANAR GRAPHS

A graph is *planar* if it can be drawn in the plane in such a way that no two edges intersect or, in other words, a graph is planar if it can be embedded in the plane. Illustrative examples of a planar graph and a nonplanar graph

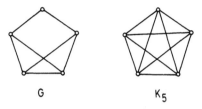

FIGURE 14. Examples of a planar graph (G) and nonplanar graph (K_5).

FIGURE 15. A planar representation of a graph G from Figure 14.

are given in Figure 14. K_5 is a complete graph of degree four and is made famous by Kuratowski[24] in his work on planar graphs. At first glance graph G in Figure 14 is not a planar graph. However, if G is given in a different representation (see Figure 15), it is clear that G is a planar graph.

The planar graphs were introduced by Euler in his investigation of polyhedra.[16] He found a fundamental relationship between vertices (V), edges (E), and faces (F) of a polyhedron, which can be expressed as the following formula, often called[7] the Euler (polyhedral) formula

$$V - E + F = 2 \tag{14}$$

The Euler formula also holds for planar graphs. However, the following should be pointed out. A planar graph partitions the plane into one infinite region and one or more finite regions. The unbounded (infinite) region is called infinite face. The face count in Equation 14, when applied to planar graphs also includes the infinite face. Thus, Formula 14 may be rewritten in the case of planar graphs as,

$$V(G) - E(G) + F(G) + 1 = 2 \tag{15}$$

or

$$V(G) - E(G) + F(G) = 1 \tag{16}$$

As an example, the application of Formulae 14 and 16 to the cube and the corresponding graph G is given in Figure 16. A graph G corresponding to the cube belongs to a class of graphs named *polyhedral graphs*. Note that

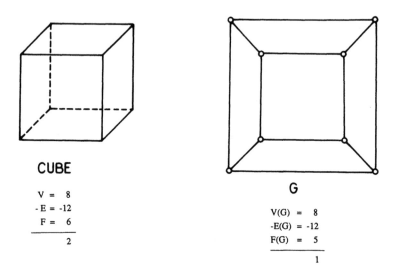

CUBE

$$V = 8$$
$$- E = -12$$
$$F = 6$$
$$\overline{}$$
$$2$$

G

$$V(G) = 8$$
$$-E(G) = -12$$
$$F(G) = 5$$
$$\overline{}$$
$$1$$

FIGURE 16. Diagrams of the cube and the corresponding polyhedral graph G.

the polyhedral graphs are planar graphs. Another name for polyhedral graphs is *Schlegel graphs.*[25]

The Euler formula can also extend to disconnected graphs. If G is a planar disconnected graph with $V(G)$ vertices, $E(G)$ edges, $F(G)$ faces, and $k(G)$ components, then

$$V(G) - E(G) + F(G) + 1 = k(G) + 1 \qquad (17)$$

or

$$V(G) - E(G) + F(G) = k(G) \qquad (18)$$

Rudel,[26] Levin,[27] and others[28-31] have noticed a similarity between the Euler formula (14) and the Gibbs phase rule,[32]

$$P - C + F = 2 \qquad (19)$$

where P, C, and F denote, respectively, the number of phases, components of the system in equilibrium between phases, and degrees of freedom. A phase diagram may be regarded as a topological simplex,[33] i.e., as the *n*-dimensional analog of a tetrahedron, where $n > 3$. A phase diagram can be represented by a *dual.*[34] Given a planar graph G, its dual graph G* is constructed as follows:[4] Place a vertex in each region of G (including the infinite region) and, if two regions have an edge *e* in common, join the corresponding vertices by an edge *e** crossing only *e*. A dual of a phase diagram is constructed similarly: each field of the diagram is represented as a vertex, including the field outside the diagram. If two fields of phase diagram have a common

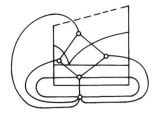

FIGURE 17.   A phase diagram and the corresponding dual.

G G_1 G_2

FIGURE 18.   A pair of homeomorphic graphs (G_1 and G_2) obtained from a graph G.

edge, then the corresponding vertices are connected in the dual. An example of a phase diagram and the corresponding dual is given in Figure 17. The exact connection between the Euler formula (14) and the Gibbs phase rule (19) has not yet been established.[35]

If new vertices are added to the edges of a graph G, a *homeomorph* of G is obtained. Two graphs are *homeomorphic* (or identical to within vertices of valency two) if they can both be obtained from the same graph by inserting new vertices of valency two into its edges.[6] The construction of homeomorphic graphs is illustrated in Figure 18.

Kuratowski[24] proved that a graph is planar, if and only if, it has no subgraphs homeomorphic to K_5 or $K_{3,3}$ (given in Figure 19). (For an explanation of the $K_{3,3}$ notation see Section XI of this chapter.)

VIII. THE STORY OF THE KÖNIGSBERG BRIDGE PROBLEM AND EULERIAN GRAPHS

Before introducing the concept of Eulerian graphs, it is worthwhile to take a short historical excursion. The Königsberg bridge problem,[15,36,37] which motivated Euler to invent graph theory in order to solve it, will be briefly presented. Besides, this is a good example to illustrate the usefulness of graphs as abstract models.

In Figure 20 a part of the map of the park in the Prussian city of Königsberg is shown as it appeared in the early 18th century. The figure displays a part of the Pregel River including two islands. Also depicted are the seven bridges connecting the islands to each other and to the banks of the Pregel. On August 30, 1944, Königsberg was destroyed in a bombing raid. Today Königsberg

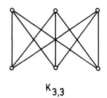

$K_{3,3}$

FIGURE 19.   A diagram of the Kuratowski graph $K_{3,3}$.

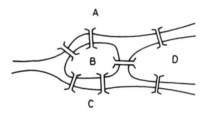

FIGURE 20.   The seven bridges of Königsberg.

is a naval base called Kaliningrad and is in the U.S.S.R. The river Pregel is now called the Pregolya. The seven bridges no longer exist.

It was the pastime of some of the burghers of Königsberg to attempt the following: one was to start at any point on the banks of the Pregel or on one of the islands and walk over all the bridges once and only once, returning to the starting point. The trial and error method of Königsberg's burghers did not produce the required result. No one succeeded in solving the above problem, but it remained unknown whether or not the Königsberg bridge problem had a solution. The problem was presented to Léonard (or Leonhard) Euler (Basle: April 15, 1707 — St. Petersburg: September 18, 1783), the Swiss mathematician who spent a large part of his life in St. Petersburg (now Leningrad).[38] Euler approached the Königsberg bridge problem by replacing the diagram in Figure 20 by another, more general, diagram — a graph using the following conversion rules for its construction: banks of the river and islands were replaced by points (vertices) and bridges by lines (edges). This is a graph G which is already given in Figure 3.

In order to solve the Königsberg bridge problem, one must find a closed walk which includes every edge of the graph G only once. Such a walk is called an *Eulerian trail*.[4] Euler found that the Eulerian trail is possible only when the valency of each vertex in a graph is *even*. Since the valencies of vertices in G in Figure 3 are all odd, this graph does not possess the Eulerian trail. Thus, the solution of the Königsberg bridge problem was negative. It is *not* possible to walk over each bridge once and return to the starting point. However, this result of Euler is a very general result valid for all problems in the same genre.

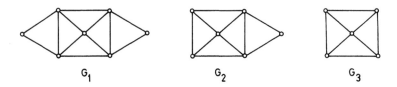

FIGURE 21. Examples of an Eulerian graph (G_1), a semi-Eulerian graph (G_2), and a non-Eulerian graph (G_3).

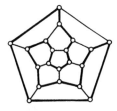

FIGURE 22. The dodecahedral graph and one of its Hamiltonian circuits.

An *Eulerian graph* is a graph which has an Eulerian trail, a closed trail containing all its vertices and edges. An Eulerian graph G may be simply detected by inspecting the valencies of its vertices. If every vertex in G has *even* valency, then G is the Eulerian graph. Graph G in Figure 3 has all vertices with *odd* valencies and thus, it is a representative of *non-Eulerian graphs*. A connected graph G is a *semi-Eulerian graph* if, and only if, it does not contain more than two vertices of *odd* valency. A semi-Eulerian graph contains a semi-Eulerian trail which starts at one vertex of odd valency and ends at the other. Every Eulerian graph is also semi-Eulerian. Graphs which are Eulerian, semi-Eulerian, and non-Eulerian, respectively, are given in Figure 21.

Problems on Eulerian graphs frequently appear in books on recreation mathematics[6] and they occur even in the primitive geometric art.[39]

IX. HAMILTONIAN GRAPHS

A connected graph G is a *Hamiltonian graph* if it has a circuit which includes every vertex of G only once. Such a circuit is called a *Hamiltonian circuit*.

The original Hamiltonian graph is connected with the icosian game invented by Hamilton.[15,40] This game uses the graph of the regular dodecahedron (the dodecahedral graph has 20 vertices and icosian is a term based on the Greek word *eikosi* = 20) labeled with names of famous cities of the world and the player is challenged to "travel around the world" along the edges through each vertex exactly once. Thus, he should find a closed path, i.e., a Hamiltonian circuit, if he plays correctly. The dodecahedral graph and one of its Hamiltonian circuits are given in Figure 22.

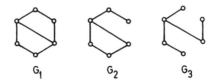

FIGURE 23. Examples of a Hamiltonian graph (G_1), a semi-Hamiltonian graph (G_2), and a non-Hamiltonian graph (G_3).

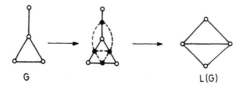

FIGURE 24. The line graph L(G) of a graph G.

The characterization of Hamiltonian graphs has been shown to be one of the major unsolved problems of formal graph theory.[6] In fact, little is known in general about Hamiltonian graphs. However, for simple graphs Dirac[41] has produced a sufficient criterion for detecting whether they are Hamiltonian: if G is a simple graph with N vertices and $D(i) \geq N/2$ for every vertex i, then G is Hamiltonian.

A connected graph G which contains a chain which passes through every vertex of G is called a *semi-Hamiltonian graph*. Every Hamiltonian graph is also a semi-Hamiltonian graph. If a connected graph G does not contain a circuit or a chain which includes every vertex of G, then G is called a *non-Hamiltonian graph*. In Figure 23, graphs are given which are Hamiltonian, semi-Hamiltonian, and non-Hamiltonian, respectively.

X. LINE GRAPHS

The *line graphs* L(G) of a simple graph G is the graph derived from G in such a way that the edges in G are replaced by vertices in L(G). Two vertices in L(G) are connected whenever the corresponding edges in G are adjacent. An illustrative example of the line graph is shown in Figure 24.

The number of vertices V(L) and the number of edges E(L) in the line graph L(G) of G are given by the following relationship:

$$V(L) = E(G) \tag{20}$$

$$E(L) = (1/2) \sum_i D^2(i) - E(G) \tag{21}$$

where D(i) is the valency of a vertex i in G.

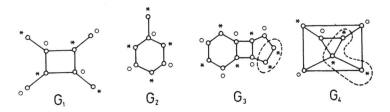

FIGURE 25.   Examples of bipartite graphs (G_1 and G_2) and nonbipartite graphs (G_3 and G_4).

XI. VERTEX COLORING OF A GRAPH

A vertex coloring of a graph G is an assignment of colors to its vertices so that no two adjacent vertices have the same color.[4,42] If N colors are used, the N coloring of G is obtained. It partitions V(G) into N color classes. It is clear that N coloring of a graph G with N vertices can always be carried out. If the number of colors used is gradually decreased, one finally arrives at a number k, called the *chromatic number*, such that k coloring of vertices of G is possible, but not $(k - 1)$ coloring. Then, G is k *chromatic*.

For chemistry, the important class of vertex colored graphs is the class of *bichromatic* or *bicolorable graphs*. These graphs are also called[4] *bipartite graphs* or *bigraphs*. A bipartite graph G is a graph whose vertex-set V(G) can be partitioned into two nonempty subsets V_1 and V_2 such that every edge in G joins V_1 with V_2. Therefore, the first neighbors of vertices V_1 are contained in V_2 and *vice versa*, i.e., the vertices of the same color are never adjacent if the graph is to be bipartite. This is not the case in the *nonbipartite graphs;* there will always appear an edge connecting two vertices of the same color.

The coloring of bipartite graphs will be indicated by stars (*) and circles (o). Vertices of different ''colors'' in bipartite graphs will be therefore called *starred* and *unstarred*. Conventionally,

$$s \geq u \qquad\qquad (22)$$

where s and u stand for the number of starred and unstarred vertices, respectively. If a bipartite graph G is a *complete bipartite graph*, it is denoted as $K_{s,u}$. A complete bipartite graph $K_{s,u}$ is a bigraph in which every vertex of V_1 is joined to every vertex of V_2.[6] An example of a complete bigraph is the graph $K_{3,3}$ (see Figure 19).

The following theorem is given by König:[43] A graph is bipartite if, and only if, all its cycles are even. Therefore, one can decide by a simple inspection whether a graph is bipartite or nonbipartite. In Figure 25 examples of bipartite graphs and nonbipartite graphs are given.

REFERENCES

1. **Berge, C.**, *The Theory of Graphs and Its Applications*, Methuen, London, 1962.
2. **Ore, O.**, *Theory of Graphs*, Vol. 38, American Mathematical Society Colloquium Publications, Providence, RI, 1962.
3. **Bursaker, R. G. and Saaty, T. L.**, *Finite Graphs and Networks*, McGraw-Hill, New York, 1965.
4. **Harary, F.**, *Graph Theory*, Addison-Wesley, Reading, MA, 1971, 2nd printing.
5. **Behzad, M. and Chartrand, G.**, *Introduction to the Theory of Graphs*, Allyn & Bacon, Boston, 1971.
6. **Wilson, R. J.**, *Introduction to Graph Theory*, Oliver & Boyd, Edinburgh, 1972.
7. **Johnson, D. E. and Johnson, J. R.**, *Graph Theory with Engineering Applications*, Ronald Press, New York, 1972.
8. **Malkevitch, J. and Meyer, W.**, *Graphs, Models and Finite Mathematics*, Prentice-Hall, Englewood Cliffs, NJ, 1974.
9. **Chartrand, G.**, *Graphs as Mathematical Models*, Prindle, Weber & Schmidt, Boston, MA, 1977.
10. **Turro, N. J.**, *Angew. Chem. Ind. Ed. Engl.*, 25, 882, 1986.
11. **Trinajstić, N.**, in *MATH/CHEM/COMP 1987*, Lacher, R. C., Ed., Elsevier, Amsterdam, 1988, 83.
12. **Sylvester, J. J.**, *Nature*, 17, 284, 1877–1878.
13. **Trindle, C.**, *Croat. Chem. Acta*, 57, 1231, 1984.
14. **Rouvray, D. H.**, *J. Mol. Struct. (Theochem.)*, 185, 1, 1989.
15. **Biggs, N. L., Lloyd, E. K., and Wilson, R. J.**, *Graph Theory 1736–1936*, Clarendon Press, Oxford, 1976.
16. **Euler, L.**, *Comment Acad. Sci. Imp. Petropolitanae*, 8, 128, 1736.
17. **Zykov, A. A.**, *Theory of Finite Graphs*, Nauka, Novosibirsk, 1969 (in Russian).
18. **Read, R. C. and Corneil, D. G.**, *J. Graph Theory*, 1, 339, 1977.
19. **Ulam, S. M.**, *A Collection of Mathematical Problems*, John Wiley & Sons, New York, 1960.
20. **Kelly, P. J.**, *Pac. J. Math.*, 7, 961, 1957.
21. **Manvel, B.**, in *Proof Techniques in Graph Theory*, Harary, F., Ed., Academic Press, New York, 1969, 103.
22. **Harary, F. and Palmer, E. M.**, *Graphical Enumeration*, Academic Press, New York, 1973.
23. **Essam, J. W. and Fisher, M. E.**, *Rev. Mod. Phys.*, 42, 272, 1970.
24. **Kuratowski, K.**, *Fund. Math.*, 15, 271, 1930.
25. **Coxeter, H. S. M.**, *Regular Polytopes*, 3rd ed., Dover, New York, 1973.
26. **Rudel, O.**, *Z. Electrochem.*, 35, 54, 1929.
27. **Levin, I.**, *J. Chem. Educ.*, 23, 183, 1946.
28. **Mindel, J.**, *J. Chem. Educ.*, 39, 512, 1962.
29. **Mallion, R. B.**, *Chem. Br.*, 8, 446, 1972.
30. **Rouvray, D. H.**, *Chem. Br.*, 10, 11, 1974.
31. **Wilson, R. J.**, *Colloquia Mathematica Societatis János Bolyai*, Vol. 18. *Combinatorics*, Keszthely, Hungary, 1976, 1147.
32. **Moore, W. J.**, *Physical Chemistry*, 5th ed., Longman, London, 1972, 207.
33. **Kurnakov, N. S.**, *Z. Anorg. Allg. Chem.*, 169, 113, 1928.
34. **Seifer, A. L. and Stein, V. S.**, *Zh. Neorg. Khim.*, 6, 2719, 1961.
35. **Radhakrishnan, T. P.**, *J. Math. Chem.*, 5, 381, 1990.
36. **Wilson, R. J.**, *J. Graph Theory*, 10, 265, 1986.
37. **Fowler, P. A.**, *Am. Math. Mon.*, 95, 42, 1988.
38. **Bell, E. T.**, *Men of Mathematics*, Simon & Schuster, New York, 1965, chap. 9.
39. **Hage, P. and Harary, F.**, *J. Graph Theory*, 10, 353, 1986.

40. **Gardner, M.**, *Mathematical Puzzles and Diversions,* Penguin Books, Harmondsworth, Middlesex, 1985, chap. 6, 12th printing.
41. **Dirac, G. A.**, *Proc. London Math. Soc.,* 2, 69, 1952.
42. **Bondy, J. A. and Murty, U. S. R.**, *Graph Theory with Applications,* North Holland/ Elsevier, Amsterdam, 1976, chap. 8.
43. **König, D.**, *Theorie der endlichen und unendlichen Graphen,* Akademische Verlagsgesellschaft, Leipzig, 1936.

Chapter 3

CHEMICAL GRAPHS

I. THE CONCEPT OF A CHEMICAL GRAPH

In chemistry graphs can represent different chemical objects: molecules, reactions, crystals, polymers, clusters, etc. The common feature of chemical systems is the presence of *sites* and *connections* between them. Sites may be atoms, electrons, molecules, molecular fragments, groups of atoms, intermediates, orbitals, etc. The connections between sites may represent bonds of any kind, bonded and nonbonded interactions, elementary reaction steps, rearrangements, van der Waals forces, etc. Chemical systems may be depicted by *chemical graphs* using a simple conversion rule:

site $\leftrightarrow$ vertex

connection $\leftrightarrow$ edge

A special class of chemical graphs are *molecular graphs*. Molecular graphs are chemical graphs which represent the *constitution* of molecules.[1-3] They are also called *constitutional graphs*.[4,5] In these graphs vertices correspond to individual atoms and edges to chemical bonds between them. Molecular graphs are necessarily connected graphs. As examples the molecular graphs corresponding to propane and cyclopropane are shown in Figure 1.

In order to simplify the handling of molecular graphs, *hydrogen-suppressed graphs*,[6] i.e., graphs depicting only molecular skeletons without hydrogen atoms and their bonds, are often used. They are also called *skeleton graphs*.[7] The hydrogen-suppressed graphs are almost universally used in chemical graph theory, because the neglect of the hydrogen atoms and their bonds in most cases cannot be the cause of any ambiguity. The hydrogen-suppressed graphs corresponding to butane and cyclobutane are given in Figure 2.

The molecular graph grossly simplifies the complex picture of a molecule by depicting only its constitution (i.e., the chemical bonds between the various pairs of atoms in the molecule) and neglecting other structural features (e.g., geometry, stereochemistry, chirality). Even so, a simple picture of a molecule as the molecular graph can enable one to make useful predictions about physical and chemical properties of molecules.[5,7-15] Since the predictions of properties and reactivities of molecules are of prime interest to chemists, the development of chemical graph theory is, thus, justified.

II. MOLECULAR TOPOLOGY

Molecular graphs depicting constitutional formulae of molecules represent their *topology*.[1-4,7,13,14,16] This is a chemist's view of molecular topology.

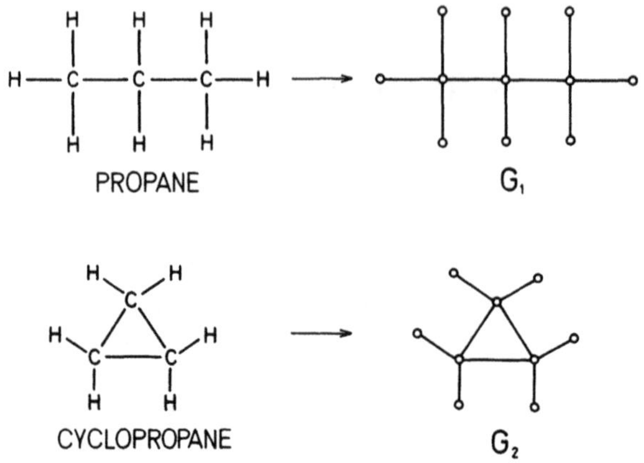

FIGURE 1. The molecular graphs corresponding to propane and cyclopropane.

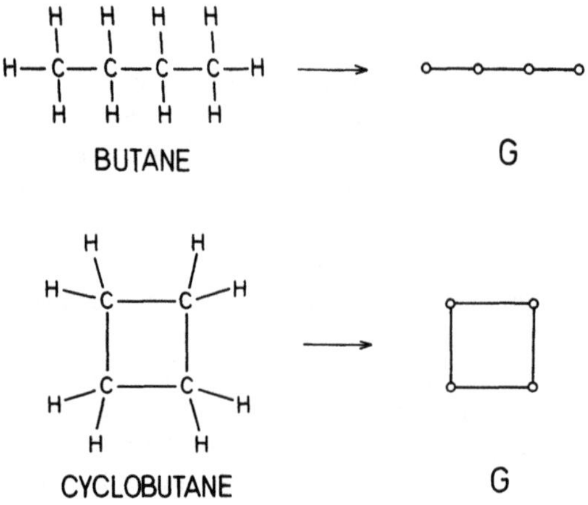

FIGURE 2. The hydrogen-suppressed molecular graphs depicting butane and cyclobutane.

However, a more precise definition of molecular topology may also be given using the concept of the molecular graph.[7,13,14]

A *topological space* is formed by a set and the topological structure defined upon the set.[17,18] A simple connected (molecular) graph can be associated with a topological space if it can be shown that a toplogical structure is defined upon its vertex-set. There are several equivalent methods available for proving the above. The Merrifield-Simmons method[19-22] is based on an *open set topology* or *O-topology formalism.*[18]

Consider a set $\Re$ and its subsets $\Re_i \subseteq \Re$ ($i = 1,2, \ldots$). An open set topology is defined on the set $\Re$ to be a collection of subsets $\mathbf{T_O} = \{\Re_1, \Re_2, \ldots\}$ which obeys the following axioms:

1. Each union of members of $\mathbf{T_O}$ is a member of $\mathbf{T_O}$
2. Each intersection of a finite number of members of $\mathbf{T_O}$ is a member of $\mathbf{T_O}$
3. The set $\Re$ belongs to $\mathbf{T_O}$
4. The empty set $\emptyset$ belongs to $\mathbf{T_O}$

Merrifield and Simmons[19-22] have shown that the open set topology formalism can be applied directly to the vertex-set of bipartite graphs, but it needs to undergo some modification via paraspectral duplex[22,23] in the case of nonbipartite graphs.

A simpler alternative to the Merrifield-Simmons method appears to be a method developed by Polansky[7,13,14,24-27] which is based on a *neighborhood topology* or *U-topology formalism*.[28] It may be outlined briefly as follows.

Consider a set $\Re$ upon which a metric is defined and let i and j be elements of $\Re$. For each $i \in \Re$ and $\epsilon > 0$ a *ball-neighborhood* $U_\epsilon(i)$ of i is defined as the set of all elements j of $\Re$ whose distance to i, $d(i,j)$, is smaller than ϵ,

$$U_\epsilon(i) = \{j | d(i,j) < \epsilon\} \tag{1}$$

ϵ is called the radius of $U_\epsilon(i)$. Each ball-neighborhood $U_\epsilon(i)$ contains at least one element, namely i itself.

Any subset of $\Re$ is called a neighborhood U of i if for some value $\epsilon > 0$ it contains a ball-neighborhood $U_\epsilon(i)$ of i,

$$U \supseteq U_\epsilon(i) \tag{2}$$

Let $U(i)$ be the collection of all neighborhoods of i. Then, upon the set $\Re$ a neighborhood topology or a U-topology $\mathbf{T_U}$ is defined if each element i of $\Re$ is associated with a collection $U(i)$ of neighborhoods of i, with the proviso that the following conditions are fulfilled:

1. $i \in U$ for any $U \in U(i)$
2. If $U \in U(i)$ and $U' \supseteq U$, then also $U' \in U(i)$
3. If $U', U'' \in U(i)$, then also $U' \cap U'' \in U(i)$
4. For any $U \in U(i)$ there exists a $U' \in U(i)$, such that $U' \in U(j)$ for all $j \in U'$

The set $\Re$ together with its U-topology forms a topological space. If the above definitions are applied to the vertex-set V(G) of a simple connected graph G with the "standard" distance function on G, then it is easy to verify

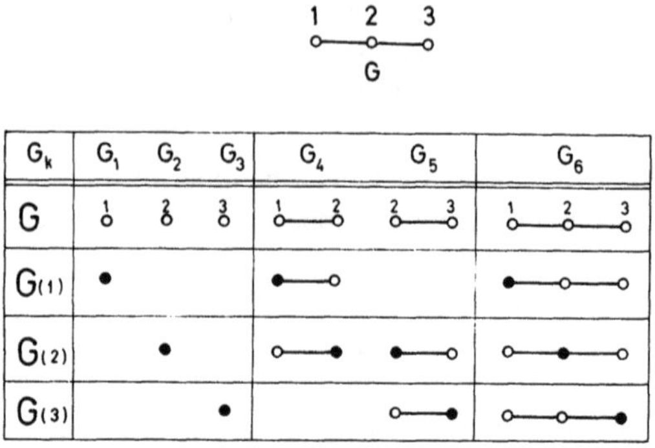

$$\underset{G}{\overset{\overset{\textstyle 1 \quad 2 \quad 3}{\circ\!-\!\!-\!\circ\!-\!\!-\!\circ}}{}}$$

G_k	G_1	G_2	G_3	G_4	G_5	G_6
G	$\overset{1}{\circ}$	$\overset{2}{\circ}$	$\overset{3}{\circ}$	$\overset{1}{\circ}\!-\!\overset{2}{\circ}$	$\overset{2}{\circ}\!-\!\overset{3}{\circ}$	$\overset{1}{\circ}\!-\!\overset{2}{\circ}\!-\!\overset{3}{\circ}$
$G_{(1)}$	•			●—○		●—○—○
$G_{(2)}$		•		○—●	●—○	○—●—○
$G_{(3)}$			•		○—●	○—○—●

FIGURE 3.   Neighborhoods and neighborhood collections of the vertices of a chain with three vertices.

that the collection $U(v)$ of all neighborhoods of a vertex v satisfies conditions (1) to (4) above. This is illustrated in Figure 3. The graph G considered is a chain with 3 vertices and 2 edges. The set $\mathcal{R}$ is identified by the vertex-set $V(G) = \{1,2,3\}$. The neighborhoods of the vertices are represented by the corresponding subgraphs $G_k \subseteq G$. There are, in total, six such subgraphs; together with the empty graph G_0 they form the collection of subgraphs **G**. The neighborhood collections $G(i)$ of the vertices i ($i = 1,2,3$) are subsets of **G**. They are shown in Figure 3. The vertex i in question is denoted by a black dot. The neighborhoods and the neighborhood collections of all vertices of G obey the axioms (1) to (4). This is easy to show. Therefore, they represent the U-topology of G.

From the above, one may conclude that each connected simple graph G is associated with a topological space $T(G)$. Since the molecular graph is uniquely derived from the molecular constitution, each molecule is associated with a topological space via its molecular graph. Since it has been shown[18] that the topological structure of a given set does not depend on the method used for its definition, the topological spaces derived for a given molecule by either the open set topology formalism.[19-23] or the neighborhood topology formalism[7,13,14,24-27] are equivalent. Therefore, the topology of a molecule can be described as an open set topology or a neighborhood topology defined upon the vertex-set of a molecular graph.

III. HÜCKEL GRAPHS

The hydrogen-suppressed undirected planar connected graphs depicting conjugated molecules are called *Hückel graphs*,[29] because they were first used by Hückel.[30] They describe the first neighbor interactions between π-electrons in conjugated systems. A Hückel graph corresponding to benzene is given in Figure 4.

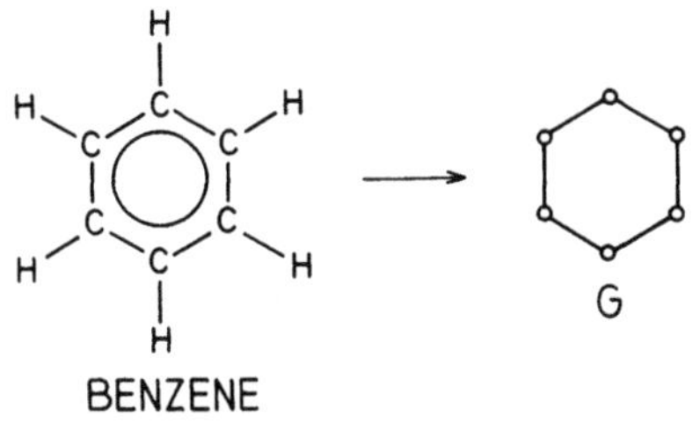

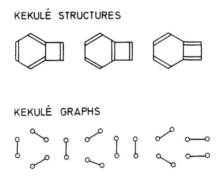

FIGURE 4. A Hückel graph G depicting benzene.

KEKULÉ STRUCTURES

KEKULÉ GRAPHS

FIGURE 5. Kekulé structures and Kekulé graphs of benzocyclobutadiene.

A. KEKULÉ GRAPHS

Conjugated molecules may be classified into two groups: *Kekuléan* and *non-Kekuléan* molecules, depending on whether or not they possess Kekulé structure(s). Kekulé structures may be represented by Kekulé graphs[31] or Kekuléan graphs as Sylvester called them.[32] A Kekulé graph is a disconnected graph consisting of two or more K_2 components. The number of K_2 components equals the number of double bonds in a Kekulé structure. In Figure 5 Kekulé structures and the corresponding Kekulé graphs belonging to benzocyclobutadiene are given.

In formal graph theory, a Kekulé graph is called a *1-factor*.[33] A 1-factor of a graph G is a spanning subgraph of G whose components are only K_2. An *n*-factor of G is a spanning subgraph of G whose components are only regular graphs of valency n. One says G is *n-factorable* if there exists at least one *n-factor* of G. Thus, Hückel graphs are 1-factorable, when they possess Kekulé graphs. A problem of interest in chemistry is the enumeration of Kekulé structures (1-factors).[34] Later on in this book (Chapter 8) the enumeration of Kekulé structures will be discussed.

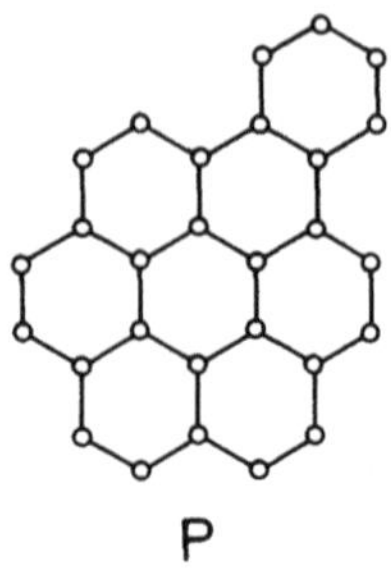

P

FIGURE 6.   A polyhex depicting benzo[a]coronene.

IV. POLYHEXES AND BENZENOID GRAPHS

Polyhexes are graphs which may be obtained by any combination of regular hexagons such that any two of its hexagons have exactly one common edge or are disjointed.[35] A polyhex P is planar if, and only if, it can be embedded in the plane without crossing bonds. A planar polyhex divides the plane into one infinite and a number of finite regions.[36] All vertices and edges which lie on the boundary of the infinite region form a *perimeter* of a polyhex. The *internal vertices* are those which do not belong to the perimeter. Their number is N_i and they all have the valency of 3. Therefore, the internal vertices are shared by three hexagons. The following relationships between N_i, N (the number of vertices in the polyhex P), M (the number of edges in P), and R (the number of rings in P) hold:

$$M - N = R - 1 \tag{3}$$

$$N + N_i = 4R + 2 \tag{4}$$

$$M + N_i = 5R + 1 \tag{5}$$

The term *polyhex hydrocarbons* is adopted for polycyclic conjugated hydrocarbons whose carbon skeletons may be depicted by polyhexes.[37] The simplest polyhex hydrocarbon is, of course, benzene. In Figure 6 a polyhex P corresponding to benzo[a]coronene is shown.

An important subset of the polyhex family are benzenoid graphs.[38] A polyhex is a benzenoid graph if, and only if, it is a 1-factorable graph.[39] A benzenoid graph B is a graph-theoretical representation of the carbon skeleton of a benzenoid hydrocarbon. The term 1-factorization signifies that the benzenoid hydrocarbon in question possesses Kekulé structure(s),[34] because only such polyhex hydrocarbons are expected to show similarity in their chemical behavior with benzene. For example, polyhex hydrocarbons without Kekulé structures are extremely unstable.[40] A polyhex P in Figure 6 is also an example

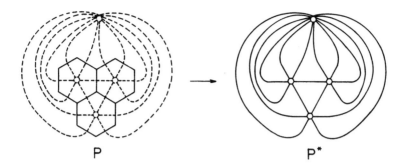

FIGURE 7.   The construction of the dual for a trihex representing phenalene.

of a benzenoid graph because it is a 1-factorable graph (this polyhex possesses 34 1-factors and consequently the corresponding hydrocarbon benzo[a]coronene 34 Kekulé structures).[41]

A. THE DUAL AND THE INNER DUAL OF A POLYHEX

The *dual* and its subgraph *inner dual* are very special types of graphs and play an important role in characterizing polyhexes.

Given a planar polyhex $P = (V, E)$, its *dual* $P^* = (V^*, E^*)$ can be constructed in the following way:[7,42,43] Place a vertex in the center of each hexagon of P and one vertex on the plane outside the perimeter of P and, if two hexagons have an edge e in common, join the corresponding vertices by an edge e^* crossing only e. The construction of the dual for a trihex depicting phenalene is given in Figure 7.

The number of vertices in P^* is given by $V^* = R + 1$, where R is the number of hexagons in P and the number of edges in P^* is equal to the number of edges in P, i.e., $E^* = E$. It should be noted that the dual of the dual of P is a graph isomorphic to P.

The *inner dual*[43] (sometimes also called *bual*)[44] is a special subgraph of the dual, because it does not contain the vertex corresponding to the infinite region of the plane. It is also a much simpler graph than the dual. The inner dual of polyhex $\mathscr{P} = \mathscr{P}(\mathscr{V}, \mathscr{E})$ can be obtained by connecting the centers of individual hexagons through the edge common to two hexagons.[7,35,41,43] The inner duals corresponding to two pentahexes: dibenzo[a,c]anthracene and benzo[e]pyrene, are shown in Figure 8.

The number of vertices in $\mathscr{P}$ is equal to the number of hexagons in P, i.e., $\mathscr{V} = R$, while the number of edges in $\mathscr{P}$ is equal to the number of adjacent hexagons in P, i.e., $\mathscr{E} = R - 1 + N_i$.

Polyhexes may be partitioned into two subsets: *catacondensed polyhexes* and *pericondensed polyhexes*. In catacondensed polyhexes (catahexes) there is no vertex at which three hexagons meet ($N_i = 0$), whereas in pericondensed polyhexes (perihexes) there is at least one such vertex ($N_i > 0$). Inner duals belonging to catahexes and those belonging to perihexes differ considerably.

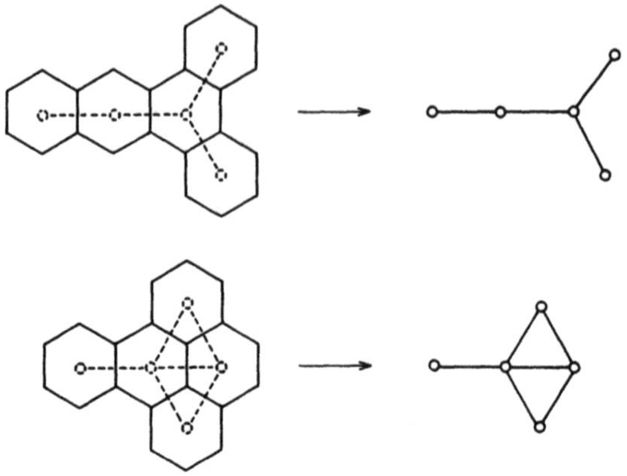

FIGURE 8. The inner duals of polyhexes depicting dibenzo[a,c] anthracene and benzo[e]pyrene.

In the former case the corresponding inner dual *always* has the form of a tree and in the latter case it contains cycles. This can also be seen in two examples shown in Figure 8: the inner dual of dibenzo[a,c]anthracene, which is a catacondensed polyhex (benzenoid) hydrocarbon, is a branched tree and the inner dual of benzo[e]pyrene, which is a pericondensed polyhex (benzenoid) hydrocarbon, contains cycles. The above property of inner duals was used for differentiating between cata- and pericondensed polyhex hydrocarbons.[45] However, the inner duals are not very useful for differentiating isomeric polyhexes, because different isomeric polyhexes may possess the same inner dual. This fact led to the introduction of the concept of *dualist*.[45]

B. THE DUALIST OF A POLYHEX

The concept of dualist has been introduced in an attempt to accommodate the structural differences between isomeric polyhexes with the isomorphic inner duals.[45] Before the concept of dualist was introduced, the same concept, albeit not identified as such, was used as a shorthand notation for benzenoid hydrocarbons.[46] The dualist $\mathscr{D}$* of a polyhex P can be simply constructed by placing a vertex in the center of each hexagon of P and then connecting those vertices which are in the adjacent hexagons. To this point the construction of the inner dual and the dualist is identical. However, the dualist unlike the inner dual preserves the geometric information on the direction of ring annelation in P. The dualists of four unbranched catacondensed tetrahexes are shown in Figure 9.

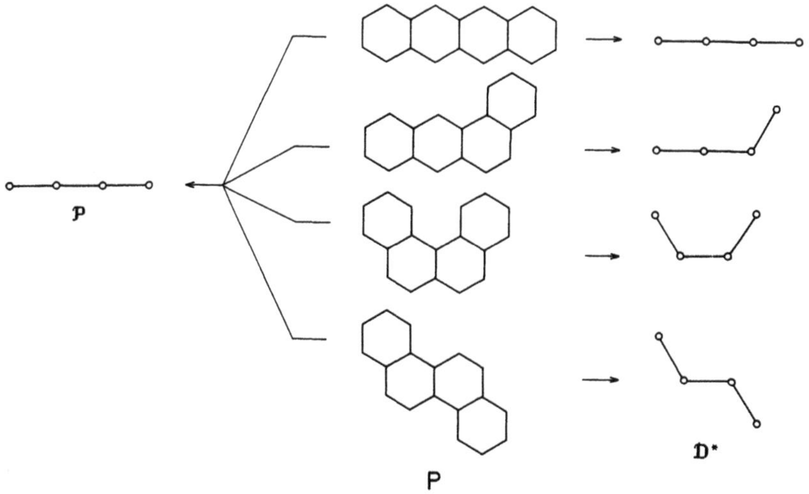

FIGURE 9. The inner dual and dualists corresponding to unbranched catacondensed tetrahexes.

A dualist is *not* a graph in the strict graph-theoretical sense,[33] because in the case of dualist, angles are also important. Therefore, the dualist can be defined[47] as $\mathcal{D}^* = \mathcal{D}^* (V^*, \mathcal{E}^*, \Theta)$, where the cardinality of the vertex-set V^* is R, the cardinality of the edge-set $\mathcal{E}^*$ is $R - 1 + N_i$ and Θ is the set of angles between incident edges.

The dualist is a unique representation of a polyhex. In other words, two or more nonisomorphic polyhexes cannot possess the same dualist. This property of dualist is a basis for its use in characterization and classification of polyhexes.[5,7,35,41,43,45-49] Planar polyhexes may be classified into the following four classes:

1. *Catahexes,* when the dualists are trees
2. *Perihexes,* when the dualists contain 3-membered cycles and no cycles of other sizes
3. *Coronahexes,* when the dualists contain a single cycle or several cycles equal to or larger than an 8-membered cycle
4. *Complex polyhexes,* whose constituent parts are built up from at least two classes from the above.

One example from each class and the corresponding dualists are shown in Figure 10.

C. FACTOR GRAPHS

Factor graphs[7,50] (which are also called Kekulé subspace graphs) are related to benzenoid hydrocarbons (and other Kekuléan hydrocarbons) by depicting the adjacency relationship of the formal CC double bonds in their

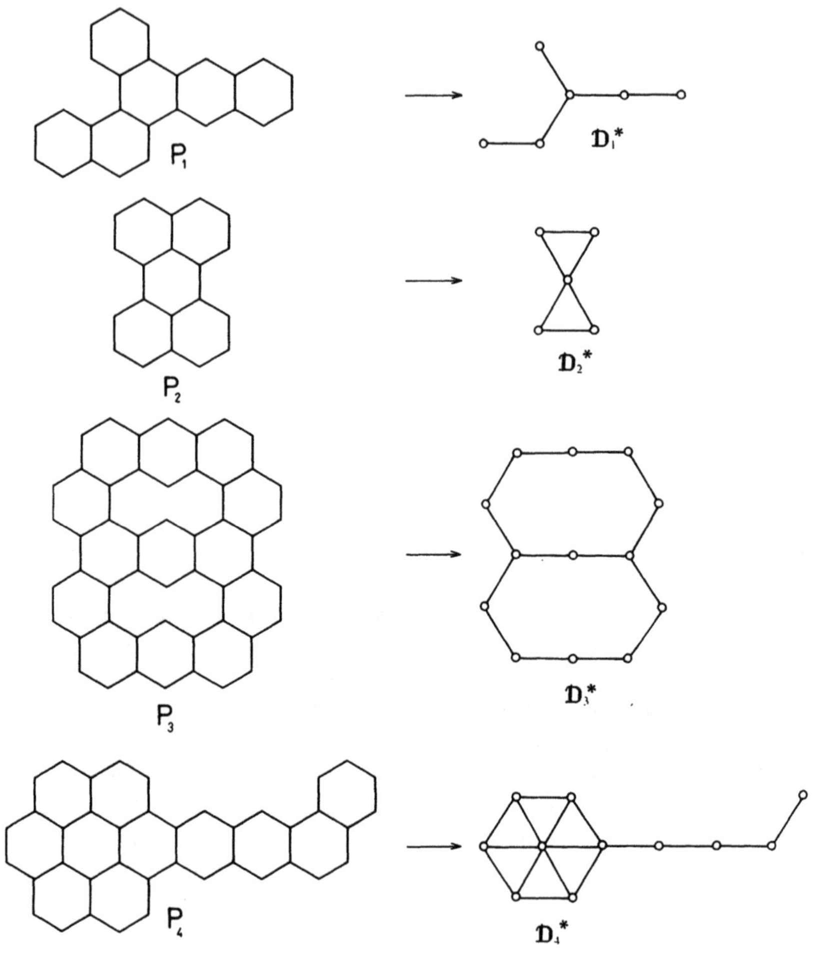

FIGURE 10. Examples of catahexes (dibenzo[c,f]tetraphene, P_1), perihexes (perylene, P_2), coronahexes (double coronoid with 13 hexagons, P_3), and complex polyhexes (tetrapheno[9,10-a]coronene, P_4), and the corresponding dualists ($\mathcal{D}_1^*$, $\mathcal{D}_2^*$, $\mathcal{D}_3^*$, $\mathcal{D}_4^*$).

Kekulé valence structures.[51] In factor graphs vertices represent double bonds and edges the couplings of double bonds separated by a single bond in a Kekulé valence structure. Factor graphs corresponding to Kekulé structures of pyrene are shown in Figure 11.

Factor graphs dramatically illustrate the differences between individual Kekulé valence structures. Symmetry-related Kekulé structures, of course, produce isomorphic factor graphs (see Figure 11).

A practical aspect of factor graphs is their use in the interpretation of electronic spectra of ions of benzenoid hydrocarbons.[51] Various other applications of factor graphs are described by El-Basil and co-workers.[50,52-54]

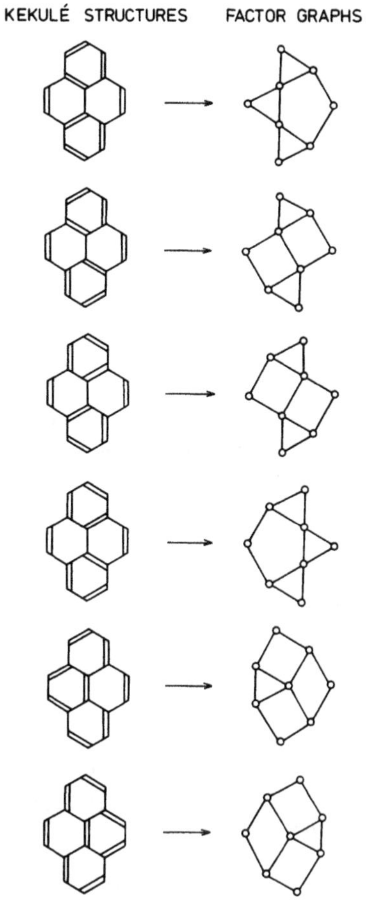

KEKULÉ STRUCTURES   FACTOR GRAPHS

FIGURE 11. Factor graphs (Kekulé subspace graphs) corresponding to Kekulé valence structures of pyrene.

V. WEIGHTED GRAPHS

A. VERTEX- AND EDGE-WEIGHTED GRAPHS

Heteroconjugated molecules may be represented by *vertex-* and *edge-weighted graphs*.[7,27,55-60] A vertex- and edge-weighted graph G_{VEW} is a graph which has one or more of its vertices and edges distinguished in some way from the rest of the vertices and edges. These vertices and edges of different "type" are weighted. Their weights are identified by parameters h (weighted vertices) and k (weighted edges) for heteroatoms and heterobonds, respectively. Weighted vertices are visually identified by loops (one-cycles) of weight h, and weighted edges by heavy lines of weight k. In Figure 12 a vertex- and edge-weighted graph G_{VEW} corresponding to thiophene is given.

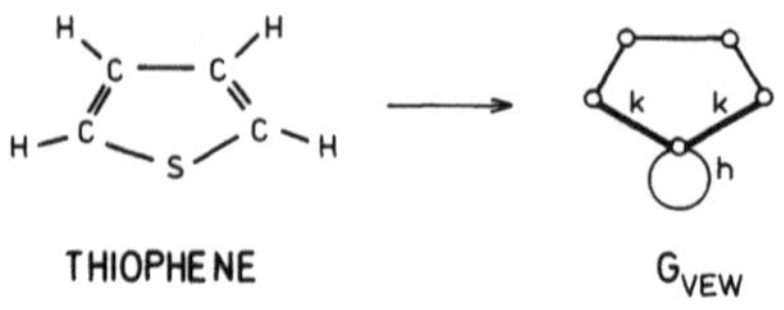

THIOPHENE G_{VEW}

FIGURE 12.   A vertex- and edge-weighted graph G_{VEW} depicting thio-phene.

Simpler classes of weighted graphs are the *vertex-weighted graphs* G_{VW} and the *edge-weighted graphs* G_{EW}. A vertex-weighted graph G_{VW} is a graph which has one or more of its vertices weighted.[55-57] Originally these graphs were named *rooted graphs*,[55] a name which is not quite appropriate for this particular use. Besides the term *rooted* is already reserved for a class of trees (e.g., rooted trees).[61] An edge-weighted graph G_{EW} is a graph in which one or more of its edges are weighted.[58] Although weighted graphs are *not* graphs in a strict sense they are of use in applied graph theory.[62-64]

Among the weighted graphs one may include the *quantum-chemical graphs*[27,65] and *reaction graphs*.[66-75] Quantum-chemical graphs consist of vertices representing atomic orbitals and edges representing various orbital interactions. In reaction graphs, the vertices symbolize chemical species and the edges represent chemical transformations involving these species.

B. MÖBIUS GRAPHS

In recent years the use of the *Hückel-Möbius concept* has been very popular in studying transition states of certain pericyclic reactions, e.g., electrocyclic closures of polyenes.[76-82] Möbius systems are defined as cyclic arrays of orbitals with *one* or, more generally, with an *odd* number of phase dislocations resulting from the negative overlap between the adjacent $2p_z$-orbitals of different signs.[83] Similarly, a more general definition of Hückel systems may also be given. Ordinarily Hückel systems are defined as those in which there is no sign inversion among the adjacent $2p_z$-orbitals. A more general definition is as follows: Hückel systems are those in which there is an *even* number of sign inversions among the adjacent $2p_z$-orbitals.

Möbius systems may be visualized by the use of a *Möbius strip*, which was introduced by Möbius.[84] Such a strip is obtained if one short edge of a narrow sheet of paper is half-rotated and then joined with the other short edge of a band. This is shown in Figure 13.

When the $2p_z$-orbitals are drawn on the narrow sheet of paper, which is twisted into a Möbius strip with one half-twist, the Möbius π-system is created in a pictorial way. Möbius systems are also called *anti-Hückel systems*,[79] because their stabilities are governed by rules opposite to those valid for Hückel systems.

Möbius systems may be depicted by Möbius graphs[29,85,86] and are denoted by $G_{Mö}$. In mathematical literature, Möbius graphs are known as the *signed*

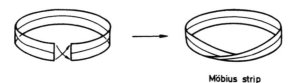

Möbius strip

FIGURE 13. A generation of a Möbius strip.

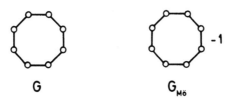

G $\qquad$ $G_{Mö}$

FIGURE 14. A Hückel graph G and a Möbius graph $G_{Mö}$ corresponding to Hückel cyclooctatetraene and Möbius cyclooctatetraene.

graphs.[87] Möbius graphs should also be viewed as kinds of edge-weighted graphs with the weight of some (odd number of) edges in these graphs being -1. A Möbius graph $G_{Mö}$ is defined as $[V(G_{Mö}), E^+(G_{Mö}), E^-(G_{Mö})]$, where $V(G_{Mö})$ is the vertex-set of $G_{Mö}$, and $E^+(G_{Mö}) = E^+$ and $E^-(G_{Mö}) = E^-$ are the edge-subsets of $G_{Mö}$ containing as elements edges representing either "$+1$" weights (E^+) or "-1" weights (E^-) between the pairs of adjacent vertices of $G_{Mö}$. The union of subsets E^+ and E^- represents, of course, the edge-set of $G_{Mö}$. A Hückel graph G and a Möbius graph $G_{Mö}$ corresponding to Hückel cyclooctatetraene and Möbius cyclooctatetraene are given in Figure 14.

REFERENCES

1. **Lederberg, J.**, *Proc. Natl. Acad. Sci. U.S.A.*, 53, 134, 1965.
2. **Prelog, V.**, Nobel Lecture, Stockholm, December 12, 1975.
3. **Turro, N. J.**, *Angew. Chem. Int. Ed., Engl.*, 25, 882, 1986.
4. **Prelog, V.**, *Science*, 193, 17, 1976.
5. **Balaban, A. T.**, *J. Chem. Inf. Comput. Sci.*, 25, 334, 1985.
6. **Spialter, L.**, *J. Chem. Doc.*, 4, 261, 1964; 4, 269, 1964.
7. **Gutman, I. and Polansky, O. E.**, *Mathematical Concepts in Organic Chemistry*, Springer-Verlag, Berlin, 1986.
8. **Balaban, A. T., Ed.**, *Chemical Applications of Graph Theory*, Academic Press, London, 1976.
9. **Kvasnička, V.**, *Coll. Czech. Chem. Commun.*, 48, 2097, 1983; 48, 2119, 1983.
10. **King, R. B., Ed.**, *Chemical Applications of Topology and Graph Theory*, Elsevier, Amsterdam, 1983.

11. **Balasubramanian K.**, *Chem. Rev.*, 85, 599, 1985.
12. **King, R. B. and Rouvray, D. H., Eds.**, *Graph Theory and Topology in Chemistry*, Elsevier, Amsterdam, 1987.
13. **Polansky, O. E.**, in *Atomic Concept and the Concept of Molecular Structure*, Maksić, Z. B., Ed., Springer-Verlag, Berlin, 1990, 29.
14. **Zander, M.**, *Z. Naturforsch.*, 45a, 1041, 1990; *Topics Curr. Chem.*, 153, 101, 1990.
15. **Trinajstić, N.**, *Rep. Mol. Theory*, 1, 185, 1990.
16. **Prelog, V.**, *J. Mol. Catalysis*, 1, 163, 1975.
17. **Alexandroff, P.**, *Elementary Concepts of Topology*, Dover, New York, 1961.
18. **Dugundji, J.**, *Topology*, Allyn & Bacon, Boston, 1966.
19. **Simmons, H. E. and Merrifield, R. E.**, *Proc. Natl. Acad. Sci. U.S.A.*, 74, 2616, 1977. *Chem. Phys. Lett.*, 62, 235, 1979.
20. **Merrifield, R. E. and Simmons, H. E.**, *Theoret. Chim. Acta*, 55, 55, 1980.
21. **Merrifield, R. E. and Simmons, H. E.**, in *Chemical Applications of Topology and Graph Theory*, King, R. B., Ed., Elsevier, Amsterdam, 1983, 1.
22. **Merrifield, R. E. and Simmons, H. E.**, *Topological Methods in Chemistry*, Wiley-Interscience, New York, 1989.
23. **Merrifield, R. E. and Simmons, H. E.**, *Chem. Phys. Lett.*, 62, 235, 1979.
24. **Polansky, O. E.**, *Z. Naturforsch.*, 41a, 560, 1986.
25. **Polansky, O. E.**, in *Mathematics and Computational Concepts in Chemistry*, Trinajstić, N., Ed., Horwood, Chichester, 1986, 262.
26. **Polansky, O. E.**, in *MATH/CHEM/COMP 1988*, Graovac, A., Ed., Elsevier, Amsterdam, 1989, 65.
27. **Polansky, O. E.**, in *Chemical Graph Theory: Introduction and Fundamentals*, Bonchev, D. and Rouvray, D. H., Eds., Abacus Press/Gordon & Breach, New York, 1991, 41.
28. **Hausdorff, F.**, *Grundzüge der Mengenlehre*, Veit, Leipzig, 1914.
29. **Graovac, A. and Trinajstić, N.**, *Croat. Chem. Acta*, 47, 95, 1975.
30. **Hückel, E.**, *Z. Phys.*, 60, 204, 1931.
31. **Cvetković, D. M., Gutman, I., and Trinajstić, N.**, *Chem. Phys. Lett.*, 16, 614, 1972.
32. **Sylvester, J. J.**, *Am. J. Math.*, 1, 64, 1878.
33. **Harary, F.**, *Graph Theory*, Addison-Wesley, Reading, MA, 1971, 2nd printing.
34. **Cyvin, S. J. and Gutman, I.**, *Kekulé Structures in Benzenoid Hydrocarbons*, Springer-Verlag, Berlin, 1988.
35. **Balaban, A. T.**, in *Chemical Applications of Graph Theory*, Balaban, A. T., Ed., Academic Press, London, 1976, 63.
36. **Cvetković, D. M., Gutman, I., and Trinajstić, N.**, *J. Chem. Phys.*, 61, 2700, 1974.
37. **Trinajstić, N.**, *J. Math. Chem.*, 5, 171, 1990.
38. **Graovac, A., Gutman, I., and Trinajstić, N.**, *Topological Approach to the Chemistry of Conjugated Molecules*, Springer-Verlag, Berlin, 1977.
39. **Knop, J. V., Müller, W. R., Szymanski, K., and Trinajstić, N.**, *J. Comput. Chem.*, 7, 547, 1986.
40. **Clar, E., Kemp, W., and Stewart, D. G.**, *Tetrahedron*, 3, 325, 1958.
41. **Knop, J. V., Müller, W. R., Szymanski, K., and Trinajstić, N.**, *Computer Generation of Certain Classes of Molecules*, SKTH, Zagreb, 1985.
42. **Gutman, I., Mallion, R. B., and Essam, J. V.**, *Mol. Phys.*, 50, 859, 1983.
43. **Nikolić, S., Trinajstić, N., Knop, J. V., Müller, W. R., and Szymanski, K.**, *J. Math. Chem.*, 4, 357, 1990.
44. **Hall, G. G.**, *Int. J. Math. Educ. Sci. Technol.*, 4, 233, 1973.
45. **Balaban, A. T. and Harary, F.**, *Tetrahedron*, 24, 2505, 1968.
46. **Smith, F. T.**, *J. Chem. Phys.*, 34, 793, 1961.
47. **Polansky, O. E. and Rouvray, D. H.**, *Math. Chem. (Mülheim/Ruhr)*, 2, 63, 1976.
48. **Balaban, A. T.**, *Tetrahedron*, 25, 2949, 1969.
49. **Balaban, A. T.**, *Pure Appl. Chem.*, 54, 1075, 1982.

50. **El-Basil, S.,** *Int. J. Quantum Chem.,* 21, 771, 1982; 21, 779, 1982; 21, 793, 1982.
51. **Joela, H.,** *Theoret. Chim. Acta,* 39, 241, 1975.
52. **El-Basil, S. and Osman, A. N.,** *Int. J. Quantum Chem.,* 24, 571, 1983.
53. **El-Basil, S. and Shalabi, A. S.,** *Math. Chem. (Mülheim/Ruhr),* 17, 11, 1985.
54. **El-Basil, S., Botros, S., and Ismail, M.,** *Math. Chem. (Mülheim/Ruhr),* 17, 45, 1985.
55. **Mallion, R. B., Schwenk, A. J., and Trinajstić, N.,** *Croat. Chem. Acta,* 46, 71, 1974.
56. **Mallion, R. B., Trinajstić, N., and Schwenk, A. J.,** *Z. Naturforsch.,* 29a, 1481, 1974.
57. **Mallion, R. B., Schwenk, A. J., and Trinajstić, N.,** in *Recent Advances in Graph Theory,* Fiedler, M., Ed., Academia, Prague, 1975, 345.
58. **Graovac, A., Polansky, O. E., Trinajstić, N., and Tyutyulkov, N.,** *Z. Natuforsch.,* 30a, 1696, 1975.
59. **Rigby, M. J., Mallion, R. B., and Day, A. C.,** *Chem. Phys. Lett.,* 51, 178, 1977; erratum *Chem. Phys. Lett.,* 53, 418, 1978.
60. **Mallion, R. B.,** in *Applications of Combinatorics,* Wilson, R. J., Ed., Shiva Publishing, Nantwich, Cheshire, U.K., 1982, 87.
61. **Harary, F. and Palmer, E. M.,** *Graphical Enumeration,* Academic Press, New York, 1973, 51.
62. **Hakimi, S. L. and Yau, S. S.,** *Q. Appl. Math.,* 22, 305, 1964.
63. **Mayeda, W.,** *Graph Theory,* John Wiley & Sons, New York, 1972.
64. **Johnson, D. E. and Johnson, J. R.,** *Graph Theory with Engineering Applications,* Ronald Press, New York, 1972.
65. **Polansky, O. E.,** *Math. Chem. (Mülheim/Ruhr),* 1, 183, 1975.
66. **Balaban, A. T. and Fărcaşiu, D.,** *J. Am. Chem. Soc.,* 89, 967, 1958.
67. **Balaban, A. T., Fărcaşiu, D., and Bănică, R.,** *Rev. Roum. Chim.,* 11, 1025, 1966.
68. **Corey, E. J. and Wipke, W. T.,** *Science,* 166, 178, 1969.
69. **Bonchev, D., Kamenski, D., and Temkin, O. N.,** *J. Comput. Chem.,* 3, 95, 1982.
70. **Kvasnička, V., Kratochvíl, M., and Koča, J.,** *Coll. Czech. Chem. Commun.,* 48, 2284, 1983.
71. **Randić, M., Klein, D. J., Katović, V., Oakland, D. O., Seitz, W. A., and Balaban, A. T.,** in *Graph Theory and Topology in Chemistry,* King, R. B. and Rouvray, D. H., Eds., Elsevier, Amsterdam, 1987, 266.
72. **Bonchev, D., Kamenski, D., and Temkin, O. N.,** *J. Math. Chem.,* 1, 345, 1987.
73. **Gimarc, B. M. and Ott, J. J.,** in *Graph Theory and Topology in Chemistry,* King, R. B. and Rouvray, D. H., Eds., Elsevier, Amsterdam, 1987, 285.
74. **Temkin, O. N., Bruk, L. G. and Bonchev, D.,** *Theoret. Exp. Khim.,* 24, 275, 1988.
75. **Gimarc, B. M., Dai, B., Warren, S. C., and Ott, J. J.,** *J. Am. Chem. Soc.,* 112, 2597, 1990.
76. **Dewar, M. J. S.,** *Angew. Chem. Int. Ed. Engl.,* 10, 761, 1971.
77. **Zimmerman, H. E.,** *Acc. Chem. Res.,* 4, 272, 1971.
78. **Shen, K.-W.,** *J. Chem. Educ.,* 50, 238, 1973.
79. **Smith, W. B.,** *Molecular Orbital Methods in Organic Chemistry: HMO and PMO,* Marcel Dekker, New York, 1974.
80. **Zimmerman, H. E.,** *Quantum Mechanics for Organic Chemists,* Academic Press, New York, 1975.
81. **Yates, K.,** *Hückel Molecular Orbital Theory,* Academic Press, New York, 1978.
82. **Klein, D. J. and Trinajstić, N.,** *J. Am. Chem. Soc.,* 106, 8050, 1984.
83. **Heilbronner, E.,** *Tetrahedron Lett.,* 1923, 1964.
84. **Möbius, A. F.,** *Ber. K. Säches. Wiss. Leipzig Math.-Phys. Cl.,* 17, 31, 1865.
85. **Graovac, A. and Trinajstić, N.,** *J. Mol. Struct.,* 30, 416, 1976.
86. **Gutman, I.,** *Z. Naturforsch.,* 33a, 214, 1978.
87. **Roberts, F. S.,** in *Applications of Graph Theory,* Wilson, R. J. and Beineke, L. W., Eds., Academic Press, London, 1979, 255.

Chapter 4

GRAPH-THEORETICAL MATRICES

Graphs, adequately labeled, may be associated with several matrices.[1] A graph G is *labeled* if a certain numbering of vertices of G is introduced. Here two graph-theoretical matrices, i.e., the adjacency matrix and the distance matrix will be discussed. They are also sometimes referred to as topological matrices.[2] These matrices may be used for identifying certain properties of graphs, which would not otherwise easily emerge.

I. THE ADJACENCY MATRIX

The most important matrix representation of a graph G is the *vertex-adjacency matrix* $\mathbf{A} = \mathbf{A}(G)$. This matrix is also of importance in chemistry and physics.[2-7]

The vertex-adjacency matrix $\mathbf{A}(G)$ of a labeled connected graph G with N vertices is the square $N \times N$ symmetric matrix which contains information about the internal connectivity of vertices in G. It is defined as,

$$(\mathbf{A})_{ij} = \begin{cases} 1 & \text{if vertices } v_i \text{ and } v_j \text{ are adjacent} \\ 0 & \text{otherwise} \end{cases} \tag{1}$$

$$(\mathbf{A})_{ii} = 0 \tag{2}$$

Therefore, a nonzero entry appears in $\mathbf{A}(G)$ only if an edge connects vertices *i* and *j*. For example, the following vertex-adjacency matrix can be constructed for a labeled graph G (see Figure 1).

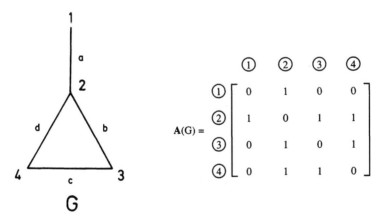

FIGURE 1.   A vertex- and edge-labeled graph G.

The adjacency matrix is symmetrical about the principal diagonal. Therefore, the transpose of the adjacency matrix **A** leaves the adjacency matrix unchanged,

$$A^T(G) = A(G) \tag{3}$$

This transpose A^T is formed by interchanging rows and columns of the matrix **A**.

For the (vertex- and edge-) weighted graphs G_{VEW}, Equations 1 and 2 should be modified,[8-10]

$$(\mathbf{A})_{ij} = \begin{cases} 1 & \text{if vertices } v_i \text{ and } v_j \text{ are adjacent} \\ k & \text{if vertices } v_i \text{ and } v_j \text{ are adjacent} \\ & \text{and edge } (v_i, v_j) \text{ is } k\text{-weighted} \\ 0 & \text{otherwise} \end{cases} \tag{4}$$

$$(\mathbf{A})_{ii} = \begin{cases} h & \text{if there is a loop of the weight } h \text{ at vertex } i \text{ in } G_{VEW} \\ 0 & \text{otherwise} \end{cases} \tag{5}$$

For example, the following vertex-adjacency matrix can be constructed for a labeled vertex- and edge-weighted graph G_{VEW} (see Figure 2).

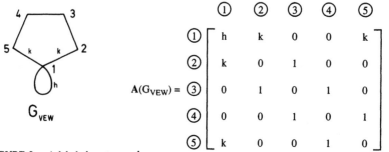

FIGURE 2.   A labeled vertex- and edge-weighted graph G_{VEW}.

Equations 1 and 2 should also be modified for Möbius graphs,[11-13]

$$(\mathbf{A})_{ij} = \begin{cases} 1 & \text{if vertices } v_i \text{ and } v_j \text{ are adjacent} \\ -1 & \text{if vertices } v_i \text{ and } v_j \text{ are adjacent} \\ & \text{and edge } (v_i, v_j) \text{ is ``}-1\text{''-weighted} \\ 0 & \text{otherwise} \end{cases} \tag{6}$$

$$(\mathbf{A})_{ii} = 0 \tag{7}$$

For example, the following vertex-adjacency matrix can be constructed for a four-membered Möbius cycle $C_4{}^{Mö}$ (see Figure 3).

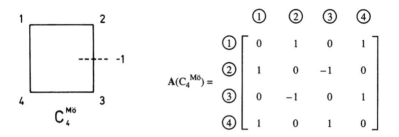

FIGURE 3.   A labeled four-membered Möbius cycle $C_4{}^{Mö}$.

The *edge-adjacency matrix* of a graph G, $^E A = {}^E A(G)$, is determined by the adjacencies of edges in G. It is very rarely used.[2] The edge-adjacency matrix is defined as

$$(^E A)_{ij} = \begin{cases} 1 & \text{if edges } e_i \text{ and } e_j \text{ are adjacent} \\ 0 & \text{otherwise} \end{cases} \tag{8}$$

$$(^E_A)_{ii} = 0 \tag{9}$$

For example, the following edge-adjacency matrix can be constructed for a labeled graph G in Figure 1:

$$^E A(G) = \begin{array}{c} \\ \text{ⓐ} \\ \text{ⓑ} \\ \text{ⓒ} \\ \text{ⓓ} \end{array} \begin{array}{cccc} \text{ⓐ} & \text{ⓑ} & \text{ⓒ} & \text{ⓓ} \\ \begin{bmatrix} 0 & 1 & 0 & 1 \\ 1 & 0 & 1 & 1 \\ 0 & 1 & 0 & 1 \\ 1 & 1 & 1 & 0 \end{bmatrix} \end{array}$$

Although both the vertex-adjacency matrix and the edge-adjacency matrix reflect the topology of a molecule, they differ in their structure. However, it should be noted while the vertex-adjacency matrix uniquely determines the graph, the edge-adjacency matrix does not. In other words, there are known nonisomorphic graphs with identical edge-adjacency matrices. A pair of such nonisomorphic graphs is shown in Figure 4. The corresponding edge-adjacency matrix is given by

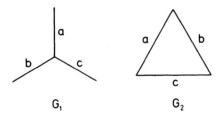

FIGURE 4. A pair of nonisomorphic graphs (G_1 and G_2) which possess the identical edge-adjacency matrices.

$$^EA(G_1) = \begin{bmatrix} 0 & 1 & 1 \\ 1 & 0 & 1 \\ 0 & 1 & 0 \end{bmatrix} = {}^EA(G_2) \tag{10}$$

Graphs G_1 and G_2 have obviously different vertex-adjacency matrices.

The edge-adjacency matrix of a graph G is identical to the vertex-adjacency matrix of the line graph L(G) of G. This must be so because the edges in G are replaced by vertices in L(G) as discussed in Chapter 2, Section X.

For some graphs these matrices are even identical. For example, $A(G) \equiv {}^EA(G)$ for cycles. Note, from now on when the term adjacency matrix will be used, it will always refer to the vertex-adjacency matrix.

The actual structure of the adjacency matrix depends on the numbering of vertices in a graph G. For example, though the differently labeled graphs G_1 and G_2 are clearly identical (see Figure 5), the corresponding adjacency matrices, $A(G_1)$ and $A(G_2)$, are not:

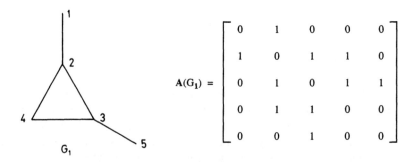

$$A(G_1) = \begin{bmatrix} 0 & 1 & 0 & 0 & 0 \\ 1 & 0 & 1 & 1 & 0 \\ 0 & 1 & 0 & 1 & 1 \\ 0 & 1 & 1 & 0 & 0 \\ 0 & 0 & 1 & 0 & 0 \end{bmatrix}$$

FIGURE 5. Two different labelings of vertices in a graph G.

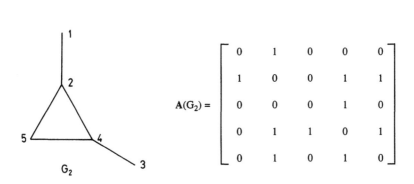

$$A(G_2) = \begin{bmatrix} 0 & 1 & 0 & 0 & 0 \\ 1 & 0 & 0 & 1 & 1 \\ 0 & 0 & 0 & 1 & 0 \\ 0 & 1 & 1 & 0 & 1 \\ 0 & 1 & 0 & 1 & 0 \end{bmatrix}$$

FIGURE 5 (continued).

Graphs G_1 and G_2 can easily be recognized as identical. However, a question of how in practice to recognize whether two complex graphs are identical or not, remains as one of the grand unsolved problems of formal graph theory.[14]

The diverse numberings of vertices of a given graph can be described by means of *permutations*. If G_1 is identified to the identity permutation, then G_2 is identified to that permutation P which carries 1,2,3,4,5 to 1,2,4,5,3. If the permutation P is acting on the columns of $A(G_1)$, i.e., the third column is placed on the position of the fifth, the fourth on the place of the third, and the fifth on the fourth, while the first and the second do not change their places, and if this is followed by the same permutation procedure on the rows, then $A(G_2)$ is obtained. This procedure may be condensed in the matrix form by use of the permutation matrices. A *permutation matrix* P is a square $N \times N$ matrix defined as

$$(\mathbf{P})_{ij} = \begin{cases} 1 & \text{if a mapping of vertices } i \rightarrow j \text{ is induced by the permutation} \\ 0 & \text{otherwise} \end{cases}$$

(11)

Therefore, in each column and in each row of **P** there is one element equal to 1, and all the other elements are equal to zero. In the example considered above, the permutation matrix **P** is given by

$$\mathbf{P} = \begin{bmatrix} 1 & 0 & 0 & 0 & 0 \\ 0 & 1 & 0 & 0 & 0 \\ 0 & 0 & 0 & 1 & 0 \\ 0 & 0 & 0 & 0 & 1 \\ 0 & 0 & 1 & 0 & 0 \end{bmatrix}$$

If $A(G_1)$ is multiplied by P from the right side, the columns of $A(G_1)$ will be permuted:

$$A(G_1) \cdot P = \begin{bmatrix} 0 & 1 & 0 & 0 & 0 \\ 1 & 0 & 0 & 1 & 1 \\ 0 & 1 & 1 & 0 & 1 \\ 0 & 1 & 0 & 1 & 0 \\ 0 & 0 & 0 & 1 & 0 \end{bmatrix}$$

The multiplication of the matrix $A(G_1) \cdot P$ by P^T from the left side leads to the permutation of the rows with the final result summarized as

$$P^T \cdot A(G_1) \cdot P = A(G_2) \tag{12}$$

where P^T is the transpose of P. Since the permutation matrix is a real unitary matrix,

$$P^T \cdot P = I \tag{13}$$

where I is the *unit matrix*. Thence Equation 12 may be rewritten as,

$$P^{-1} \cdot A(G_1) \cdot P = A(G_2) \tag{14}$$

where P^{-1} is the inverse of P.

The result stated as Equation 14 holds generally: matrices $A(G_1)$ and $A(G_2)$, which correspond to two labelings of the same graph G, are related by a similarity transformation.

Transformation (14) is called an *isomorphism* between graph G_1 [given by $V(G_1) = \{1,2,3,4,5\}$ and $E(G_1) = \{(1,2),(2,3),(2,4),(3,4),(3,5)\}$] and graph G_2 [given by $V(G_2) = V(G_1)$ and $E(G_2) = \{(1,2),(2,4),(2,5),(3,4),(4,5)\}$]. Note that G_1 and G_2 are isomorphic graphs, but not identical graphs because $E(G_1) \neq E(G_2)$.

If there exists such a transformation (14) between one and the same graph, i.e., an isomorphism of the graph onto itself,

$$P^{-1} \cdot A(G) \cdot P = A(G) \tag{15}$$

this transformation is called an *automorphism* of the graph G (actually at least one exists always: the identical one). Graph G_1 has only one nonidentical automorphism, namely $P_1 : 4 \leftrightarrow 4, 2 \leftrightarrow 3, 1 \leftrightarrow 5$. Similarly, graph G_2 has also only one nonidentical automorphism, namely $P_2 : 5 \leftrightarrow 5, 2 \leftrightarrow 4, 1 \leftrightarrow 3$.

Clearly in this case, P_i^2 = identity. In general the set of all automorphisms of a graph forms a group.

A. THE ADJACENCY MATRIX OF A BIPARTITE GRAPH

If a bipartite graph G is so labeled that 1,2, . . . ,s are starred and s + 1, s + 2, . . . , s + u (= N) unstarred vertices, then

$$(\mathbf{A})_{ij} = 0 \quad \text{for } 1 \le i,j \le s \text{ and } s + 1 \le i,j \le s + u \tag{16}$$

because two vertices of the same color are never connected. The consequence of the above is the *block-form* of the adjacency matrix of a conveniently labeled bipartite graph,[15]

$$
A(G) =
\begin{array}{c c}
 & * \quad\quad 0 \\
\begin{array}{c} * \\[12pt] 0 \end{array} &
\begin{bmatrix} \mathbf{0} & \mathbf{B} \\ \mathbf{B}^\mathrm{T} & \mathbf{0} \end{bmatrix}
\end{array}
\tag{17}
$$

where **B** is a submatrix of **A**(G) with dimensions $s \times u$ and $\mathbf{B}^\mathrm{T}$ is its transpose. For example, the following adjacency matrix in the block-form can be obtained for a labeled bipartite graph G (see Figure 6).

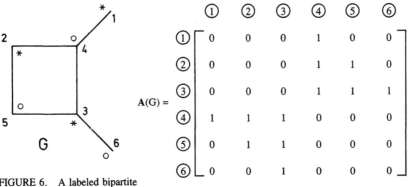

$$
A(G) =
\begin{array}{c}
\begin{array}{cccccc} ① & ② & ③ & ④ & ⑤ & ⑥ \end{array} \\
\begin{array}{c} ① \\ ② \\ ③ \\ ④ \\ ⑤ \\ ⑥ \end{array}
\begin{bmatrix}
0 & 0 & 0 & 1 & 0 & 0 \\
0 & 0 & 0 & 1 & 1 & 0 \\
0 & 0 & 0 & 1 & 1 & 1 \\
1 & 1 & 1 & 0 & 0 & 0 \\
0 & 1 & 1 & 0 & 0 & 0 \\
0 & 0 & 1 & 0 & 0 & 0
\end{bmatrix}
\end{array}
$$

FIGURE 6. A labeled bipartite graph G.

where the submatrix **B** is given by

$$
B(G) =
\begin{array}{c}
\begin{array}{ccc} 0 & 0 & 0 \end{array} \\
\begin{array}{c} * \\ * \\ * \end{array}
\begin{bmatrix}
1 & 0 & 0 \\
1 & 1 & 0 \\
1 & 1 & 1
\end{bmatrix}
\end{array}
$$

B. THE RELATIONSHIP BETWEEN THE ADJACENCY MATRIX AND THE NUMBER OF WALKS IN A GRAPH

A nonvanishing element of A(G), $(\mathbf{A})_{ij} = 1$, when the vertices v_i and v_j are connected, represents also a walk of the length *one* between the vertices v_i and v_j. Therefore, in general,[16]

$$(\mathbf{A})_{ij} = \begin{cases} 1 & \text{if there is a walk of length one between vertices } v_i \text{ and } v_j \\ 0 & \text{otherwise} \end{cases} \tag{18}$$

However, there are walks of various lengths which can be found in a given graph. Thus,

$$(\mathbf{A})_{ir} \cdot (\mathbf{A})_{rj} = \begin{cases} 1 & \text{if there is a walk of length two between} \\ & \text{vertices } v_i \text{ and } v_j \text{ passing through vertex } v_r \\ 0 & \text{otherwise} \end{cases} \tag{19}$$

Therefore, the expression,

$$(\mathbf{A}^2)_{ij} = \sum_{r=1}^{N} (\mathbf{A})_{ir} \cdot (\mathbf{A})_{rj} \tag{20}$$

represents the total number of walks of the length two in G between vertices v_i and v_j. $(\mathbf{A}^2)_{ij}$ is an element of the matrix $\mathbf{A} \cdot \mathbf{A}$.

For a walk of an arbitrary length ℓ,

$$(\mathbf{A})_{ir} \cdot (\mathbf{A})_{rs} \cdot \ldots \cdot (\mathbf{A})_{zj} = \begin{cases} 1 & \text{if there is a walk of length } \ell \text{ between} \\ & \text{vertices } v_i \text{ and } v_j \text{ passing through} \\ & \text{vertices } v_r, v_s, \ldots, v_z \\ 0 & \text{otherwise} \end{cases} \tag{21}$$

or

$$(\mathbf{A}^\ell)_{ij} = \text{the total number of walks of length } \ell \text{ between vertices } v_i \text{ and } v_j \tag{22}$$

$(\mathbf{A}^\ell)_{ij}$ is an element of the matrix $\mathbf{A}^\ell$.

C. THE DETERMINANT OF THE ADJACENCY MATRIX

The *determinant* of the adjacency matrix, det $\mathbf{A}$, is defined by:[17-19]

$$\det \mathbf{A} = \sum_{P} (-1)^{I(k_1, k_2, \ldots, k_N)} (\mathbf{A})_{1k_1} (\mathbf{A})_{1k_2} \ldots (\mathbf{A})_{1k_N} \tag{23}$$

where the summation is over all permutations of $k_i s$ and $I(k_1,k_2, \ldots ,k_N)$ is the number of inversions in a given permutation P.

The value of a determinant can be found using *the method of minors*: each of the elements in a row (or column) of a matrix is multiplied by its minor. If a given element is in the ith row and jth column, the sign associated with the product is $(-1)^{i+j}$. The summation of all products gives the value of the determinant. For example, the evaluation of the determinant of the adjacency matrix for the labeled complete graph K_4 (see Figure 7), using the method of minors, proceeds in the following way:

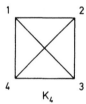

$$A(K_4) = \begin{bmatrix} 0 & 1 & 1 & 1 \\ 1 & 0 & 1 & 1 \\ 1 & 1 & 0 & 1 \\ 1 & 1 & 1 & 0 \end{bmatrix}$$

FIGURE 7.   The labeled complete graph K_4.

$$\det A(K_4) = \begin{vmatrix} 0 & 1 & 1 & 1 \\ 1 & 0 & 1 & 1 \\ 1 & 1 & 0 & 1 \\ 1 & 1 & 1 & 0 \end{vmatrix} = - \begin{vmatrix} 1 & 1 & 1 \\ 1 & 0 & 1 \\ 1 & 1 & 0 \end{vmatrix} +$$

$$\begin{vmatrix} 1 & 1 & 1 \\ 0 & 1 & 1 \\ 1 & 1 & 0 \end{vmatrix} - \begin{vmatrix} 1 & 1 & 1 \\ 0 & 1 & 1 \\ 1 & 0 & 1 \end{vmatrix} =$$

$$- \begin{vmatrix} 0 & 1 \\ 1 & 0 \end{vmatrix} + \begin{vmatrix} 1 & 1 \\ 1 & 0 \end{vmatrix} - \begin{vmatrix} 1 & 1 \\ 0 & 1 \end{vmatrix} +$$

$$\begin{vmatrix} 1 & 1 \\ 1 & 0 \end{vmatrix} + \begin{vmatrix} 1 & 1 \\ 1 & 1 \end{vmatrix} - \begin{vmatrix} 1 & 1 \\ 0 & 1 \end{vmatrix} -$$

$$\begin{vmatrix} 1 & 1 \\ 1 & 1 \end{vmatrix} = -3$$

The evaluation and manipulations of determinants can be greatly simplified because of the following properties of determinants:[17-19]

1. The value of the determinant changes sign when two rows (or two columns) are interchanged
2. If every element in a row (or column) is zero, the value of the determinant is zero
3. If two rows (or columns) are equal, the determinant is zero
4. If all elements in a row (or column) are multiplied by a constant, the value of the determinant is multiplied by the same constant
5. If a multiple of one row (or column) is added to any other, the value of the determinant remains unchanged
6. If the elements of the determinant are arranged as follows:

$$\begin{vmatrix} 1 & 2 & 3 & \ldots & N \\ 2 & 3 & 4 & \ldots & 1 \\ 3 & 4 & 5 & \ldots & 2 \\ \ldots & \ldots & \ldots & \ldots & \ldots \\ N & 1 & 2 & \ldots & N\text{-}1 \end{vmatrix} \tag{24}$$

Then the value of the determinant is given by

$$\det = 1/2(-1)^{N(N-1)/2} (N + 1)N^{N-1} \tag{25}$$

7. The transposed matrix M^T has the same value of the determinant as the matrix M itself,

$$\det M^T = \det M \tag{26}$$

D. THE PERMANENT OF THE ADJACENCY MATRIX

The *permanent* of the adjacency matrix, per A, is defined by[20-22]

$$\text{per } A = \sum_P (A)_{1k_1}(A)_{2k_2}\ldots(A)_{Nk_N} \tag{27}$$

where the summation is over all permutations of k_is.

There are several methods available for the evaluation of the permanent of a square matrix.[22] One way to compute the permanent of the adjacency matrix is also by the method of minors described in the preceding section. However, in this case the sign associated with every product is always positive. For example, evaluation of the permanent of the adjacency matrix for the labeled complete graph K_4 (see Figure 6), using the method of minors, proceeds as follows (the adjacency matrix of K_4 is also given above):

$$
\text{per } A(K_4) = \begin{vmatrix} 0 & 1 & 1 & 1 \\ 1 & 0 & 1 & 1 \\ 1 & 1 & 0 & 1 \\ 1 & 1 & 1 & 0 \end{vmatrix} = \begin{vmatrix} 1 & 1 & 1 \\ 1 & 0 & 1 \\ 1 & 1 & 0 \end{vmatrix} +
$$

$$
\begin{vmatrix} 1 & 1 & 1 \\ 0 & 1 & 1 \\ 1 & 1 & 0 \end{vmatrix} + \begin{vmatrix} 1 & 1 & 1 \\ 0 & 1 & 1 \\ 1 & 0 & 1 \end{vmatrix} =
$$

$$
\begin{vmatrix} 0 & 1 \\ 1 & 0 \end{vmatrix} + \begin{vmatrix} 1 & 1 \\ 1 & 0 \end{vmatrix} + \begin{vmatrix} 1 & 1 \\ 0 & 1 \end{vmatrix} +
$$

$$
\begin{vmatrix} 1 & 1 \\ 1 & 0 \end{vmatrix} + \begin{vmatrix} 1 & 1 \\ 1 & 1 \end{vmatrix} + \begin{vmatrix} 1 & 1 \\ 0 & 1 \end{vmatrix} +
$$

$$
\begin{vmatrix} 1 & 1 \\ 1 & 1 \end{vmatrix} = 9
$$

The permanent of the matrix has a number of useful properties.[20-22] They will be illustrated below for the permanent of the adjacency matrix.

1. The transposed adjacency matrix A^T has the same value of the permanent as the adjacency matrix $A(G)$,

$$\text{per } A^T = \text{per } A \tag{28}$$

2. The permanent of the adjacency matrix remains constant when any two rows (or columns) of $A(G)$ are interchanged. The interchange of any two (or more) rows and the corresponding columns of $A(G)$ correspond to relabeling of two (or more) vertices of G.
3. The permanent of the adjacency matrix with a zero row and/or column is equal to zero. The adjacency matrix with a zero row and/or column corresponds to a disconnected graph.

4. If the adjacency matrix **A** has all elements equal to one, then the permanent of **A** is given by

$$\text{per } \mathbf{A} = N! \tag{29}$$

5. If the adjacency matrix **A** has all elements equal to one, but one being zero, the permanent of **A** is given by

$$\text{per}^{(1)}\mathbf{A} = N! - (N - 1)! = (N - 1)[(N - 1)!] \tag{30}$$

where the superscript (1) on the left of **A** denotes a single zero in the adjacency matrix **A**.

6. Expression 30 may be generalized for the case of the adjacency matrix $^{(k)}\mathbf{A}$ with k zero entries each in a different row and column

$$\text{per}^{(k)}\mathbf{A} = N! - \binom{k}{1}(N - 1)! + \binom{k}{2}(N - 2)! - \dots$$

$$+ (-1)^k\binom{k}{k}(N - k)! \tag{31}$$

E. THE INVERSE OF THE ADJACENCY MATRIX

The *inverse* of the adjacency matrix $\mathbf{A}^{-1}$ is defined by

$$\mathbf{A}^{-1} \cdot \mathbf{A} = \mathbf{A} \cdot \mathbf{A}^{-1} = \mathbf{I} \tag{32}$$

It can be obtained by

$$\mathbf{A}^{-1} = \frac{\text{adj } \mathbf{A}}{\det \mathbf{A}} \tag{33}$$

where adj **A** is the *matrix adjoint* to **A**. The matrix adjoint to **A** can be obtained by first replacing each matrix element $(\mathbf{A})_{ij}$ by its cofactor in det **A**, and then transposing rows and columns. If the adjacency matrix **A** is to have an inverse $\mathbf{A}^{-1}$, the determinant of **A** must not vanish:

$$\det \mathbf{A} \neq 0 \tag{34}$$

Note, the *(i,j)*th minor with the sign $(-1)^{i+j}$ is the *(i,j)*th *cofactor*.

II. THE DISTANCE MATRIX

The *distance matrix* (which is also sometimes called the metrics matrix) is, in a sense, a more complicated and also a richer structure than the adjacency matrix.[1,2,4] It is a graph-theoretical (topological) matrix less common than the adjacency matrix, but it has been increasingly used in the last two decades in many different areas of chemistry and physics, as has been well documented by Rouvray.[23] This author has pointed out an interesting fact that the distance matrix has also found considerable use in the areas of research which are relatively remote from chemistry and physics and to a great extent nonmathematical, such as anthropology, geography, geology, ornithology, philology, and psychology.

The distance matrix $\mathbf{D} = \mathbf{D}(G)$ of a labeled connected graph G is a real symmetric $N \times N$ matrix whose elements $(\mathbf{D})_{ij}$ are defined as follows:[1,2,3,23]

$$(\mathbf{D})_{ij} = \begin{cases} \ell_{ij} & \text{if } i \neq j \\ 0 & \text{if } i = j \end{cases} \qquad (35)$$

where ℓ_{ij} is the length of the shortest path (i.e., the minimum number of edges) between the vertices v_i and v_j. The length ℓ_{ij} is also called the *distance* between the vertices v_i and v_j, thence the term distance matrix. For example, the following distance matrix can be constructed for a labeled graph G (see Figure 8):

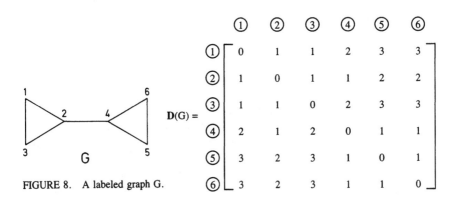

FIGURE 8.   A labeled graph G.

The distance matrix $\mathbf{D}(G_{EW})$ of a labeled edge-weighted graph G_{EW} is a real symmetric $N \times N$ matrix whose elements $(\mathbf{D})_{ij}$ are defined as follows:[24]

$$(\mathbf{D})_{ij} = \begin{cases} w_{ij} & \text{if } i \neq j \\ 0 & \text{if } i = j \end{cases} \qquad (36)$$

where w_{ij} is the minimum sum of weights of edges along the path between the vertices v_i and v_j which is not necessarily the shortest possible path between these two vertices. This is so because the shortest possible path between the vertices v_i and v_j in G_{EW} may have a greater value of w_{ij} than a longer path.

For example, the following distance matrix can be constructed for an edge-weighted graph G_{EW} (see Figure 9):

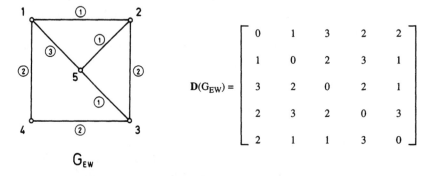

$$D(G_{EW}) = \begin{bmatrix} 0 & 1 & 3 & 2 & 2 \\ 1 & 0 & 2 & 3 & 1 \\ 3 & 2 & 0 & 2 & 1 \\ 2 & 3 & 2 & 0 & 3 \\ 2 & 1 & 1 & 3 & 0 \end{bmatrix}$$

FIGURE 9.   An edge-weighted graph G_{EW}.
The edge-weights are encircled.

The distance matrix $D(G_{VEW})$ of a labeled vertex- and edge-weighted graph G_{VEW} is a real symmetric $N \times N$ matrix whose elements $(D)_{ij}$ are defined as[25-27]

$$(D)_{ij} = \begin{cases} w_{ij} & \text{if } i \neq j \\ w_{ii} & \text{if } i = j \end{cases} \tag{37}$$

where w_{ii} is the weight of a vertex v_i and w_{ij} is the minimum sum of the edge-weights k_{ij} along the path between the vertices v_i and v_j which is not necessarily the shortest possible. One way to define the vertex-weights and the edge-weights is as follows. Since in chemical graph theory the vertex- and edge-weighted graphs represent molecules containing heteroatoms (see Chapter 3), the vertex- and edge-weights w_{ii} and k_{ij} may be defined using a comparison between carbon and noncarbon atoms via the corresponding atomic numbers. Thus, the vertex-weight w_{ii} is defined as:[25]

$$w_{ii} = 1 - \frac{Z_C}{Z_i} \tag{38}$$

where Z_i is the atomic number of the element and $Z_C = 6$. Values of w_{ii} for common atoms are given in Table 1. The edge-weight (bond parameter) k_{ij} is similarly defined as:[25]

$$k_{ij} = \frac{1}{b_{ij}} \cdot \frac{Z_C^2}{Z_i Z_j} \tag{39}$$

where b_{ij} is the bond multiplicity parameter with values 1, 1.5, 2, and 3 for a single, an aromatic, a double, and a triple bond, respectively. The off-diagonal matrix element w_{ij} is given by

$$w_{ij} = \sum_{i,j} k_{ij} \tag{40}$$

where the summation goes over i,j bonds.

The values of k_{ij} parameters for several common bonds are given in Table 2.

For example, the following distance matrix can be constructed for a vertex- and edge-weighted graph G_{VEW} depicting a specified heterosystem (see Figure 10).

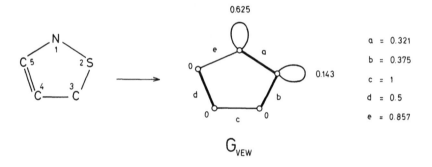

FIGURE 10. A vertex- and edge-weighted graph G_{VEW} representing a specified heterocycle. The requisite vertex- and edge-weights are taken from Table 1 and Table 2, respectively.

$$\mathbf{D}(G_{VEW}) = \begin{bmatrix} 0.625 & 0.321 & 1.071 & 1.357 & 0.857 \\ 0.321 & 0.143 & 0.375 & 1.375 & 1.178 \\ 1.071 & 0.375 & 0 & 1 & 1.5 \\ 1.357 & 1.375 & 1 & 0 & 0.5 \\ 0.857 & 1.178 & 1.5 & 0.5 & 0 \end{bmatrix}$$

TABLE 1
Values of w_{ii} for
Common Atoms

Atom i	w_{ii}
Carbon	0
Nitrogen	0.143
Oxygen	0.250
Sulfur	0.625
Fluorine	0.333
Chlorine	0.647
Phosphorus	0.600

TABLE 2
k_{ij} Parameters for Various
Types of Bonds

Bond type	Bond parameter k_{ij}
C–C	1
C=C	0.500
C≡C	0.333
Aromatic CC bond	0.670
C–N	0.857
C=N	0.429
Aromatic CN bond	0.571
C–O	0.750
C=O	0.375
C–S	0.375
C=S	0.188
N–N	0.735
N=N	0.368
N–S	0.321
C–F	0.667
C–Cl	0.353

There were also other attempts to define the distance matrix of vertex-and/or edge-weighted graphs.[28-31]

The construction of the distance matrix for large complex graphs is a nontrivial task and for its execution the computer must be employed. Traditionally, the distance matrix of a graph G has been generated using powers of the corresponding adjacency matrix of G.[32,33] However, there are other algorithms and computer programs in the literature which are faster than matrix power construction.[33-37] For example, the Bersohn algorithm[34] is about 30 times faster than the method based on powers of the adjacency matrix, when used to construct the distance matrices for steroids on an IBM 3033 computer. The weak point of the matrix power method is the number of matrix multiplications required by this procedure, since computers multiply much more slowly than they subtract, for example. The method of Bersohn does not involve any multiplication at all. Recently, Tratch et al.[38] have compared several algorithms from the literature for the construction of the distance matrix and have found that the algorithm developed by the Düsseldorf-Zagreb group[36] is the fastest algorithm on an ISKRA-226 microcomputer.

The distance matrix has found a widespread application in chemistry in both explicit and implicit forms.[23] The first explicit use of the distance matrix was made by Clark and Kettle.[39] They have employed the distance matrix for studying the permutational isomers of stereochemically nonrigid molecules. These authors distinguished between the various interconversion mechanisms for pairs of permutational isomers by determining the shortest path sequence necessary to effect the rearrangement. These shortest paths were then used as the ℓ_{ij} entries in the construction of an appropriate distance matrix. The distance matrix in explicit form is also used[40-43] to generate the *distance polynomial* and the *distance spectrum*.

The earliest application of the distance matrix in implicit form was made, albeit unknowingly, by Wiener.[44] He was studying additive physical parameters of alkanes and introduced a *path number*, which is defined as the sum of the distances between every pair of carbon atoms in an alkane with distances measured in terms of the number of intervening carbon-carbon bonds. This is the origin of the first topological (graph-theoretical) index, named in honor of Wiener, the *Wiener number*.[45] The connection between the distance matrix and the Wiener number was first pointed out by Hosoya:[46] The Wiener number is simply equal to the half-sum of the elements of the distance matrix. A discussion about the Wiener number and its uses will be given in Chapter 10.

The distance matrix has been used continuously as a source for deriving novel topological indices.[47-53] This appears to date to be one of the most important uses of this matrix in chemistry.[45,54,55] However, another use of the distance matrix is becoming increasingly important.[56-59] This particular use is based on the distance matrix whose elements no longer represent graph-theoretical distances, but geometric (Euclidean) distances.[60] In order to distinguish between the two types of the distance matrices, they are named the *graph-theoretical (topological)* distance matrix and the *geometric (topographic)* distance matrix. There are already several proposals in the literature for the construction of the geometric distance matrix.[56,59,61,62] This matrix, among other applications, serves as a generator for *topographic indices* which have also found a use in structure-property-activity studies.[60-64,67] Topographic indices will also be mentioned briefly in Chapter 10.

REFERENCES

1. **Harary, F.,** *Graph Theory,* Addison-Wesley, Reading, MA, 1971, 2nd printing.
2. **Rouvray, D. H.,** in *Chemical Applications of Graph Theory,* Balaban, A. T., Ed., Academic Press, London, 1976, 175.

3. **Trinajstić, N.,** in *Semiempirical Methods of Electronic Structure Calculation. Part A. Techniques,* Vol. 7, *Modern Theoretical Chemistry,* Segal, G. A., Ed., Plenum Press, New York, 1977, 1.

4. **Cvetković, D. M., Doob, M., and Sachs, H.,** *Spectra of Graphs: Theory and Application,* Academic Press, New York, 1980, chap. 8.

5. **Cvetković, D. M., Doob, M., Gutman, I., and Torgašev, A.,** *Recent Results in the Theory of Graph Spectra,* North-Holland, Amsterdam, 1988, chap. 5.

6. **Trinajstić, N.,** in *Graph Theory and Applications in Chemistry,* Tyutyulkov, N. and Bonchev, D., Eds., Nauka i iskustvo, Sofia, 1987, 86 (in Bulgarian).

7. **Stankevich, I. V.,** in *Applications of Graph Theory in Chemistry,* Zefirov, N. S. and Kuchanov, C. I., Eds., Nauka, Novosibirsk, 1988, 7 (in Russian).

8. **Schmidtke, H. H.,** *J. Chem. Phys.,* 45, 3920, 1966.

9. **Schmidtke, H. H.,** *Coord. Chem. Rev.,* 2, 3, 1967.

10. **Graovac, A. and Trinajstić, N.,** *Math. Chem. (Mülheim/Ruhr),* 1, 159, 1975; erratum *Math. Chem. (Mülheim/Ruhr),* 5, 290, 1979.

11. **Graovac, A. and Trinajstić, N.,** *Croat. Chem. Acta,* 47, 95, 1975.

12. **Graovac, A. and Trinajstić, N.,** *J. Mol. Struct.,* 30, 416, 1976.

13. **Gutman, I.,** *Z. Naturforsch.,* 33a, 214, 1976.

14. **Read, R. C. and Corneil, D. G.,** *J. Graph Theory,* 1, 339, 1977.

15. **Ham, N. S.,** *J. Chem. Phys.,* 29, 1229, 1958.

16. **Marcus, R. A.,** *J. Chem. Phys.,* 43, 2643, 1965.

17. **Kowalewski, G.,** *Determinantentheorie,* 3rd ed., Chelsea, New York, 1943.

18. **Finkbeiner, D. T., II,** *Introduction to Matrices and Linear Transformations,* W. H. Freeman, San Francisco, 1960, chap. 5.

19. **Kurosh, A.,** *Higher Algebra,* Mir, Moscow, 1980, chap. 1, 3rd printing.

20. **Muir, T.,** *Proc. R. Soc. (Edinburgh),* 11, 409, 1882.

21. **Marcus, M. and Minc, H.,** *Am. Math. Mon.,* 72, 577, 1965.

22. **Minc, H.,** *Permanents,* Addison-Wesley, Reading, MA, 1978.

23. **Rouvray, D. H.,** in *Mathematics and Computational Concepts in Chemistry,* Trinajstić, N., Ed., Horwood, Chichester, 1986, 295.

24. **Hakimi, S. L. and Yau, S. S.,** *Q. Appl. Math.,* 22, 305, 1964.

25. **Barysz, M., Jashari, G., Lall, R. S., Srivastava, V. K., and Trinajstić, N.,** in *Chemical Applications of Topology and Graph Theory,* King, R. B., Ed., Elsevier, Amsterdam, 1983, 222.

26. **Trinajstić, N.,** *Kem. Ind. (Zagreb),* 33, 311, 1984.

27. **Randić, M., Sabljić, A., Nikolić, S., and Trinajstić, N.,** *Int. J. Quantum Chem.: Quantum Chem. Symp.,* 15, 267, 1988.

28. **Balaban, A. T.,** *Chem. Phys. Lett.,* 80, 399, 1982.

29. **Balaban, A. T.,** *Pure Appl. Chem.,* 55, 189, 1983.

30. **Balaban, A. T.,** *Math. Chem. (Mülheim/Ruhr),* 21, 115, 1986.

31. **Balaban, A. T. and Ivanciuc, O.,** in *MATH/CHEM/COMP 1988,* Graovac, A., Ed., Elsevier, Amsterdam, 1989, 193.

32. **Roberts, F. S.,** *Discrete Mathematical Models,* Prentice-Hall, Englewood Cliffs, 1976, 58.

33. **Mohar, B. and Pisanski, T.,** *J. Math. Chem.,* 2, 267, 1988.

34. **Bersohn, M.,** *J. Comput. Chem.,* 4, 110, 1983.

35. **Deo, N. and Pang, C.,** *Networks,* 14, 275, 1984.

36. **Müller, W. R., Szymanski, K., Knop, J. V., and Trinajstić, N.,** *J. Comput. Chem.,* 8, 170, 1987.

37. **Senn, P.,** *Comput. Chem.,* 12, 219, 1988.

38. **Tratch, S. S., Stankevich, M. I., and Zefirov, N. S.,** *J. Comput. Chem.,* 11, 899, 1990.

39. **Clark, M. J. and Kettle, S. F. A.,** *Inorg. Chim. Acta,* 14, 201, 1975.

40. **Hosoya, H., Murakami, M., and Gotoh, M.,** *Natl. Sci. Rep. Ochanomizu Univ. (Tokyo)*, 24, 27, 1973.
41. **Edelberg, M., Garey, M. R., and Graham, R. L.,** *Discrete Math.*, 14, 23, 1976.
42. **Křivka, P. and Trinajstić, N.,** *Appl. Math. (Prague)*, 28, 357, 1983.
43. **Balasubramanian, K.,** *J. Comput. Chem.*, 11, 829, 1990.
44. **Wiener, H.,** *J. Am. Chem. Soc.*, 69, 17, 1947.
45. **Rouvray, D. H.,** *Sci. Am.*, 254, 40, 1986.
46. **Hosoya, H.,** *Bull. Chem. Soc. Jpn.*, 44, 2332, 1971.
47. **Bonchev, D. and Trinajstić, N.,** *J. Chem. Phys.*, 67, 4517, 1977.
48. **Bonchev, D. and Trinajstić, N.,** *Int. J. Quantum Chem.: Quantum Chem. Symp.*, 12, 293, 1978.
49. **Needham, D. E., Wei, I-C., and Seybold, P. G.,** *J. Am. Chem. Soc.*, 110, 4186, 1988.
50. **Schultz, H. P.,** *J. Chem. Inf. Comput. Sci.*, 29, 227, 1989.
51. **Schultz, H. P., Schultz, E. B., and Schultz, T. P.,** *J. Chem. Inf. Comput. Sci.*, 30, 27, 1990.
52. **Müller, W. R., Szymanski, K., Knop, J. V., and Trinajstić, N.,** *J. Chem. Inf. Comput. Sci.*, 30, 160, 1990.
53. **Lukovits, I.,** *J. Chem. Soc. Perkin Trans. II*, 84, 1667, 1988; *Quant. Struct.-Act. Relat.*, 9, 227, 1990; *Rep. Mol. Theory*, 1, 127, 1990.
54. **Rouvray, D. H.,** in *Chemical Applications of Topology and Graph Theory*, King, R. B., Ed., Elsevier, Amsterdam, 1983, 159.
55. **Stankevich, M. I., Stankevich, I. V., and Zefirov, N. S.,** *Russ. Chem. Rev.*, 57, 191, 1988.
56. **Motoc, I., Dammkoehler, R. A., and Marshall, G. R.,** in *Mathematics and Computational Concepts in Chemistry*, Trinajstić, N., Ed., Horwood, Chichester, 1986, 222.
57. **Randić, M.,** *Int. J. Quantum Chem.: Quantum Biol. Symp.*, 15, 201, 1988.
58. **Crippen, G. M. and Havel, T. F.,** *Distance Geometry and Molecular Conformation*, John Wiley & Sons, New York, 1988; **Crippen, G. M.,** *J. Math. Chem.*, 6, 307, 1991.
59. **Balasubramanian, K.,** *Chem. Phys. Lett.*, 169, 224, 1990.
60. **Randić, M.,** in *MATH/CHEM/COMP 1987*, Lacher, R. C., Ed., Elsevier, Amsterdam, 1988, 101.
61. **Bogdanov, B., Nikolić, S., and Trinajstić, N.,** *J. Math. Chem.*, 3, 291, 1981; 5, 305, 1990.
62. **Bošnjak, N., Mihalić, Z., and Trinajstić, N.,** *J. Chromatogr.*, 540, 430, 1991.
63. **Mihalić, Z. and Trinajstić, N.,** *J. Mol. Struct. (Theochem.)*, 232, 65, 1991.
64. **Nikolić, S., Trinajstić, N., Mihalić, Z., and Carter, S.,** *Chem. Phys. Lett.*, 179, 21, 1991.

Chapter 5

THE CHARACTERISTIC POLYNOMIAL OF A GRAPH

There are a number of polynomials that one can associate with a graph.[1-9] Among the graph-theoretical (graphic) polynomials the most important appears to be the *characteristic polynomial*.[10-13]

There are, in the main, three reasons why graphic polynomials are of interest for (chemical) graph theorists:[8]

1. Graphic polynomials appear as generating functions of combinatorial graph invariants
2. Graphic polynomials can be used to introduce algebraic concepts into graph theory
3. The study of graphic polynomials is an interesting and relevant research area per se

I. THE DEFINITION OF THE CHARACTERISTIC POLYNOMIAL

The characteristic (secular, spectral) polynomial $P(G;x)$ of a graph G is the characteristic polynomial of its adjacency matrix,[10,14]

$$P(G;x) = \det |x\mathbf{I} - \mathbf{A}| \qquad (1)$$

where $\mathbf{A}$ and $\mathbf{I}$ are, respectively, the adjacency matrix of a graph G with N vertices and the $N \times N$ unit matrix. A graph eigenvalue x_i is a zero of the characteristic polynomial

$$P(G;x_i) = 0 \qquad (2)$$

for $i = 1,2, \ldots ,N$. The complete set of graph eigenvalues $\{x_1,x_2, \ldots x_N\}$ forms the spectrum of the graph.[13] The eigenvalues are all real and the interval in which they lie is bounded. According to the Frobenius theorem,[15] the limits of the graph spectrum are determined by the maximum valency of a vertex D_{max} in a graph:

$$-D_{max} \leq x_i \leq D_{max} \qquad (3)$$

Among connected graphs, equality is achieved on the right-hand side if, and only if, all vertices in the graph have the same valency, while equality is achieved on the left-hand side if, and only if, the graph is also bipartite; in either case, these extreme eigenvalues are nondegenerate.

The characteristic polynomial of a graph is most often given in the coefficient form,

$$P(G;x) = \sum_{n=0}^{N} a_n x^{N-n} \qquad (4)$$

The polynomial coefficients $a_0, a_1, \ldots, a_N$ are graph invariants and are related to the structure of a graph.[8,10,13,15-17]

The characteristic polynomial can also be given in terms of its zeros,[18]

$$P(G;x) = \prod_{i=1}^{N} (x - x_i) \qquad (5)$$

If the permanent of the adjacency matrix is introduced instead of the determinant in Equation 1, the *permanental polynomial* $\mathcal{P}(G;x)$ of G is obtained:[19]

$$\mathcal{P}(G;x) = \text{per} \, |x\mathbf{I} - \mathbf{A}| \qquad (6)$$

The characteristic polynomial is an important structural invariant although it is not always unique to a single graph, because of the fact that nonisomorphic graphs may possess identical characteristic polynomials and consequently identical spectra.[2,8,13,20-29] Nonisomorphic graphs with identical spectra are called *isospectral*[23] or *cospectral*[30] graphs.

A. THE USES OF THE CHARACTERISTIC POLYNOMIAL

The characteristic polynomial of a graph is of interest in chemistry because it appears explicitly or implicitly in numerous applications. Some of the applications are summarized below:[31]

1. The characteristic polynomial (and the related acyclic polynomial) is used in the topological theory of aromaticity (see Chapter 7).
2. The characteristic polynomial is useful in predicting the relative stabilities of conjugated hydrocarbons[32] and in the formulation of the topological effect on molecular orbitals (TEMO) concept.[14,33-36]
3. The last two coefficients (a_{N-2}, a_N) in the characteristic polynomials of benzenoids, arenecyclobutadienoids (such as benzocyclobutadienoids), azulenoids, pentalenoids, etc., can be used for counting Dewar valence structures and Kekulé valence structures;[37-40] quantities that represent the basis of resonance-theoretic approaches such as the structure-resonance theory[41-43] and the conjugated-circuit model (see Chapter 9).
4. The characteristic polynomial has found application in quantum chemistry,[8,13,14,34-36,44-48] chemical kinetics,[49] dynamics of oscillatory reac-

tions,[50] solutions of Navier-Stokes equations,[51] and in statistical mechanics because it serves as a generating function for dimer statistics on trees such as Bethe lattices.[52-54] (A *Bethe lattice* is a tree of valence D and length ℓ in which each nonterminal vertex has D neighbors (of valency D) and there are ℓ bonds from the central vertex to any terminal vertex.)[53,55]

5. The coefficients of the characteristic polynomial may be interpreted as the counts of random walks over the structural network (lattice).[56-59] Random walks on the structural network are fundamental in several areas in representing the path of a diffusing particle or in modeling the conformations of flexible macromolecules, especially in dilute solution.[60-64] (A *random walk* in a graph is a sequence of edges which can be continuously traversed, starting from any vertex and ending at any vertex, also permitting the use of the same edge several times.)

6. The characteristic polynomial is related to several other graph invariants such as spectral moments.[56,57,65,66] They are auxiliary functions for counting subgraphs of various kinds belonging to a given graph.[8,10-17,31,37,44]

7. The characteristic polynomial may be used for a partial ordering of forests.[67] (A *forest* is defined to be a graph which contains no cycles. Components of a forest are trees.)[68]

8. The characteristic polynomial may be used for counting the spanning trees of labeled planar graphs.[69,70]

9. The characteristic polynomial may be used in analysis of NMR spectra.[45,71,72]

10. The characteristic polynomial was used in the studies on the factorization of graphs.[73,74]

11. The characteristic polynomial arises in the Y-conjugation model of cyanine dyes.[75]

12. The characteristic polynomial serves as the generator for the characteristic equations of a graph.[2] (The *characteristic equations* for a graph are defined as the system of equations for the coefficients of the characteristic polynomial.)

13. The characteristic polynomial has also been found useful in areas other than mathematics,[1,10,13,17,18,20,22,30,76-78] physics,[51,52,60-63] and chemistry, for example, in computer science[79] and biology.[80]

B. COMPUTATIONAL METHODS FOR THE CHARACTERISTIC POLYNOMIAL

The computation of the characteristic polynomial is rather tedious due to the combinatorial complexity of the problem.[30] There are a number of methods available for the computation of the characteristic polynomial of a graph. Some of them are listed below:

1. The Laplace expansion of the determinant (Equation 1),[18] although this is cumbrous for larger systems.

2. The matrix diagonalization and the use for the Viete formulae.[18] There are many methods available for the diagonalization of a matrix,[81] such as the Jacobi method, the Givens method, the Householder-QR algorithm, etc.

3. Direct graphical constructions such as the method of Sachs.[17] This is a very general, elegant, and imaginative way of constructing the coefficients of the characteristic polynomial. However, the method of Sachs is not really practical for large graphs because with increasing size there is an explosion of combinatorial possibilities which even large computers cannot handle.[82,83] Nevertheless, it reveals directly the structural basis of the characteristic polynomial.

4. The characteristic polynomials of simple graphs such as chains or cycles can be obtained by recurrence formulae.[13,15,84]

5. Special methods such as the transfer matrix method,[85-87] the partition technique,[88-90] the polynomial matrix method,[91-94] the pruning technique,[53,54,95-99] the ultimate pruning technique,[100] the symmetry blocking method,[101,102] the operator technique,[103] the Chebishev expansion,[2,57,104-107] the use of the Frobenius matrix,[108] the use of the power sum symmetric functions,[109,110] the use of symmetry properties of a structure,[111] the use of powers of the adjacency matrix,[112] the use of the Ulam subgraphs,[8,113] the use of a functional-group concept,[114-119] the use of the Sturm sequence,[120] the differential operator technique,[121] the low-rank perturbation method,[122] the matrix decomposition method,[123] etc.

6. Recursive techniques such as those of Le Verrier,[124] Faddeev,[125] and Frame,[126] which were adopted for computer processing by Balasubramanian[3,45,58,127-131] and the Zagreb Group.[4,9,121,132]

In the remainder of this chapter the method of Sachs, the recurrence formulae for several classes of simple graphs, and the recursive techniques of Le Verrier, Faddeev, and Frame will be detailed.

II. THE METHOD OF SACHS FOR COMPUTING THE CHARACTERISTIC POLYNOMIAL

A lot of the research has been carried out on generating the coefficients $a_n(G)$ of $P(G;x)$ from the structure of the graph.[8,10-17,20,22,44,76,90,114-119,133-135] Most elegant, if not the most practical,[82,83] is the method of Sachs,[17] which can best be summarized in the following formula:

$$a_n(G) = \sum_{s \in S_n} (-1)^{p(s)} 2^{c(s)} \tag{7}$$

where s is a *Sachs graph,* S_n a set of all Sachs graphs with n vertices, while $p(s)$ and $c(s)$ denote, respectively, the total number of components and the

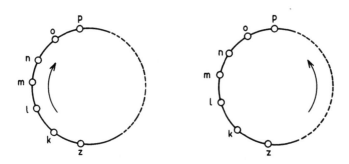

FIGURE 1. The equivalent directions in a cycle.

total number of cycles in s. A Sachs graph[37] (Grundfigure,[17] elementary figure,[13] mutation graph[22]) s is defined as such a subgraph of G whose components are K_2 and/or C_m ($m = 3,4, \ldots ,N$). Therefore, the coefficients a_n of P(G;x) are either made up of complete graphs K_2 or cycles C_m ($m = 3,4, \ldots ,n$) or combinations between $\ell\, K_2$ components and $k\, C_m$ components with the restriction $2\ell + km = n$. It should also be noted that for the case of $n = 0$, $a_0 = 1$ (by definition) and if $S_n = \emptyset$, $a_n = 0$.

Sachs graphs are related to permutations of nonzero off-diagonal elements $(\mathbf{A})_{ij}$ in the secular determinant. Taking into account K_2 components corresponds to considering only products $(\mathbf{A})_{k\ell} \cdot (\mathbf{A})_{k\ell}$ which are related to edges in G. Counting cyclic components C_m ($m = 3,4, \ldots ,N$) corresponds to taking into account only products such as $(\mathbf{A})_{k\ell} \cdot (\mathbf{A})_{\ell m} \cdot (\mathbf{A})_{mn} \cdot \ldots \cdot (\mathbf{A})_{zk}$ and $(\mathbf{A})_{kz} \cdot \ldots \cdot (\mathbf{A})_{nm} \cdot (\mathbf{A})_{m\ell} \cdot (\mathbf{A})_{\ell k}$, which are products of nonvanishing elements related to cycles in G. Two products appear because each cycle may be counted in either of two directions (see Figure 1). This is the origin of the factor 2 in Equation 1.

A. THE APPLICATION OF THE METHOD OF SACHS TO SIMPLE GRAPHS

The application of the method of Sachs to simple graphs is straightforward. It will be illustrated for the hydrogen-suppressed graph G corresponding to methylenecyclopropene (see Figure 2). In Figure 3 all Sachs graphs that can be associated with G in Figure 2 are given.

The coefficients $a_n(G)$ of the corresponding characteristic polynomial P(G;x) can be directly computed by means of the Sachs formula (7) as follows:

$$a_0 = 1 \text{ (by definition)}$$
$$a_1 = 0$$
$$a_2 = 4(-1)^1\, 2^0 = -4$$
$$a_3 = (-1)^1\, 2^1 = -2$$
$$a_4 = (-1)^2\, 2^0 = 1$$

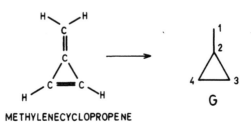

METHYLENECYCLOPROPENE

FIGURE 2.   The hydrogen-suppressed graph G corresponding to meth-ylenecyclopropene.

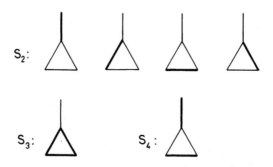

FIGURE 3.   Sachs graphs derived from G in Figure 2. Thicker line(s) denote a particular Sachs graph.

Then, the characteristic polynomial of methylenecyclopropene is given by

$$P(G;x) = x^4 - 4x^2 - 2x + 1 \qquad (8)$$

with the following spectrum: $\{2.1701, 0.3111, -1.0000, -1.4812\}$.

B. THE EXTENSION OF THE SACHS FORMULA TO MÖBIUS SYSTEMS

The extended Sachs formula which embraces the Möbius structures is given by[136,137]

$$a_n(G_{Mö}) = \sum_{s \in S_n} (-1)^{p(s) + r(s)} 2^{c(s)} \qquad (9)$$

where $r(s)$ is the number of "-1" connectivities in the cycles of a Sachs graph s. Other symbols have their previous meaning.

It should be noted that a Sachs graph s of a Möbius graph $G_{Mö}$ is defined as such a subgraph of $G_{Mö}$ whose components are K_2 and/or C_m ($m = 3, 4, \ldots, N$) and/or Möbius cycles $C_m^{Mö}$ ($m = 3, 4, \ldots, N$). In Figure 4 Sachs graphs belonging to a Möbius cycle $C_3^{Mö}$ depicting Möbius cyclopro-

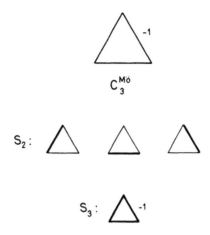

FIGURE 4. The Möbius cycle $C_3^{M\ddot{o}}$ corresponding to Möbius cyclopropenyl. The corresponding Sachs graphs are given in thicker lines.

penyl are given. The coefficients a_n $(C_3^{M\ddot{o}})$ of the corresponding characteristic polynomial $P(C_3^{M\ddot{o}};x)$ can be obtained by the direct use of the extended Sachs formula (9) in the following way:

$$a_0 = 1 \text{ (by definition)}$$
$$a_1 = 0$$
$$a_2 = 3(-1)^{1+0}\, 2^0 = -3$$
$$a_3 = (-1)^{1+1}\, 2^1 = 2$$

The explicit form of the corresponding characteristic polynomial is given by

$$P(C_3^{M\ddot{o}}) = x^3 - 3x + 2 \tag{10}$$

with the following spectrum: $\{1,1,-2\}$.

C. THE EXTENSION OF THE SACHS FORMULA TO WEIGHTED GRAPHS

In this section will be demonstrated how the structure of a vertex- and edge-weighted graph G_{VEW} is related to the corresponding characteristic polynomial $P(G_{VEW};x)$. This will be done by means of the extension of the Sachs formula to weighted graphs.[12,31,138,139] Before doing that, a new type of the Sachs graph containing weighted vertices and/or weighted edges needs to be defined. It is also immediately evident that a graph G_{VEW} may contain two types of Sachs graphs: *unweighted* and *weighted* Sachs graphs. A weighted Sachs graph is defined as such a subgraph of G_{VEW} whose components are loops and/or (weighted) complete graphs K_2 and/or (weighted) cycles C_m ($m = 3,4,\ldots,N$). A complete set of Sachs graphs of G_{VEW}, corresponding to cyclopropenthione (see Figure 5), is given in Figure 6.

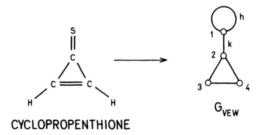

CYCLOPROPENTHIONE

FIGURE 5. The vertex- and edge-weighted graph G_{VEW} depicting cyclopropenthione.

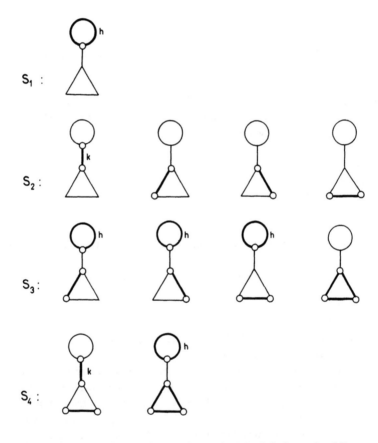

FIGURE 6. A complete set of (unweighted and weighted) Sachs graphs of G_{VEW} (see Figure 5). Thicker line(s) denote a particular Sachs graph.

The Sachs formula adopted for computing coefficients of the characteristic polynomial of a vertex- and edge-weighted graphs G_{VEW} is given by[31]

$$a_n(G_{VEW}) = \sum_{s \in S_n} (-1)^{p(s)} 2^{c(s)} \prod_i^{\ell(s)} h_i \prod_j^{K_2 \text{ in } s} k_j^2 \prod_{j'}^{C_m \text{ in } s} k_{j'} \qquad (11)$$

where the symbols have the following meaning: h_i is the weight of the ith vertex depicted by a loop, $\ell(s)$ is the number of loops in a vertex-weighted Sachs graph, the first product gives the contribution from all the weighted vertices i with weights h_i in s, k_j is the weight of the jth edge, which may be either the K_2 component of s or an edge in the C_m component of s, the second product gives the contribution from all the weighted K_2 components in s and the third product gives the contribution from all the weighted edges in the C_m component of s. Other symbols in Equation 3 have their previous meaning.

The coefficient $a_n(G_{VEW})$ of the characteristic polynomial $P(G_{VEW};x)$ for a weighted graph G_{VEW}, depicting cyclopropenthione, can be computed using Sachs formula (11) and the set of Sachs graphs from Figure 6 as follows:

$a_0 = 1$ (by definition)
$a_1 = (-1)^1 \, 2^0 \, h = -h$
$a_2 = (-1)^1 \, 2^0 \, k^2 + 3(-1)^1 \, 2^0 = -k^2 - 3$
$a_3 = 3(-1)^2 \, 2^0 \, h + (-1)^1 \, 2^1 = 3h - 2$
$a_4 = (-1)^2 \, 2^0 \, k^2 + (-1)^2 \, 2^1 \, h = k^2 + 2h$

Then, the characteristic polynomial $P(G_{VEW};x)$ for G_{VEW} is given by,

$$P(G_{VEW};x) = x^4 - hx^3 - (k^2 + 3)x^2 + (3h - 2)x + (k^2 + 2h) \qquad (12)$$

Arbitrary selection of parameters $h = 0.5$ and $k = 1.5$, leads to the polynomial:

$$P(G_{VEW};x) = x^4 - 0.5x^3 - 5.25x^2 - 0.5x + 3.25 \qquad (13)$$

with the following spectrum: $\{2.4831, 0.7536, -1.0000, -1.7367\}$. The polynomial (12) for $h = 0$ and $k = 1$ becomes the characteristic polynomial of methylenecyclopropene (see polynomial [8]).

When the vertex-weighted G_{VW} is considered only, then $k = 1$ and Equation 11 reduces to

$$a_n(G_{VW}) = \sum_{s \in S_n} (-1)^{p(s)} 2^{c(s)} \prod_i^{\ell(s)} h_i \qquad (14)$$

Similarly, when the edge-weighted graph G_{EW} is considered only, then Equation 11 contracts to

$$a_n(G_{EW}) = \sum_{s \in S_n} (-1)^{p(s)} 2^{c(s)} \overset{K_2 \text{ in } s}{\underset{j}{\prod}} k_j^2 \overset{C_m \text{ in } s}{\underset{j'}{\prod}} k_{j'} \tag{15}$$

Obviously, when there are no weighted vertices and edges, Equation 11 converts into Equation 7.

D. SUMMARY OF SOME RESULTS OBTAINED BY THE USE OF THE SACHS FORMULA

The characteristic polynomials of molecular graphs can be constructed in a manner described by the use of the Sachs formula or its adaptations. Some general results arising from the application of the Sachs formula are summarized below:

1. For unweighted graphs, the sole Sachs graphs are K_2 or C_m ($m \geq 3$). Therefore, there cannot be a Sachs graph with only *one* vertex,

$$S_1 = \emptyset \tag{16}$$

and

$$a_1(G) = 0 \tag{17}$$

Note that the symbol $\emptyset$ stands for empty set, i.e., a set which contains no elements. The implication of Equations 16 and 17 is that the sum of the roots of $P(G;x)$ is always zero:

$$a_1(G) = \sum_{i=1}^{N} x_i = 0 \tag{18}$$

which is, in fact, expected because the trace of the adjacency matrix belonging to a simple graph G is zero.

The nonvanishing value of a_1 coefficient appears only when the vertex-weighted graphs G_{VW} and the vertex- and edge-weighted graphs G_{VEW} are considered. Then

$$a_1(G_{VW}) = a_1(G_{VEW}) = \sum_{j} h_j \tag{19}$$

where h_j stands for the "weight" of the jth vertex of G_{VW} or G_{VEW}. In addition, the sum of the whole spectrum of G_{VW} (or G_{VEW}) is equal to the sum of the selected parameters for the "weighted" vertices appearing in it

$$a_1(G_{VW}) = a_1(G_{VEW}) = \sum_{i=1}^{N} x_i = \sum_{j} h_j \tag{20}$$

2. The construction of the characteristic polynomial via the formula of Sachs, reveals that the a_2 and a_3 coefficients are related to the number of edges M and the three-membered cycles C_3, respectively, in G

$$a_2(G) = -M \tag{21}$$
$$a_3(G) = -2C_3 \tag{22}$$

In addition, the half-sum of squares of the polynomial zeros is related to the a_2 coefficient and, consequently, to the number of edges in a graph G

$$a_2(G) = \frac{1}{2} \sum_{i=1}^{N} x_i^2 = -M \tag{23}$$

3. The explicit formulae for the a_4, a_6, a_8, and a_{10} coefficients of the characteristic polynomials of benzenoids are given by Dias,[117] but they are rather unwieldy. However, an interesting result is that the a_6 coefficient is related to the number of bay regions in the benzenoid hydrocarbon.

4. The absolute value of the last coefficient $|a_N(G)|$ of the characteristic polynomial of a graph G is related to the number of 1-factors K of G

$$|a_N(G)| = K^2 \tag{24}$$

This is a consequence of the following relationships between $a_N(G)$, det(A), and K:[137,140,141]

$$\det(\mathbf{A}) = (-1)^{N/2} K^2 \tag{25}$$
$$\prod_{i=1}^{N} x_i = \det(\mathbf{A}) \tag{26}$$
$$a_N(G) = \prod_{i=1}^{N} x_i \tag{27}$$

5. Since, by definition, there are no odd-membered cycles in bipartite graphs, it follows that for alternant structures $S_n = \emptyset$ and $a_n = 0$ when n = odd. Therefore, the characteristic polynomial of a bipartite graph G must be of the following general form:

$$P(G;x) = x_N + a_2 x^{N-2} + a_4 x^{N-4} + \dots$$
$$+ \begin{cases} a_{N-2}x & N = \text{odd} \\ a_{N-2}x^2 + a_N & N = \text{even} \end{cases} \tag{28}$$

Such a structure of P(G;x) implies that the corresponding graph spectrum must be *symmetric* with respect to $x = 0$, or in other words, if x is a

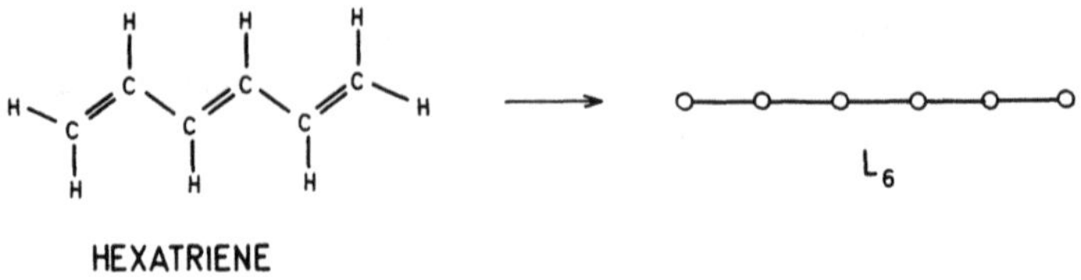

HEXATRIENE

FIGURE 7.   Hexatriene and the corresponding chain with six vertices L_6.

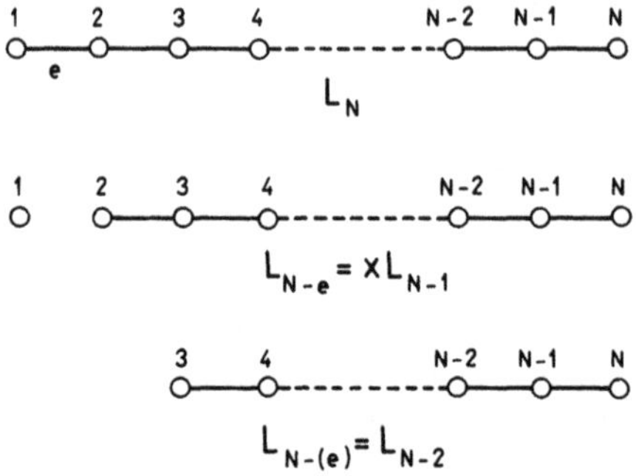

FIGURE 8.   Chains L_N, L_{N-e}, and $L_{N-(e)}$.

root of $P(G;x)$ then $-x$ must also be a root of $P(G;x)$. However, when N is odd there will be one zero element in the spectrum, because N − 1 roots will be symmetrically arranged around $x = 0$, and this leaves a single root which must be zero. Note that only bipartite graphs have symmetric spectra.

III. THE CHARACTERISTIC POLYNOMIALS OF SOME CLASSES OF SIMPLE GRAPHS

A. CHAINS

Linear polyenes or *n*-alkanes with N vertices may be depicted by chains, denoted by L_N, whose characteristic polynomials are symbolized by $P(L_N;x)$. In Figure 7 hexatriene and the corresponding chain L_6 are-shown.

The recurrence relation for computing $P(L_N;x)$ is given by[30,84]

$$P(L_N;x) = P(L_{N-e};x) - P(L_{N-(e)};x) \tag{29}$$

where L_{N-e} is obtained by deletion of the end-edge e from L_N, while $L_{N-(e)}$ is generated by the removal of the end-edge e and both vertices incident to it. Hence, L_{N-e} and $L_{N-(e)}$ possess N and N − 2 vertices, respectively. Chains L_N, L_{N-e}, and $L_{N-(e)}$ are given in Figure 8.

If $xP(L_{N-1};x)$ and $P(L_{N-2};x)$ are substituted for $P(L_{N-e};x)$ and $P(L_{N-(e)};x)$ in Equation 29, the following recurrence relation is obtained:[11,30,76,84]

TABLE 1
Characteristic Polynomials for Chains with up to 20 Vertices

$P(L_0;x) = 1$

$P(L_1;x) = x$

$P(L_2;x) = x^2 - 1$

$P(L_3;x) = x^3 - 2x$

$P(L_4;x) = x^4 - 3x^2 + 1$

$P(L_5;x) = x^5 - 4x^3 + 3x$

$P(L_6;x) = x^6 - 5x^4 + 6x^2 - 1$

$P(L_7;x) = x^7 - 6x^5 + 10x^3 - 4x$

$P(L_8;x) = x^8 - 7x^6 + 15x^4 - 10x^2 + 1$

$P(L_9;x) = x^9 - 8x^7 + 21x^5 - 20x^3 + 5x$

$P(L_{10};x) = x^{10} - 9x^8 + 28x^6 - 35x^4 + 15x^2 - 1$

$P(L_{11};x) = x^{11} - 10x^9 + 36x^7 - 56x^5 + 35x^3 - 6x$

$P(L_{12};x) = x^{12} - 11x^{10} + 45x^8 - 84x^6 + 70x^4 - 21x^2 + 1$

$P(L_{13};x) = x^{13} - 12x^{11} + 55x^9 - 120x^7 + 126x^5 - 56x^3 + 7x$

$P(L_{14};x) = x^{14} - 13x^{12} + 66x^{10} - 165x^8 + 210x^6 - 126x^4 + 28x^2 - 1$

$P(L_{15};x) = x^{15} - 14x^{13} + 78x^{11} - 220x^9 + 330x^7 - 252x^5 + 84x^3 - 8x$

$P(L_{16};x) = x^{16} - 15x^{14} + 91x^{12} - 286x^{10} + 495x^8 - 462x^6 + 210x^4 - 36x^2 + 1$

$P(L_{17};x) = x^{17} - 16x^{15} + 105x^{13} - 364x^{11} + 715x^9 - 792x^7 + 462x^5 - 120x^3 + 9x$

$P(L_{18};x) = x^{18} - 17x^{16} + 120x^{14} - 445x^{12} + 1001x^{10} - 1287x^8 + 924x^6 - 330x^4 + 45x^2 - 1$

$P(L_{19};x) = x^{19} - 18x^{17} + 136x^{15} - 560x^{13} + 1365x^{11} - 2002x^9 + 1716x^7 - 792x^5 + 165x^3 - 10x$

$P(L_{20};x) = x^{20} - 19x^{18} + 153x^{16} - 680x^{14} + 1820x^{12} - 3003x^{10} + 3003x^8 - 1716x^6 + 495x^4 - 55x^2 + 1$

$$P(L_N;x) = xP(L_{N-1};x) + P(L_{N-2};x) \qquad (30)$$

which enables the calculation of the characteristic polynomials of chains starting with

$$P(L_0;x) = 1 \qquad (31)$$

$$P(L_1;x) = x \qquad (32)$$

In Table 1 the characteristic polynomials for chains with up to 20 vertices are tabulated.

B. TREES

Trees with N vertices, denoted by T_N, may be used to depict acyclic structures such as branched alkanes. Their characteristic polynomials are denoted by $P(T_N;x)$. The generating formula for the characteristic polynomial $P(T_N;x)$ of a tree T_N with N vertices is identical to Equation 29,[30,84]

$$P(T_N;x) = P(T_{N-e};x) - P(T_{N-(e)};x) \qquad (33)$$

The strategy of using Equation 33 is to break up the tree into constituting chains in the smallest possible number of steps. Note that the chain is the unbranched tree. The computation of the characteristic polynomial for a given tree is shown in Figure 9.

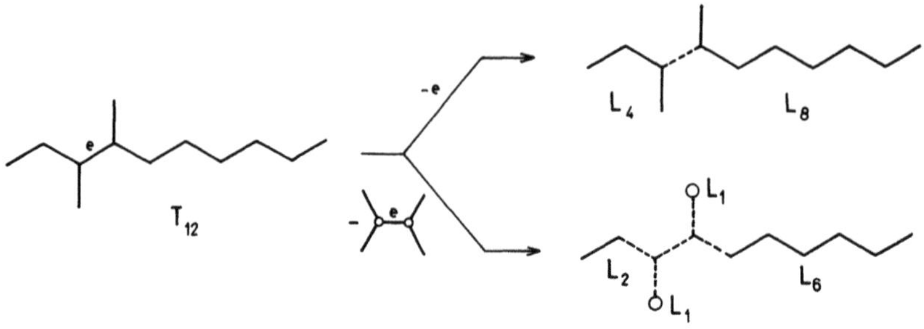

$$P(T_{12};x) = L_4 \cdot L_8 - L_1^2 \cdot L_2 \cdot L_6 = x^{12} - 11x^{10} + 43x^8 - 73x^6 + 53x^4 - 14x^2 + 1$$

FIGURE 9. The computation of the characteristic polynomial for the tree T_{12} depicting 3,4-dimethyldecane.

For large and highly branched trees, the application of Equation 33 becomes rather involved. In that case the pruning technique of Balasubramanian and Randić[53,54,95-99] or the computational procedure of Mohar[142] appear to be the convenient methods to use.

C. CYCLES

Cycles C_N (N $\geq$3) may be used to represent the carbon skeletons of cycloalkanes or [N]annulenes. The characteristic polynomial $P(C_N;x)$ of a cycle C_N can be computed by relating it to the characteristic polynomials $P(L_N;x)$ and $P(L_{N-2};x)$ of [N]chain and [N − 2]chain, respectively, which can be obtained by dissecting the [N]cycle[143] (see Figure 10).

$$P(C_N;x) = P(L_N;x) - P(L_{N-2};x) - 2 \qquad (34)$$

The idea behind Equation 34 is to reduce the [N]cycle into constituting chains in two steps: the first step is to remove an edge e from C_N and the second step is to remove this edge, incident vertices, and adjacent edges from C_N. In the first case a chain L_N with N vertices is obtained whose characteristic polynomial is $P(L_N;x)$ and in the second case a chain L_{N-2} with N − 2 vertices is generated whose characteristic polynomial is $P(L_{N-2};x)$. The term "− 2" is the contribution for the cycle closure.[84] How this term arises can be seen, for example, if one applies the Sachs formula to the [N]cycle. Since $P(L_N;x)$ and $P(L_{N-2};x)$ are known polynomials (see Table 1), one can easily compute the characteristic polynomial of any cycle by means of Equation 34.

D. MÖBIUS CYCLES

Möbius cycles $C_N^{M\ddot{o}}$ (N $\geq$3) may be used to depict Möbius annulenes.[136,144-146] A Möbius cycle $C_N^{M\ddot{o}}$ is shown in Figure 11.

The characteristic polynomial $P(C_N^{M\ddot{o}};x)$ of a Möbius cycle $C_N^{M\ddot{o}}$ can be obtained in the same way as the characteristic polynomial $P(C_N;x)$ of a cycle

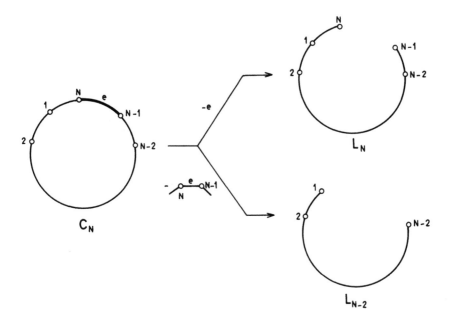

FIGURE 10. A dissection of an [N]cycle into two chains L_N and L_{N-2}.

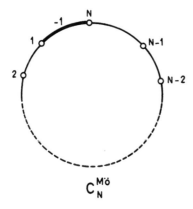

FIGURE 11. A diagram of the Möbius cycle $C_N^{M\ddot{o}}$.

C_N by dissecting $C_N^{M\ddot{o}}$ into two chains L_N and L_{N-2}, except that the constant factor is now "+2".

$$P(C_N^{M\ddot{o}};x) = P(L_N;x) - P(L_{N-2};x) + 2 \tag{35}$$

The term "+2" is the contribution for the Möbius cycle closure. How this term arises can also be seen, for example, if one applies the Sachs formula to the Möbius [N]Cycle.

An interesting footnote to this section is the fact that the characteristic polynomial of a cycle with an even number of vertices can be represented as the product of the characteristic polynomials of a cycle and a Möbius cycle, each half size.[147]

IV. THE LE VERRIER-FADDEEV-FRAME METHOD FOR COMPUTING THE CHARACTERISTIC POLYNOMIAL

A very efficient method for the computation of the characteristic polynomial of any real and symmetric square matrix is the recursive approach of Le Verrier, Faddeev, and Frame.[124-126] This approach was introduced by Le Verrier in 1840 and was used in the perturbative computation of the orbits of planets in our solar system. The approaches developed by Faddeev and by Frame essentially reduce to the Le Verrier method.[148] The *Le Verrier-Faddeev-Frame method* has been shown by Balasubramanian[3,45,58,127-131] and by the Zagreb group[4,9,132] to be a computationally convenient procedure for generating the characteristic polynomial for any graph.

The recursive formulae for computing the coefficients $a_n(G)$ of the characteristic polynomial for G, given in the following form:

$$P(G;x) = x^N - a_1 x^{N-1} - a_2 x^{N-2} - \dots - a_{N-1} x - a_N \qquad (36)$$

are presented for each of the three procedures as follows:

1. *The Le Verrier procedure*

$$a_n = (1/n)(\text{tr } \mathbf{A}^n - a_1 \text{ tr } \mathbf{A}^{n-1} - \dots - a_{n-1} \text{ tr } \mathbf{A}) \qquad (37)$$

where tr $\mathbf{A}$ is the trace of the adjacency matrix $\mathbf{A}$.

2. *The Faddeev procedure*

$$a_n = (1/n) \text{ tr } \mathbf{A}_n \qquad (38)$$

3. *The Frame procedure*

$$a_n = (1/n) \text{ tr } \mathbf{A} \mathbf{B}_{n-1} \qquad (39)$$

Equations 37 and 38 are related by

$$\mathbf{A}_n = \mathbf{A}^n - a_1 \mathbf{A}^{n-1} - \dots - a_{n-1} \mathbf{A} \qquad (40)$$

and Equations 38 and 39 by

$$\mathbf{A}_n = \mathbf{A} \mathbf{B}_{n-1} \qquad (41)$$

where B_{n-1} are N × N matrices which can be obtained from the following recursive relationship:

$$B_{n-1} = A_{n-1} - a_{n-1} I \tag{42}$$

It should be noted that

$$B_N = A_N - a_N I = 0 \tag{43}$$

Since Equations 37 to 39 can easily be transformed among themselves, all three of the above procedures reduce to the same method for computing the characteristic polynomial of a graph. Hence, the term Le Verrier-Faddeev-Frame method.

The algorithm for computing the coefficients of the characteristic polynomial for a given graph G by the Le Verrier-Faddeev-Frame method may be schematized as follows:

$$G \rightarrow A = A_1 \rightarrow a_1 \rightarrow B_1 \rightarrow A_2 \rightarrow a_2 \rightarrow \ldots$$

$$\ldots \rightarrow a_{N-1} \rightarrow B_{N-1} \rightarrow A_{N-1} \rightarrow a_N \rightarrow B_N = 0 \tag{44}$$

An example of computing the characteristic polynomial by the above algorithm is given for a weighted graph G_{VEW}, depicting cyclopropenthione (see Figure 5), in Table 2.

The Le Verrier-Faddeev-Frame algorithm was modified in the following way:[121] The adjacency matrix is first diagonalized (for example, by means of the Householder-QL algorithm)[81] and then the recursive approach of the Le Verrier-Faddeev-Frame is carried out with A_n and B_n matrices in the diagonal form. However, Balasubramanian[149] has shown that the Le Verrier-Faddeev-Frame method is more universal than this modification in its applicability to direct graphs, signed graphs, and complex graphs.

TABLE 2
The Computation of the Characteristic Polynomial for the Weighted Graph G_{VEW} Depicting Cyclopropenthione[a] by the Le Verrier-Faddeev-Frame Method

(1) $a_0 = 1$ (by definition)

$$A = \begin{bmatrix} h & k & 0 & 0 \\ k & 0 & 1 & 1 \\ 0 & 1 & 0 & 1 \\ 0 & 1 & 1 & 0 \end{bmatrix} = A_1$$

(2) $a_1 = \text{tr } A_1 = h$

$$B_1 = A_1 - a_1 I = \begin{bmatrix} 0 & k & 0 & 0 \\ k & -h & 1 & 1 \\ 0 & 1 & -h & 1 \\ 0 & 1 & 1 & -h \end{bmatrix}$$

$$A_2 = AB_1 = \begin{bmatrix} k^2 & 0 & k & k \\ 0 & k^2+2 & -h+1 & -h+1 \\ k & -h+1 & 2 & -h+1 \\ k & -h+1 & -h+1 & 2 \end{bmatrix}$$

(3) $a_2 = (1/2) \text{ tr } A_2 = k^2 + 3$

$$B_2 = A_2 - a_2 I = \begin{bmatrix} -3 & 0 & k & k \\ 0 & -1 & -h+1 & -h+1 \\ k & -h+1 & -(k^2+1) & -h+1 \\ k & -h+1 & -h+1 & -(k^2+1) \end{bmatrix}$$

TABLE 2 (continued)
The Computation of the Characteristic Polynomial for the
Weighted Graph G_{VEW} Depicting Cyclopropenthione[a]
by the Le Verrier-Faddeev-Frame Method

$$A_3 = AB_2 = \begin{bmatrix} -3h & -k & k & k \\ -k & -2(h-1) & -h & -h \\ k & -h & -2(h-1) & -(k^2+h) \\ k & -h & -(k+h) & -2(h-1) \end{bmatrix}$$

(4) $a_3 = (1/3)\,\mathrm{tr}\,A_3 = -3h + 2$

$$B_3 = A_3 - a_3\,I = \begin{bmatrix} -2 & -k & k & k \\ -k & h & -h & -h \\ k & -h & h & -(k^2+h) \\ k & -h & -(k^2+h) & h \end{bmatrix}$$

$$A_4 = AB_3 = \begin{bmatrix} -(k^2+2h) & 0 & 0 & 0 \\ 0 & -(k^2+2h) & 0 & 0 \\ 0 & 0 & -(k^2+2h) & 0 \\ 0 & 0 & 0 & -(k^2+2h) \end{bmatrix}$$

(5) $a_4 = (1/4)\,\mathrm{tr}\,A_4 = -(k^2+2h)$

$$B_4 = A_4 - a_4\,I = \begin{bmatrix} 0 & 0 & 0 & 0 \\ 0 & 0 & 0 & 0 \\ 0 & 0 & 0 & 0 \\ 0 & 0 & 0 & 0 \end{bmatrix}$$

(6) $P(G_{VEW};x) = x^4 - hx^3 - (k^2+3)\,x^2 + (3h-2)x - (k^2+2h)$

REFERENCES

1. **Knop, J. V. and Trinajstić, N.**, *Int. J. Quantum Chem.: Quantum Chem. Symp.*, 14, 503, 1980.
2. **Randić, M.**, *SIAM J. Alg. Disc. Math.*, 6, 145, 1985.
3. **Balasubramanian, K.**, in *Mathematics and Computational Concepts in Chemistry*, Trinajstić, N., Ed., Horwood, Chichester, 1986, 20.
4. **Trinajstić, N., Klein, D. J., and Randić, M.**, *Int. J. Quantum Chem.: Quantum Chem. Symp.*, 20, 699, 1986.
5. **Randić, M., Hosoya, H., Ohkami, N., and Trinajstić, N.**, *J. Math. Chem.*, 1, 97, 1987.
6. **Berman, D. M. and Holladay, K. W.**, *J. Math. Chem.*, 1, 405, 1987.
7. **King, R. B.**, *J. Math. Chem.*, 2, 89, 1988.
8. **Gutman, I.**, in *Chemical Graph Theory: Introduction and Fundamentals*, Bonchev, D. and Rouvray, D. H., Eds., Abacus Press/Gordon & Breach, New York, 1991, 133.
9. **Trinajstić, N.**, *Rep. Mol. Theory*, 1, 185, 1990.
10. **Mowshowitz, A.**, *J. Comb. Theory (B)*, 17, 177, 1972.
11. **Hosoya, H.**, *Theoret. Chim. Acta*, 25, 215, 1972.
12. **Trinajstić, N.**, *Croat. Chem. Acta*, 49, 593, 1977.
13. **Cvetković, D. M., Doob, M., and Sachs, H.**, *Spectra of Graphs: Theory and Application*, Academic Press, New York, 1980.
14. **Gutman, I. and Polansky, O. E.**, *Mathematical Concepts in Organic Chemistry*, Springer-Verlag, Berlin, 1986.
15. **Coulson, C. A.**, *Proc. Cambridge Philos. Soc.*, 46, 202, 1950; *Proc. Phys. Soc. (London)*, 60, 257, 1948.
16. **Spialter, L.**, *J. Chem. Doc.*, 4, 269, 1964.
17. **Sachs, H.**, *Publ. Math. (Debrecen)*, 11, 119, 1964.
18. **Kurosh, A. G.**, *Higher Algebra*, Mir Moscow, 1980, 154, 3rd printing.
19. **Kasum, D., Trinajstić, N., and Gutman, I.**, *Croat. Chem. Acta*, 54, 321, 1981.
20. **Collatz, L. and Sinogowitz, U.**, *Abh. Math. Semin. Univ. Hamburg*, 21, 63, 1967.
21. **Balaban, A. T. and Harary, F.**, *J. Chem. Doc.*, 11, 258, 1971.
22. **Schwenk, A. J.**, in *New Directions in the Theory of Graphs*, Harary, F., Ed., Academic Press, New York, 1973, 275.
23. **Gutman, I. and Trinajstić, N.**, *Topics. Curr. Chem.*, 42, 49, 1973.
24. **Herndon, W. C.**, *Tetrahedron Lett.*, 671, 1974.
25. **Živković, T., Trinajstić, N., and Randić, M.**, *Mol. Phys.*, 30, 517, 1975.
26. **D'Amato, S. S., Gimarc, B. M., and Trinajstić, N.**, *Croat. Chem. Acta*, 54, 1, 1981.
27. **Knop, J. V., Müller, W. R., Szymanski, K., Trinajstić, N., Kleiner, A., and Randić, M.**, *J. Math. Phys.*, 27, 2601, 1986.
28. **Randić, M., Barysz, M., Nowakowski, J., Nikolić, S., and Trinajstić, N.**, *J. Mol. Struct. (Theochem.)*, 185, 95, 1989.
29. **Lowe, J. P. and Davis, M. V.**, *J. Math. Chem.*, 5, 275, 1990.
30. **Harary, F., King, C., Mowshowitz, A., and Read, R. C.**, *Bull. London Math. Soc.*, 3, 321, 1971.
31. **Trinajstić, N.**, *J. Math. Chem.*, 2, 197, 1988.
32. **Gutman, I. and Polansky, O. E.**, *Theoret. Chim. Acta*, 60, 203, 1981.
33. **Polansky, O. E. and Zander, M.**, *J. Mol. Struct.*, 84, 361, 1982.
34. **Polansky, O. E.**, in *Mathematics and Computational Concepts in Chemistry*, Horwood, Chichester, 1986, 262.
35. **Polansky, O. E.**, in *MATH/CHEM/COMP 1988*, Graovac, A., Ed., Elsevier, Amsterdam, 1989, 65.
36. **Polansky, O. E.**, in *Atomic Hypothesis and the Concept of Molecular Structure*, Maksić, Z. B., Ed., Springer-Verlag, Berlin, 1990, 29.
37. **Graovac, A., Gutman, I., Trinajstić, N., and Živković, T.**, *Theoret. Chim. Acta*, 26, 67, 1972.

38. Randić, M., Ruščić, B., and Trinajstić, N., *Croat. Chem. Acta*, 54, 295, 1981.
39. Eilfeld, P. and Schmidt, W., *J. Electron Spectr. Rel. Phenom.*, 24, 101, 1981.
40. Ruščić, B., Trinajstić, N., and Křivka, P., *Theoret. Chim. Acta*, 69, 107, 1986.
41. Herndon, W. C., *J. Am. Chem. Soc.*, 95, 2404, 1973.
42. Herndon, W. C., *J. Am. Chem. Soc.*, 98, 887, 1976.
43. Herndon, W. C., *Isr. J. Chem.*, 20, 270, 1980.
44. Trinajstić, N., in *Modern Theoretical Chemistry — Semiempirical Methods of Electronic Structure Calculation, Part A. Techniques*, Vol. 7, Segal, G. A., Plenum Press, New York, 1977, 1.
45. Balasubramanian, K., *Chem. Rev.*, 85, 599, 1985.
46. Balaban, A. T., *J. Mol. Struct. (Theochem)*, 120, 117, 1985.
47. Wang, Y., George, T. F., Lindsay, D. M., and Beri, A. C., *J. Chem. Phys.*, 86, 3493, 1987.
48. Lindsay, D. M., Wang, Y., and George, T. F., *J. Chem. Phys.*, 86, 3500, 1987.
49. Glass, L., *J. Chem. Phys.*, 63, 1325, 1975.
50. King, R. B., *Theoret. Chim. Acta*, 56, 269, 1980.
51. Amit, R., Hall, C. A., and Porsching, T. A., *J. Comput. Phys.*, 40, 183, 1981.
52. Perugi, E., Liberto, F., and Monroy, G., *J. Phys. A*, 16, 811, 1983.
53. Balasubramanian, K., *J. Math. Chem.*, 2, 69, 1988.
54. Balasubramanian, K., *J. Math. Chem.*, 4, 89, 1990.
55. Essam, J. W. and Fisher, M. E., *Rev. Mod. Phys.*, 42, 271, 1970.
56. Marcus, R. A., *J. Chem. Phys.*, 43, 2643, 1965.
57. Randić, M., *J. Comput. Chem.*, 1, 386, 1980.
58. Balasubramanian, K., *J. Comput. Chem.*, 9, 43, 1985.
59. Balasubramanian, K., *Comput. Chem.*, 12, 106, 1991.
60. Sykes, M. F. and Fisher, M. E., *Adv. Phys.*, 9, 315, 1960.
61. Baker, G. K., *J. Math. Phys.*, 7, 2238, 1966.
62. Brastow, W. and Schinitzel, A., *J. Stat. Phys.*, 4, 103, 1972.
63. Wall, F. T. and Klein, D. J., *Proc. Natl. Acad. Sci. U.S.A.*, 76, 1529, 1979.
64. Klein, D. J. and Seitz, W. A., in *Chemical Applications of Topology and Graph Theory*, King, R. B., Ed., Elsevier, Amsterdam, 1983, 430.
65. Jiang, Y., Tang, A., and Hoffmann, R., *Theoret. Chim. Acta*, 66, 183, 1984.
66. Jiang, Y. and Zhang, H., *J. Math. Chem.*, 3, 357, 1989.
67. Gutman, I., *Colloq. Math. Soc. Jànos Bolyai*, 18, 429, 1976.
68. Wilson, R. J., *Introduction to Graph Theory*, Oliver & Boyd, Edinburgh, 1972, 44.
69. Gutman, I., Mallion, R. B., and Essam, J. W., *Mol. Phys.*, 50, 859, 1983.
70. O'Leary, B. and Mallion, R. B., in *Graph Theory and Topology in Chemistry*, King, R. B. and Rouvray, D. H., Eds., Elsevier, Amsterdam, 1987, 544.
71. Balasubramanian, K., *J. Chem. Phys.*, 73, 3321, 1980; *J. Math. Chem.*, 3, 227, 1989.
72. Balasubramanian, K., in *Computational Chemical Graph Theory*, Rouvray, D. H., Ed., Nova Science Publishers, New York, 1990, 67.
73. Kirby, E. C., in *Graph Theory and Topology in Chemistry*, King, R. B. and Rouvray, D. H., Eds., Elsevier, Amsterdam, 1987, 529.
74. Kirby, E. C., *Croat. Chem. Acta*, 59, 635, 1986; *J. Math. Chem.*, 1, 175, 1987.
75. Grajcar, L., Berthier, G., Faure, J., and Fleury, J.-P., *Theoret. Chim. Acta*, 71, 299, 1987.
76. Schwenk, A. J., in *Graphs and Combinatorics*, Bari, R. and Harary, F., Eds., Springer-Verlag, Berlin, 1974, 153.
77. Rovnyak, J., *Am. Math. Mon.*, 94, 289, 1987.
78. Cvetković, D. M., Doob, M., Gutman, I., and Torgašev, A., *Recent Results in the Theory of Graph Spectra*, North-Holland, Amsterdam, 1988.
79. Rao, T. M., *Comput. Math. Appl.*, 4, 61, 1978.
80. Goldstein, B. N. and Shevelev, E. L., *J. Theoret. Biol.*, 112, 493, 1985.

81. **Wilkinson, J. H.**, *The Algebraic Eigenvalue Problem*, Clarendon Press, Oxford, 1965; **Fröberg, C. E.**, *Introduction to Numerical Analysis*, 2nd ed., Addison-Wesley, Reading, MA, 1970.

82. **Hess, B. A., Jr., Schaad, L. J., and Agranat, I.**, *J. Am. Chem. Soc.*, 100, 5268, 1978.

83. **Džonova-Jerman-Blažič, B., Mohar, B., and Trinajstić, N.**, in *Applications of Information and Control Systems*, Lainiotis, D. G. and Tzannes, N. S., Eds., Reidel, Dordrecht, 1980, 395.

84. **Heilbronner, E.**, *Helv. Chim. Acta*, 36, 170, 1953.

85. **Hori, J.-i. and Asahi, T.**, *Prog. Theoret. Phys.*, 17, 523, 1957.

86. **Jido, Y., Inagaki, T., and Fukutome, H.**, *Prog. Theoret. Phys.*, 48, 808, 1972.

87. **Tesár, A. and Fillo, L.**, *Transfer Matrix Method*, Kluwer Academic Publishers, Dordrecht, 1988.

88. **Tang, A.-C. and Kiang, Y.-S.**, *Sci. Sin.*, 19, 208, 1976.

89. **Tang, A.-C. and Kiang, Y.-S.**, *Sci. Sin.*, 20, 595, 1977.

90. **Kiang, Y.-S.**, *Int. Quantum Chem.: Quantum Chem. Symp.*, 15, 293, 1981.

91. **Kaulgud, M. V. and Chitgopkar, V. H.**, *J. Chem. Soc. Faraday Trans. II*, 73, 1385, 1977.

92. **Kaulgud, M. V. and Chitgopkar, V. H.**, *J. Chem. Soc. Faraday Trans. II*, 74, 951, 1978.

93. **Gutman, I.**, *J. Chem. Soc. Faraday Trans. II*, 76, 1161, 1980.

94. **Graovac, A., Polansky, O. E., and Tyutyulkov, N.**, *Croat. Chem. Acta*, 56, 325, 1983.

95. **Balasubramanian, K. and Randić, M.**, *Theoret. Chim. Acta*, 61, 307, 1982.

96. **Balasubramanian, K.**, *Int. Quantum Chem.*, 21, 581, 1982.

97. **Balasubramanian, K.**, in *Chemical Applications of Topology and Graph Theory*, King, R. B., Ed., Elsevier, Amsterdam, 1983, 243.

98. **Balasubramanian, K. and Randić, M.**, *Int. J. Quantum Chem.*, 28, 481, 1985.

99. **Balasubramanian, K.**, *J. Math. Chem.*, 3, 147, 1989.

100. **Randić, M., Baker, B., and Kleiner, A. F.**, *Int. J. Quantum Chem.: Quantum Chem. Symp.*, 19, 107, 1986.

101. **McClelland, B. J.**, *J. Chem. Soc. Faraday Trans. II*, 78, 911, 1982.

102. **McClelland, B. J.**, *Mol. Phys.*, 45, 189, 1982.

103. **Hosoya, H. and Ohkami, N.**, *J. Comput. Chem.*, 4, 585, 1983.

104. **Randić, M.**, *Theoret. Chim. Acta*, 62, 485, 1983.

105. **Hosoya, H. and Randić, M.**, *Theoret. Chim. Acta*, 63, 473, 1983.

106. **Kassman, A. J.**, *Theoret. Chim. Acta*, 67, 255, 1985.

107. **He, WH. and He, WJ.**, *Theoret. Chim. Acta*, 70, 35, 1986.

108. **McWorter, W. A., Jr.**, *Math. Mag.*, 56, 158, 1983.

109. **Barakat, R.**, *Theoret. Chim. Acta*, 69, 35, 1986.

110. **Randić, M.**, *J. Math. Chem.*, 1, 145, 1987.

111. **Brocas, J.**, *Theoret. Chim. Acta*, 68, 155, 1985.

112. **Krylov, A. N.**, *Izv. Akad. Nauk S.S.S.R., Otd. Mat. Estest. Nauk*, Ser. 7, 491, 1931.

113. **Křivka, P., Mallion, R. B., and Trinajstić, N.**, *J. Mol. Struct. (Theochem.)*, 164, 363, 1988.

114. **Dias, J. R.**, *Can. J. Chem.*, 65, 734, 1987.

115. **Dias, J. R.**, in *Graph Theory and Topology in Chemistry*, King, R. B. and Rouvray, D. H., Eds., Elsevier, Amsterdam, 1987, 466.

116. **Dias, J. R.**, *Handbook of Polycyclic Hydrocarbons. Part A. Benzenoid Hydrocarbons*, Elsevier, Amsterdam, 1987.

117. **Dias, J. R.**, *J. Mol. Struct. (Theochem.)*, 165, 125, 1988.

118. **Dias, J. R.**, *J. Math. Chem.*, 4, 127, 1990.

119. **Dias, J. R.**, *Theoret. Chim. Acta*, 68, 107, 1985.

120. **Herndon, W. C., Radhakrishnan, T. P., and Živković, T. P.,** *Chem. Phys. Lett.,* 152, 233, 1988.
121. **Živković, T. P.,** *J. Math. Chem.,* 4, 143, 1990; *J. Comput. Chem.,* 11, 217, 1990.
122. **Rosenfeld, V. R. and Gutman, I.,** *Math. Chem. (Mülheim/Ruhr),* 24, 191, 1989.
123. **Goodwin, T. H. and Vand, V.,** *J. Chem. Soc.,* 1683, 1955; **Kirby, E. C.,** *J. Chem. Res.,* S, 4, 1984.
124. **Le Verrier, U. J. J.,** *J. Math.,* 5, 95, 1840; 5, 220, 1840.
125. **Faddeev, D. K. and Sominskii, I. S.,** *Problems in Higher Algebra,* W.H. Freeman, San Francisco, 1965.
126. **Dwyer, P. S.,** *Linear Computations,* John Wiley & Sons, New York, 1951, 225.
127. **Balasubramanian, K.,** *Theoret. Chim. Acta,* 65, 49, 1984.
128. **Balasubramanian, K.,** *J. Comput. Chem.,* 5, 387, 1984.
129. **Ramaraj, R. and Balasubramanian, K.,** *J. Comput. Chem.,* 6, 122, 1985.
130. **Balasubramanian, K.,** *J. Comput. Chem.,* 6, 656, 1985.
131. **Balasubramanian, K.,** *J. Comput. Chem.,* 9, 204, 1988.
132. **Křivka, P., Jeričević, Ž., and Trinajstić, N.,** *Int. J. Quantum Chem.: Quantum Chem. Symp.,* 19, 129, 1986.
133. **Samuel, I.,** *C.R. Acad. Sci.,* 229, 1236, 1949.
134. **Gourné, R.,** *J. Rech. C.N.R.S.,* 34, 81, 1950.
135. **Harary, F.,** *SIAM Rev.,* 4, 202, 1962.
136. **Graovac, A. and Trinajstić, N.,** *Croat. Chem. Acta,* 47, 95, 1975.
137. **Graovac, A. and Trinajstić, N.,** *J. Mol. Struct.,* 30, 316, 1975.
138. **Graovac, A., Polansky, O. E., Trinajstić, N., and Tyutyulkov, N.,** *Z. Naturforsch.,* 30a, 1696, 1975.
139. **Rigby, M. J., Mallion, R. B., and Day, A. C.,** *Chem. Phys. Lett.,* 51, 178, 1977; erratum *Chem. Phys. Lett.,* 53, 418, 1978.
140. **Dewar, M. J. S. and Longuet-Higgins, H. C.,** *Proc. R. Soc. (London), Ser. A,* 214, 482, 1952.
141. **Cvetković, D. M., Gutman, I., and Trinajstić, N.,** *J. Mol. Struct.,* 28, 289, 1975.
142. **Mohar, B.,** *J. Math. Chem.,* 3, 403, 1989.
143. **Gutman, I., Milun, M., and Trinajstić, N.,** *Croat. Chem. Acta,* 49, 441, 1977.
144. **Heilbronner, E.,** *Tetrahedron Lett.,* 1923, 1964.
145. **Zimmerman, H. E.,** *Acc. Chem. Res.,* 4, 272, 1972.
146. **Polansky, O. E.,** *Z. Naturforsch.,* 38a, 909, 1983.
147. **Kirby, E. C.,** unpublished.
148. **Barysz, M., Nikolić, S., and Trinajstić, N.,** *Math. Chem. (Mülheim/Ruhr),* 19, 117, 1986.
149. **Balasubramanian, K.,** *J. Comput. Chem.,* 12, 248, 1991.

Chapter 6

TOPOLOGICAL ASPECTS OF HÜCKEL THEORY

Hückel theory is the first introduced,[1-3] and the simplest,[4] form of the molecular orbital theory of conjugated molecules.[5] Since its inception, Hückel theory has been rather successfully used, on a qualitative level, as a guide for chemists in planning and interpreting experiments.[6-19] Attractive features of Hückel theory for experimental chemists are its simplicity and limited computational efforts. The last two decades have produced a number of results which indicate that the successful longevity of Hückel theory is based on the fact that this theory contains intrinsic information about the internal connectivity in the conjugated systems, i.e., it reflects the neighborhood of the atoms in conjugated structures.[4,15,20-25] This chapter will be concerned with the connection between Hückel theory and the topology of the molecular π-network.

I. ELEMENTS OF HÜCKEL THEORY[1-5,26,27]

In Hückel theory, only the π-electrons are considered explicitly. This is the result of the *Hückel approximation* of σ-π separability,

$$\langle \sigma | \pi \rangle = 0 \tag{1}$$

The Hückel approximation (Equation 1) is based on the following qualitative argument. A conjugated molecule is in general a planar, or near-planar, system. In this very special situation, molecular orbitals, MOs, describing the molecule may be partitioned into two orthogonal groups; σ MOs which are symmetric and π MOs which are antisymmetric to reflection in the plane of the molecule. Physically, the Hückel approximation may be viewed as one which has the π-electrons moving in a potential field due to the nuclei and a σ core, which is assumed to be rigid as the π-electrons move around. The Hückel approximation, when introduced, had a threefold justification. First, with this bold approximation, Hückel reduced an insurmountable computational problem to one which was feasible in those precomputer days. For example, benzene, instead of the formidable problem of 30 valence electrons, converts into a problem of only 6 electrons. Secondly, Hückel found that, by treating only the π-electrons explicitly, it is possible to reproduce many of the observed properties of unsaturated molecules.[1-3,28-30] Finally, organic chemists of his day have related quite correctly the physical and chemical properties of conjugated molecules (thermodynamic and structural parameters, spectroscopic features, reactivity) to the presence of their π-electrons. In addition, the subsequent work by a large number of researchers[4,5,15,26,27,31] has also supported the Hückel approximation, because it has revealed many useful correlations between the quantities obtained from Hückel theory and experimental findings.

The Hückel molecular orbitals, HMOs, of a conjugated molecule are eigenfunctions ψ_i of the effective one-electron Hamiltonian, called the Hückel Hamiltonian, $\hat{H}$ (Hückel), the precise nature of which is not specified,

$$\hat{H}(\text{Hückel})\psi_i = E_i\psi_i \qquad (i = 1,2,\ldots, N) \tag{2}$$

where E_i is the energy eigenvalue associated with ψ_i. N stands for the number of atoms in a molecule bearing π-electrons.

The individual Hückel orbital ψ_i is expressed as a linear combination of atomic orbitals (LCAO approximation),[32]

$$\psi_i = \sum_{r=1}^{N} c_{ir}\phi_r \tag{3}$$

where c_{ir} are the linear expansion coefficients, while ϕ_r is a $2p_z$ orbital on atom r. The summation is over all atoms r in a conjugated molecule.

The total π-electron energy, E_π, is given by

$$E_\pi = \sum_{i=1}^{N} g_i E_i \tag{4}$$

or more explicitly,

$$E_\pi = \sum_{i=1}^{N} g_i\left(\sum_r\sum_s c_{ir}^* H_{rs} c_{is} \Big/ \sum_r\sum_s c_{ir}^* c_{is} S_{rs}\right) \tag{5}$$

where g_i is the occupation number of ψ_i, i.e., the number of electrons that populate ψ_i, while H_{rs} and S_{rs} are shorthand notations for the integrals

$$H_{rs} = \int \phi_r^* \hat{H}(\text{Hückel})\phi_s \, d\tau \tag{6}$$

$$S_{rs} = \int \phi_r^* \phi_s \, d\tau \tag{7}$$

The coefficients c_{ir} are obtained from the requirement that E_π should be a minimum. Minimization of E_π by means of the variational procedure leads to a set of simultaneous, linear, homogeneous equations,

$$\sum_{t=1}^{N} c_{it}(H_{rt} - E_i S_{rt}) = 0) \qquad (i,r = 1,2,\ldots, N) \tag{8}$$

These equations have nontrivial solutions (the trivial solution has all $c_{it} = 0$) only if the corresponding Hückel (secular) determinant vanishes:

$$\det|H_{rt} - E_i S_{rt}| = 0 \qquad (i,r = 1,2,\ldots, N) \tag{9}$$

The Hückel determinant can be simplified by using the set of approximations originally introduced by Bloch[33] and utilized by Hückel.[1] Since the Hückel Hamiltonian is not known explicitly, the backbone of the *Bloch-Hückel approximations* is the presumption that the entries to the Hückel determinant may either be related to empirical quantities or removed entirely.

1. The diagonal elements, H_{rr}

$$H_{rr} = \int \phi_r^* \hat{H}(\text{Hückel}) \phi_r \, d\tau = \alpha \tag{10}$$

The diagonal elements H_{rr} are *Coulomb integrals* with empirical values α. It is assumed that the αs are constant for all orbitals ϕ_r centered on similar atoms r, regardless of the variations in the neighboring atoms and groups.

2. The off-diagonal elements, H_{rs}

The off-diagonal elements H_{rs} are assumed to be zero unless orbitals ϕ_r and ϕ_s are located on bonded atoms

$$H_{rs} = \int \phi_r^* \hat{H}(\text{Hückel}) \phi_s \, d\tau = \begin{cases} \beta & \text{if atom } r \text{ and } s \text{ are bonded} \\ 0 & \text{otherwise} \end{cases} \tag{11}$$

For bonded atoms the off-diagonal elements H_{rt} are called *resonance integrals* with empirical value β. It is assumed that the βs have the same value for π-bonds between the same kind of atoms, regardless of the environment.

3. The overlap integral, S_{rs}

$$S_{rs} = \int \phi_r^* \phi_s \, d\tau = \begin{cases} 1 & \text{if } r = s \\ 0 & \text{if } r \neq s \end{cases} \tag{12}$$

This is the so-called *zero-overlap approximation*, which is rather drastic because it says that there is no overlap between the atoms making up the π-network of a conjugated molecule. However, the neglect of overlap is justified empirically by the success of Hückel theory over the past 60 years. The inclusion of overlap among the bonded atoms changes the spacing of energy levels and the values of the total π-electron energies, but other quantities remain unchanged.[34]

The parameters α and β for carbon atoms (α_C and β_{CC}) are taken to be the reference points. The introduction of a heteroatom X into the conjugated system alters both of these parameters. These changes are expressed as:

$$\alpha_X = \alpha_C + h_X \beta_{CC} \tag{13}$$

$$\beta_{CX} = k_{CX} \beta_{CC} \tag{14}$$

where h_X and k_{CX} are dimensionless parameters for a given heteroatom in a specific molecular environment. Tables of the parameters h_X and k_{CX} are collected, for example, by Purcell and Singer.[35]

II. ISOMORPHISM OF HÜCKEL THEORY AND GRAPH SPECTRAL THEORY

Equation 9 may be presented in a more compact, and elegant, form if matrix notation is used. Thus, Equation 9 in the matrix form is given by

$$\det |\mathbf{H} - E_i\mathbf{S}| = 0 \quad (i = 1,2,..., N) \tag{15}$$

where $\mathbf{H}$ and $\mathbf{S}$ are the Hamiltonian and overlap matrices, respectively. As a consequence of the Bloch-Hückel approximations, the matrices $\mathbf{H}$ and $\mathbf{S}$ have the following composition:[36]

$$\mathbf{H} = \alpha\mathbf{I} + \beta\mathbf{A} \tag{16}$$

$$\mathbf{S} = \mathbf{I} \tag{17}$$

where $\mathbf{A}$ is the adjacency matrix of the Hückel graph (conjugated molecule). The matrix $[\mathbf{H} - E_i\mathbf{S}]$ is called the *Hückel matrix*.

Substitution of $\mathbf{H}$ and $\mathbf{S}$ by Equations 16 and 17 into Equation 15 and dividing each row of a determinant by β leads to

$$\det \left| \frac{E_i - \alpha}{\beta}\mathbf{I} - \mathbf{A} \right| = 0 \quad (i = 1,2,..., N) \tag{18}$$

If the normalized form of Hückel theory is used, i.e., if β is taken as the energy unit and α the zero-energy reference point ($\beta = 1$ and $\alpha = 0$), Equation 18 becomes

$$\det |E_i\mathbf{I} - \mathbf{A}| = 0 \quad (i = 1,2,..., N) \tag{19}$$

The comparison between the secular determinant (Equation 1, Chapter 5) and the Hückel determinant (Equation 19) reveals that the numbers E_i, representing the energies of individual Hückel orbitals, are *identical* to the elements of the spectrum of eigenvalues of the adjacency matrix of a Hückel graph (i.e., the graph spectrum)

$$E_i = x_i \quad (i = 1,2,..., N) \tag{20}$$

Since matrices $\mathbf{H}$ and $\mathbf{A}$ commute (this can easily be seen from Equation 16),

$$[\mathbf{H}, \mathbf{A}] = 0 \tag{21}$$

they possess the same set of eigenvectors. Therefore, the eigenvectors of the adjacency matrix are *identical* to the Hückel MOs. On account of this the

HMOs are sometimes called the *topological orbitals*. Equation 16 also discloses that the Hückel Hamiltonian is a function of the adjacency matrix,[21,37]

$$\mathbf{H} = \mathbf{H(A)} \tag{22}$$

This is due to the singular nature of the Hückel Hamiltonian, with the short-range forces being dominant in the effective potential.[38]

The above analysis leads to two important conclusions: (1) the spacing and general pattern of Hückel eigenvalues are specified by the connectivity (the neighborhood topology)[25] in the conjugated molecule; and (2) the connectivity (the neighborhood topology)[25] in the conjugated molecule, rather than its geometry, shapes up the form of Hückel orbitals. Thus, what chemists commonly refer to as Hückel theory is essentially the same thing as graph-spectral theory (in the area of planar connected undirected graphs with maximum valency 3) referred to by graph theoreticians. In fact Hückel theory and graph-spectral theory are isomorphic theories.[5,21,22,37,39-45]

At this point the reader should be reminded that in order to derive various quantities from the knowledge of Hückel orbitals and energies (total π-electron energy, charge density, bond order, free valence, etc.), one also needs to know the electronic configuration appropriate to the ground state (or to any other state, if required) of the molecule under consideration.[46,47] In order to assign the appropriate electronic configuration, the fundamental physical notion of the *Aufbau principle*[48] must be called upon. According to the Aufbau principle, each of the energy levels in a molecule can be populated by a maximum of two electrons of opposite spin in accordance with the *Pauli exclusion principle*.[48] The electrons are fed in from the lowest to the highest energy levels until all of the available electrons are used up. In cases when degeneracies occur in the energy levels the filling up process also involves the use of the *Hund rules*.[48]

The total π-electron energy E_π (the Hückel energy) of a conjugated molecule in the ground state is given by

$$E_\pi = E(HMO) = \sum_{i=1}^{N} g_i \, E_i \tag{23}$$

or by virtue of Equation 20,

$$E_\pi = \sum_{i=1}^{N} g_i \, x_i \tag{24}$$

where g_i is the occupation number of the ith molecular orbital which can take values 0, 1, or 2, respectively. The x_i values are conventionally ordered as

$$x_1 \geq x_2 \geq \ldots \geq x_N \tag{25}$$

Let the conjugated molecule have N atoms and N_e π-electrons. Then the ground-state Hückel energy is given by

$$E_\pi = \begin{cases} 2 \displaystyle\sum_{i=1}^{N_e/2} x_i & \text{if } N_e = \text{even} \\[2em] 2\left[\displaystyle\sum_{i=1}^{(N_e-1)/2} x_i + x_{(N_e+1)/2}\right] & \text{if } N_e = \text{odd} \end{cases} \tag{26}$$

Since for the majority of conjugated systems $N_e = N$, it follows

$$E_\pi = \begin{cases} 2 \displaystyle\sum_{i=1}^{N} x_i & \text{if } N = \text{even} \\[2em] 2\left[\displaystyle\sum_{i=1}^{(N-1)/2} x_i + x_{(N+1)/2}\right] & \text{if } N = \text{odd} \end{cases} \tag{27}$$

III. THE HÜCKEL SPECTRUM

The Hückel spectrum is given by an ordered sequence of eigenvalues of the Hückel matrix

$$\{x_1, x_2, \ldots, x_N\} \tag{28}$$

The extreme values of these eigenvalues are defined by the Frobenius theorem (see discussion in Chapter 5, Section I). Since the maximum valency in Hückel graphs is 3, the interval which limits all the elements in the Hückel spectrum is

$$-3 \leq x_i \leq +3 \qquad (i = 1, 2, \ldots, N) \tag{29}$$

Because the maximum graph-theoretical valency in the linear polyenes and annulenes is 2, the extreme values in the Hückel spectra for these systems are ± 2. Since the maximum valency in K_2 graph is 1, the ethylene Hückel spectrum consists only of two integers $\{1, -1\}$. There are only five other conjugated molecules whose Hückel spectra have only integers.[49] These are given in Figure 1.

There exist nonidentical conjugated molecules with identical Hückel spectra.[21,22,50] They are named *isospectral molecules*.[51-54] A classical example of isospectral molecules is provided by the following pair: 1,4-divinylbenzene and 2-phenylbutadiene. Their graphs are given in Figure 2. These two molecules possess identical Hückel spectra: $\{\pm 2.214, \pm 1.675, \pm 1.000, \pm 1.000, \pm 0.539\}$.

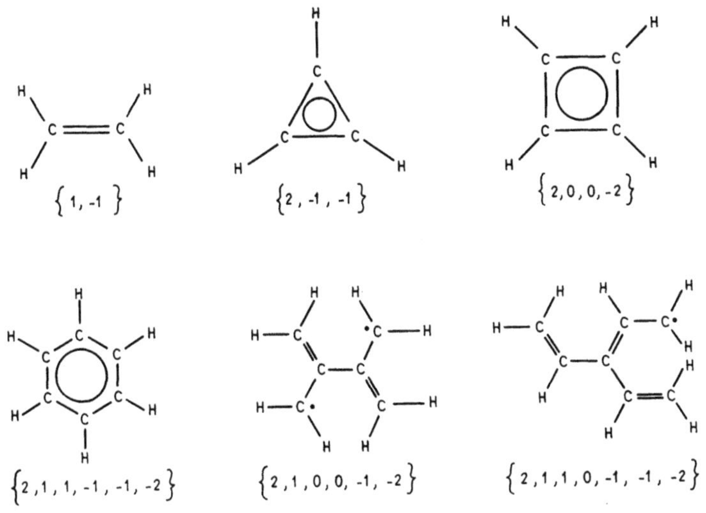

FIGURE 1. The only conjugated systems with integral Hückel spectra.

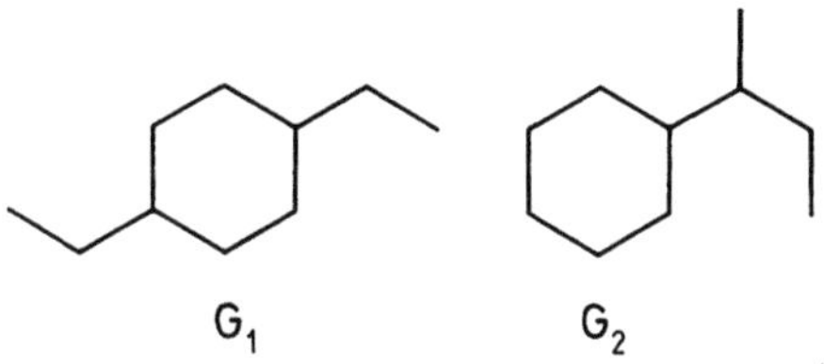

FIGURE 2. Hückel graphs G_1 and G_2 corresponding to 1,4-divinylbenzene and 2-phenylbutadiene, respectively.

There are a variety of methods for the construction of isospectral molecules and graphs available in the literature.[50-65]

A more frequent occurrence is that in which the Hückel spectra of two different conjugated molecules have one or more common eigenvalues.[34] In some cases, the spectrum of one Hückel graph contains the complete spectrum of a second, smaller Hückel graph. The two graphs are then said to be *subspectral*.[56] The corresponding molecules are called *subspectral molecules*.[56,57] The origin of subspectrality was studied in details by Dias.[50,66,67] A typical pair of subspectral molecules is naphthalene and butadiene. The Hückel spectrum of naphthalene $\{\pm 2.30278, \pm 1.61803, \pm 1.30278, \pm 1.00000, \pm 0.61803\}$ contains the complete Hückel spectrum of butadiene: $\{\pm 1.61803, \pm 0.61803\}$.

The Hückel spectrum may be partitioned into three subsets corresponding to the bonding, nonbonding, and antibonding energy levels, denoted by N_+,

N_0, and N_-, respectively. They are related to the number of atoms N in the conjugated systems:

$$N_+ + N_0 + N_- = N \qquad (30)$$

These quantities are important for the chemistry of conjugated molecules. It is especially important to establish whether the nonbonding molecular orbitals, NBMOs, are present in the Hückel spectrum, because their existence leads to the prediction[68] that such molecules should have open-shell ground states and be very reactive. Although in reality the situation is much more complicated, for example, because of the Jahn-Teller effects in the case of the triplet ground states,[69] it is an established fact[70] that the structures possessing NBMOs are rarely encountered in the chemistry of conjugated hydrocarbons and even these are obtained under the drastic conditions of low-temperature chemistry.[71]

A. A METHOD FOR THE ENUMERATION OF NBMOs

There is a simple way to determine whether or not a conjugated system has NBMOs. Since the number of NBMOs is identical with N_0, i.e., the number of zeros in the Hückel spectrum, the determinant of the adjacency matrix vanishes

$$\det \mathbf{A} = \prod_{i=1}^{N} x_i = 0 \qquad \text{if } x_m = 0 \qquad (31)$$

Therefore, the determinant of **A** will be zero if, and only if, there exists at least one zero element in the Hückel spectrum of a molecule. Hence, the problem of whether or not a conjugated system has NBMOs can be solved. The question is how to obtain the number of NBMOs without going through the procedure of diagonalization of the Hückel matrix. One way will be described below.

If $\mathbf{C} = (c_1, c_2, \ldots, c_N)$ is a NBMO (not necessarily normalized), the following equation holds:

$$\mathbf{C} \, \mathbf{A} = \mathbf{0} \qquad (32)$$

This equation in the scalar form represents a *zero-sum rule* first used by Longuet-Higgins,[68]

$$\sum_{r \to s} c_r = 0 \qquad (s = 1,2,\ldots, N) \qquad (33)$$

The summation is over all vertices r joined to the vertex s. The number of the independent parameters in an unnormalized NBMO is equal to the number

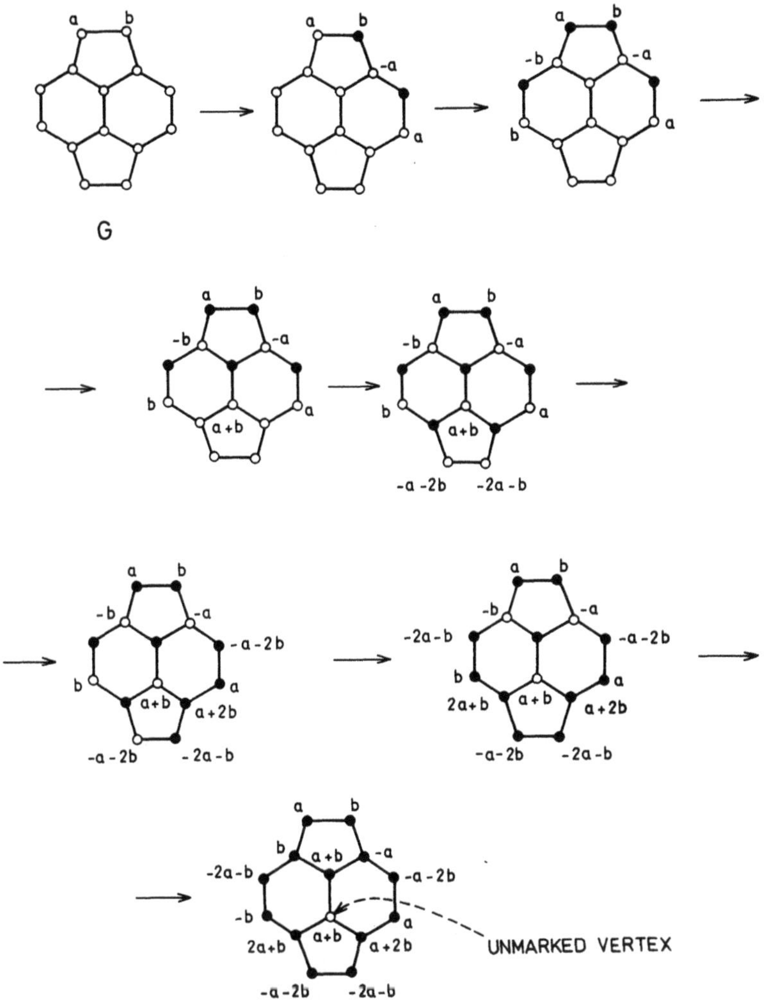

FIGURE 3. Graphical enumeration of the NBMO for pyracyclene represented by a graph G.

N_0 in the Hückel spectrum.[72] Thus, the enumeration of N_0 is reduced to a determination of the number of independent parameters in the unnormalized NBMO which satisfy the zero-sum rule (Equation 33). The application of this method is illustrated for pyracyclene in Figure 3. The procedure is as follows: Equation 33 is satisfied stepwise for each vertex of a graph G. Vertices for which the zero-sum rule is fulfilled are denoted by ●. In order that Equation 32 holds for the last (unmarked) vertex of the pyracyclene graph, the following relation must be equal to zero:

$$a + b = 0 \qquad\qquad (34)$$

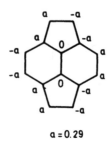

$$a = 0.29$$

FIGURE 4.   The normalized NBMO of pyracyclene.

If this is so, only *one* parameter is independent

$$a = -b \tag{35}$$

and consequently, N_0 (pyracyclene) $= 1$. The normalized NBMO of pyra-cyclene is given in Figure 4.

The presence of the NBMO in the Hückel spectrum of pyracyclene in-dicates this molecule is unstable. The chemistry of pyracyclene appears to be in accord with this prediction: pyracyclene is a very unstable molecule and it could not be isolated from solution.[73]

The procedure described in this section may be simplified by applying certain graph-theoretical transformations under which the value of N_0 remains the same, but which make the graphical enumeration much easier to carry out.[74,75]

B. THE ENUMERATION OF N_0 AND N_+ FROM THE CHARACTERISTIC POLYNOMIAL

The information of numbers N_0, N_+, and N_- is contained in the char-acteristic polynomial of a conjugated system.[76,77] N_0 and N_+ are available directly from $P(G;x)$ by using the following relations:

$$a_N = a_{N-1} = \ldots = a_{N-N_0+1} = 0 \tag{36}$$

$$a_{N-N_0} \neq 0 \tag{37}$$

$$Ch(a_0, a_1, \ldots, a_{N-1}, a_N) = N_+ \tag{38}$$

where Ch denotes the number of sign changes in the sequence of coefficients a_i (i $= 0, 1, 2, \ldots, N$). Equation 38 is known as the *Descartes theorem.*[78] The application of Equations 36 to 38 is shown in Table 1.

TABLE 1
The Enumeration of Numbers N_0, N_+, and N_- for Pentalene Depicted by Graph G

$$P(G;x) = x^8 - 9x^6 + 24x^4 - 4x^3 - 16x^2 + 8x$$
$a_8 = 0$, but $a_7 \neq 0$. Therefore
$8 = N - N_0 + 1$ or $N_0 = 1$
$N_+ = Ch(1, -9, 24, -4, -16, 8) = 4$
$N_- = N - N_+ - N_0 = 3$

G

C. A GRAPH-THEORETICAL CLASSIFICATION OF CONJUGATED HYDROCARBONS BASED ON THEIR SPECTRAL CHARACTERISTICS

The quantity

$$N_+ - N_- = \sigma \tag{39}$$

is called a *graph (molecule) signature* and when combined with N_0 can be used as a basis for a graph-theoretical classification of conjugated hydrocarbons.[79] There are *four* classes of conjugated hydrocarbons possible:

(1) Stable molecules characterized by

$$N_+ = N_- \text{ and } N_0 = 0 \tag{40}$$

Molecules in this class are expected to be stable. However, the above criterion is *necessary*, but not sufficient for a given molecule to be stable in a chemical sense. The prediction of the stability of a chemical compound is a rather complex problem involving thermodynamic and kinetic aspects. Therefore, the above may serve only as a suggested indicator for neutral molecules whose stabilities may vary from very stable molecules to rather unstable species.

(2) Polyradical molecules characterized by

$$N_+ = N_- \text{ and } N_0 > 0 \tag{41}$$

Structures belonging to this class are extremely reactive.

(3) Electron-deficient molecules characterized by

$$N_+ > N_- \text{ and } N_0 \geq 0 \tag{42}$$

These molecules tend to generate stable anions by accepting π-electrons from a suitable donor in their empty MO bonding levels.[73]

(4) Electron-excessive molecules characterized by

$$N_+ < N_- \text{ and } N_0 \geq 0 \tag{43}$$

These molecules show a tendency to generate stable cations by releasing π-electrons from the antibonding MO levels.[73]

Pentalene studied earlier in this chapter belongs to class 3. It is a reactive electron-deficient molecule whose dianion is relatively stable.[80]

Polycyclic hydrocarbons consisting of $(4m + 2)$ and/or $(4m)$ rings belong either to class 1 or class 2. The presence of a $(4m - 1)$-membered ring in a graph is a necessary topological condition for a corresponding molecule to belong to class 3, while the $(4m + 1)$ ring is required for class 4.

This classification is in fair agreement with experimental findings.[73,81-89] The above is, of course, a very simple approach and it should be used rather cautiously.

IV. CHARGE DENSITIES AND BOND ORDERS IN CONJUGATED SYSTEMS

It was Coulson who first defined charge densities and bond orders in Hückel theory.[90] An element in the Coulson charge density-bond order matrix $\mathbf{P}^C$ is given by

$$(\mathbf{P}^C)_{rs} = \sum_{i=1}^{N} g_i \, c_{ir} \, c_{is} \tag{44}$$

where c_{ir} and c_{is} are linear expansion coefficients defining the contributions of the rth and sth atomic orbitals, respectively, in the ith MO.

Alternative definitions, within the MO theory, of charge densities and bond orders were offered by Mulliken[91] and Ruedenberg.[36] Elements of the Mulliken, $\mathbf{P}^M$, and the Ruedenberg, $\mathbf{P}^R$, charge density-bond order matrices are defined as

$$(\mathbf{P}^M)_{rs} = (1 + S_{rs}) \sum_{i=1}^{N} g_i \, \frac{c_{ir} \, c_{is}}{1 + x_i \, S_{rs}} \tag{45}$$

$$(\mathbf{P})^R_{rs} = \sum_{i=1}^{N} g_i \, \frac{c_{ir} \, c_{is}}{x_i} \tag{46}$$

The formulae by Coulson, Mulliken, and Ruedenberg can be collected together in the general charge density-bond order matrix

$$(\mathbf{P})_{rs} = \sum_{i=1}^{N} g_i \, c_{ir} \, c_{is} \, f(x_i) \tag{47}$$

where $f(x_i)$ is a weighting factor. An analysis by Ham and Ruedenberg[92] indicated that there is a connection between the Coulson and Mulliken charge density-bond order matrices,

$$[(\mathbf{P}^M)_{rs} - 1] \approx 1.2[(\mathbf{P}^C)_{rs} - 1] \tag{48}$$

In addition, they have also shown that the charge density-bond order matrix **P** may be expressed in terms of the adjacency matrix of the molecule,

$$\mathbf{P} = \mathbf{P}^C\, f(\mathbf{A}) \tag{49}$$

where

$$\mathbf{P}^C = \mathbf{I} + (\mathbf{A}^2)^{1/2}\, \mathbf{A}^{-1} \tag{50}$$

Since the bond orders may be correlated with bond lengths,[90,93] the important conclusion of these considerations is that molecular topology in a subtle way influences the geometry of a conjugated system.

Pauling and co-workers[94] have defined a bond order within the resonance theory, a simple version of the valence bond theory,[95] as

$$(\mathbf{P}^P)_{rs} = \frac{K[G - (r - s)]}{K(G)} \tag{51}$$

where $K(G)$ and $K[G - (r - s)]$ are the numbers of 1-factors (Kekulé structures) for a graph G (molecule) and subgraph $G - (r - s)$ (molecular fragment) obtained by deletion of the vertices (atoms) r and s and the adjacent edges (bonds) from G (molecule). The above formulation of $\mathbf{P}^P$ is given in the graph-theoretical formalism.[96-98] The definition of the bond order within the resonance theory precedes the definitions in the framework of the MO theory.

An interesting result is obtained when $\mathbf{P}^R$ and $\mathbf{P}^P$ are compared. It is found that for alternant hydrocarbons without 4m-membered rings and NBMOs, bond orders defined by Ruedenberg are *identical* with the Pauling bond orders,[99]

$$\mathbf{P}^R - \mathbf{P}^P \tag{52}$$

This result reveals that one can use the tables of Hückel values[34] to evaluate Pauling bond orders. More important is the conclusion that both HMO and Pauling concepts of bond order are, in essence, topological quantities.

V. THE TWO-COLOR PROBLEM IN HÜCKEL THEORY

Conjugated molecules that can be represented by the bipartite graphs (the two-colored structures) are called *alternant hydrocarbons* (AHs).[100] *Nonalternant hydrocarbons* (NAHs) are depicted by nonbipartite graphs.

The fact that the adjacency matrix belonging to a bipartite graph appears in the block form (see Equation 17 in Chapter 4) may be used in discussing the *pairing theorem* of Coulson and Rushbrooke.[101] According to the pairing theorem if x_i is an element of a Hückel spectrum (with an associated eigenvector $\psi_i = \Sigma_r\, c^*_{ir}\, \phi_r + \Sigma_s\, c^0_{is}\, \phi_s$), then $-x_i$ (with an associated eigenvector $\psi_i = \Sigma_r\, c^*_{ir}\, \phi_r - \Sigma_s\, c^0_{is}\, \phi_s$) is also an eigenvalue of $\mathbf{A(G)}$. In other words, the pairing theorem states that Hückel energy levels of an alternant hydrocarbon should be symmetrically distributed about $x = 0$. This result was also observed by Hückel[3] eight years before the work by Coulson and Rushbrooke. Since the Hückel eigenvalues for alternants appear in pairs: $x_i + x_{N+1-i} = 0$, in order to construct the whole set of the HMO levels it is sufficient to obtain, in some way, only positive (or negative) eigenvalues and their corresponding eigenvectors.[102]

There are a number of proofs and demonstration of the pairing theorem available in the literature.[4,5,26,27,34,44,67,101-114] Here the matrix formulation of the spectral symmetry will be briefly given.

Let the eigenvalue equation

$$\mathbf{C_i A} = x_i \mathbf{C_i} \tag{53}$$

be satisfied for the eigenvector

$$\mathbf{C_i} = [c^*_{i1},\, c^*_{i2},\, \ldots,\, c^*_{is},\, c^0_{i,s+1},\, c^0_{i,s+2},\, \ldots,\, c^0_{i,s+u}] \tag{54}$$

There should be another eigenvalue equation, producing the other member of the pair: $x_i,\ -x_i$,

$$\mathbf{C_i^{pair}\, A} = -x_i \mathbf{C_i^{pair}} \tag{55}$$

with the eigenvector

$$\mathbf{C_i^{pair}} = [c^*_{i1},\, c^*_{i2},\, \ldots,\, c^*_{is},\, -c^0_{i,s+1},\, -c^0_{i,s+2},\, \ldots,\, -c^0_{i,s+u}] \tag{56}$$

In the matrix notation Equations 54 and 56 are given by

$$\mathbf{C} = \begin{bmatrix} \mathbf{C}_{\text{bonding}} \\ \mathbf{C}_{\text{antibonding}} \end{bmatrix} = \begin{bmatrix} \mathbf{N} & \mathbf{M} \\ \mathbf{N} & -\mathbf{M} \end{bmatrix} \tag{57}$$

where

$$\mathbf{N} = [c^*_{i1},\, c^*_{i2},\, \ldots,\, c^*_{is}] \tag{58}$$

$$\mathbf{M} = [c^0_{i,s+1},\, c^0_{i,s+2},\, \ldots,\, c^0_{i,s+u}] \tag{59}$$

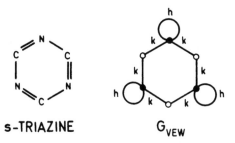

s-TRIAZINE G_{VEW}

FIGURE 5.   The skeleton of s-triazine and the corresponding vertex- and edge-weighted graph G_{VEW}.

Since the adjacency matrix of an alternant structure can, by an appropriate labeling of atoms (vertices), be given in the block form, the eigenvalue matrix **X** is expressed as follows:

$$X = C A C^\dagger = \begin{bmatrix} Y & 0 \\ 0 & -Y \end{bmatrix} \tag{60}$$

where

$$Y/2 = N B M^\dagger = M B^\dagger N^\dagger \tag{61}$$

where **B** is a submatrix of **A**, while **Y** is a diagonal matrix. Therefore, the symmetry of Hückel levels is the very consequence of the particular topology of alternant systems.

Another way of introducing the pairing theorem is by using the Sachs formula. Because the bipartite graphs, by a theorem, do not contain odd-membered cycles, it follows that for alternant structures $S_n = 0$ and $a_n(G) = 0$ for $n = 2i + 1$. Therefore, the characteristic polynomial of alternant hydrocarbons is necessarily of the form

$$P(G;x) = \sum_{i=0}^{[N/2]} (-1)^i a_{2i}(G) x^{N-2i} \tag{62}$$

where the coefficients $a_{2i}(G)$ are the nonnegative quantities for all i. The structure of the polynomial (Equation 62) is such that the corresponding spectrum must be symmetrically arranged with respect to $x = 0$, because if x is a zero of a polynomial (Equation 62), then $-x$ must also be a zero. This is possibly the simplest demonstration of the pairing theorem.

The pairing theorem may be extended to include other bicolorable structures encountered in chemistry. However, in these structures the starred and unstarred positions are populated with different types of atoms. Here is given a theorem which is valid for alternant heteroconjugated structures, like s-triazine (see Figure 5).

The extension of the pairing theorem to include bipartite graphs of the type G_{VEW} is stated as follows:[108] if G_{VEW} is a bipartite graph with the same number of vertices in each set, and exactly those in the first are weighted, then

$$x_i + x_{N+1-i} = h \quad \text{for} \quad 1 \leq i \leq N \tag{63}$$

To illustrate the use of the above theorem let us consider the *s*-triazine system. Its characteristic polynomial is given by

$$P(G_{VEW};x) = x^6 - 3hx^5 + (3h^2 - 6k^2)x^4 - (12hk^2 - h^3)x^3$$
$$+ (9k^4 - 6h^2)x^2 - 9hk^4x - 4k^6 \tag{64}$$

Taking, for example, $h = 2.00$ and $k = 1.00$, the following spectrum is obtained:

$$\{3.23607, 2.41421, 2.41421, -0.41421, -0.41421, -1.23607\} \tag{65}$$

The use of Equation 63 shows that the above spectrum is symmetric about $h = 2.00$. For example,

$$x_2 + x_5 = 2.41421 - 0.41421 = 2.00 \tag{66}$$

There are other extensions and generalizations of the pairing theorem reported in the literature. For example, Tyutyulkov and Polansky[115] and Karabunarliev and Tyutyulkov[116] have shown that the pairing theorem can be extended to some classes of nonclassical (non-Kekuléan) systems even containing heteroatoms.

A. PROPERTIES OF ALTERNANT HYDROCARBONS

There are some unique properties of AHs which can be easily rationalized in terms of the pairing theorem.[117]

If NBMOs appear in the Hückel spectrum of *even* AHs, their number will always be *even,* because all elements of the spectrum must be paired. However, in the spectrum of *odd* AHs there is always at least *one* NBMO present, because $N - 1$ levels will be symmetrically arranged around $x = 0$, and this leaves a single level which must be a NBMO. Examples of even and odd AHs are given in Figure 6.

Another consequence of the pairing theorem is the fact that the charge density distribution in alternant hydrocarbons is uniform and that each atom carries the π-electron density value equal unity.[34] This can be proved in the following way. The Coulson charge density-bond order matrix is given by

$$\mathbf{P}^C = 2\mathbf{C}^\dagger_{\text{bonding}} \mathbf{C}_{\text{bonding}} \tag{67}$$

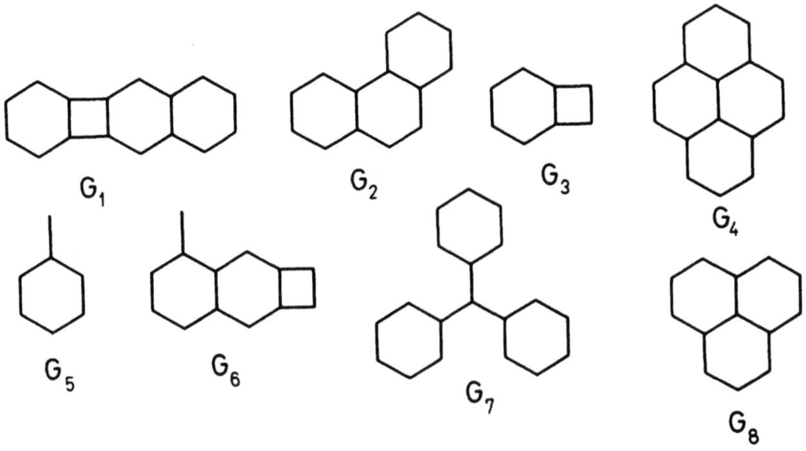

FIGURE 6.   Examples of even (G_1, G_2, G_3, G_4) and odd alternant hydrocarbons (G_5, G_6, G_7, G_8).

where

$$C_{bonding} = [N, M] \tag{68}$$

Applying the relations,

$$C^\dagger C = I \tag{69}$$

$$2N^\dagger N = 2M^\dagger M \tag{70}$$

matrix (67) may be given in the form,

$$P^C = \begin{bmatrix} I & 2N^\dagger M \\ 2M^\dagger N & I \end{bmatrix} \tag{71}$$

From this matrix is seen that the π-electron charge density on any arbitrary atom r of any AH is identically equal to unity

$$q_r = (P^C)_{rr} = 1 \tag{72}$$

The above result leads to the prediction of a zero π-component of dipole moment in AHs. Available experimental data verify this prediction: AHs have indeed either zero or negligibly small total dipole moments.[118]

The fact that the charge-density distribution is uniform is related to the self-consistency of the topological orbitals of alternant systems. It has been shown[119,120] that this self-consistency of topological MOs depends on the relations

$$(\mathbf{P}^C)_{11} = (\mathbf{P}^C)_{22} = \ldots = (\mathbf{P}^C)_{\pi} = \ldots = (\mathbf{P}^C)_{NN} \qquad (73)$$

An important property of AHs is that the bond orders between atoms of the same color are zero:

$$(\mathbf{P}^C)_{**} = (\mathbf{P}^C)_{\text{oo}} = 0 \qquad (74)$$

The nonvanishing bond orders in AHs are possible only between differently colored atoms

$$(\mathbf{P})^C)_{*\text{o}} = 2\mathbf{N}^\dagger\,\mathbf{M} \qquad (75)$$

A previously derived topological formula for bond orders by Hall,[103,105,121]

$$(\mathbf{P}^C)_{*\text{o}} = (\mathbf{BB}^\dagger)^{-1/2}\,\mathbf{B} \qquad (76)$$

is just a special case of the more general formula (Equation 50). This can be demonstrated in the following way. Equation 50 may be given in a more convenient form:

$$\mathbf{P}^C = \begin{bmatrix} \mathbf{I} & \mathbf{B}(\mathbf{B}^\dagger\,\mathbf{B})^{-1/2} \\ \mathbf{B}^\dagger(\mathbf{B}\,\mathbf{B})^{-1/2} & \mathbf{I} \end{bmatrix} \qquad (77)$$

where use is made of the matrices

$$\mathbf{A}^{-1} = \begin{bmatrix} \mathbf{0} & (\mathbf{B}^\dagger)^{-1} \\ \mathbf{B}^{-1} & \mathbf{0} \end{bmatrix} \qquad (78)$$

and

$$(\mathbf{A}^2)^{1/2} = \begin{bmatrix} (\mathbf{B}\,\mathbf{B}^\dagger)^{1/2} & \mathbf{0} \\ \mathbf{0} & (\mathbf{B}^\dagger\,\mathbf{B})^{1/2} \end{bmatrix} \qquad (79)$$

Equation 77 immediately enables one to obtain the results of Equations 72, 74, and 75. In the case of NAHs, the matrix $[(\mathbf{A}^2)^{1/2}\,\mathbf{A}^{-1}]$ has nonvanishing diagonal elements; the π-electron charge density is thus nonuniform, and dipole moments are nonzero.

Recently, several papers have appeared on *topological charge stabilization*.[23,122] The concept of topological charge stabilization is based on the observation that the pattern of charge densities in a molecule depends on the

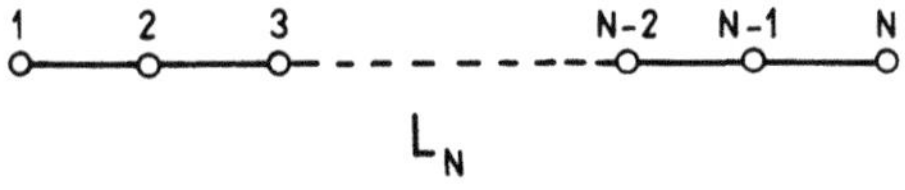

FIGURE 7.   A chain L_N of N vertices depicting [N]polyene.

topology of the molecule (molecular connectivity) and the number of electrons available. Atoms of greater electronegativity tend to be placed in those positions where the topology of the structure inclines to pile up extra charge. Since such heteroatomic systems are preferentially stabilized by charge distributions established by molecular connectivity, this effect has been called the *rule of topological charge stabilization.*[23] This rule has been applied successfully to a number of inorganic and organic planar and nonplanar systems.[123-129]

VI. EIGENVALUES OF LINEAR POLYENES

A linear polyene, [N]polyene, of the general formula $CH_2(CH)_{N-2}CH_2$, may be represented by a chain of N vertices (see Figure 7).

The corresponding adjacency matrix is of the form

$$A(L_N) = \begin{array}{cccccccc} & 1 & 2 & 3 & \cdots & N-2 & N-1 & N \\ 1 & 0 & 1 & 0 & \cdots & \cdots & & \\ 2 & 1 & 0 & 1 & & & \mathbf{0} & \\ 3 & & 1 & 0 & 1 & & & \\ \cdots & & & 1 & 0 & 1 & & \\ \cdots & & & & & & & \\ \cdots & & & & & & & \\ N-2 & & & & 1 & 0 & 1 & \\ N-1 & & \mathbf{0} & & & 1 & 0 & 1 \\ N & & & & & & 1 & 0 \end{array} \qquad (80)$$

This matrix reflects clearly the topology of the chain, nonzero entries appearing only alongside the principal diagonal. Since it is known that $|x| \leq 2$ for chains one can formally substitute $x = 2\cos\theta$ in det $|x\mathbf{I} - \mathbf{A}(L_N)|$. The expansion of the determinant gives the following characteristic polynomial:[130,131]

$$P(L_N;x) = 2\cos\theta\, P(L_{N-1};x) - P(L_{N-2};x) \qquad (81)$$

It can be shown by induction that

$$P(L_N;x) = \frac{\sin(N+1)\theta}{\sin\theta} \tag{82}$$

This polynomial vanishes when and only when

$$\theta = \frac{i\pi}{N+1} \qquad (i = 1,2,..., N) \tag{83}$$

that is, when

$$x_i = 2\cos\frac{i\pi}{N+1} \qquad (i = 1,2,..., N) \tag{84}$$

These are in fact the N distinct zeros of the polynomial $P(L_N;x)$. The eigenvalues of linear polyenes are identical to Equation 84 because of Equation 20.

A very useful mnemonic device for determining the energy level diagrams of linear polyenes has been found by Frost and Musulin.[132] They have noticed that one can derive the Hückel energies for linear polyenes by inscribing the appropriate regular hexagon in a cycle of radius 2 with one vertex down. The cycle is centered at $E_i = 0$. For every intersection of the cycle and polygon, there is an MO energy level corresponding to the vertical displacement. Since every polygon has one vertex down and intersecting at the bottom of the cycle, there always will be an Hückel level of 2, i.e., the value corresponding to the cycle radius. The horizontal projection of all other points of contact onto a vertical line yields the exact eigenvalues for the cyclic molecule. In order to obtain the energy levels of a given [N]polyene, a fictitious [N+2]polyene is added on a [2N+2]polygon in such a way that the first vertex of the polyene graph touches the second vertex of the polygon. The use of the Frost-Musulin device is illustrated in Figure 8. It should be noted that all eigenvalues appear in pairs. This is yet another demonstration of the pairing theorem. An additional interesting observation is that [10]annulene and butadiene represent a pair of subspectral molecules because the whole Hückel spectrum of butadiene is contained in the spectrum of [10]annulene. However, they represent a very special pair of subspectral molecules because the Hückel spectrum of butadiene is contained twice in the spectrum of [10]annulene, as can be seen from the Frost-Musulin diagram. Furthermore, from the knowledge of the butadiene Hückel spectrum: $\{\pm 1.618, \pm 0.618\}$ and the relationship between [2N+2]annulene and [N]polyene as demonstrated by Frost and Musulin, one can immediately construct the complete spectrum of [10]annulene: $\{\pm 2, \pm 1.618, \pm 1.618, \pm 0.618, \pm 0.618\}$. This result is valid for any pair of [2N+2]annulene and [N]polyene. The above represents one way to show the origin of subspectrality.

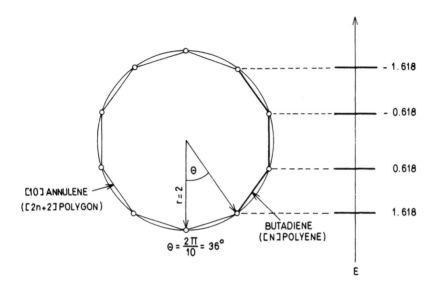

FIGURE 8. The Frost-Musulin projection diagram for the eigenvalues of butadiene.

VII. EIGENVALUES OF ANNULENES

A cyclic polyene, [N]annulene,[133] of the general formula $(CN)_N$ may be represented by a cycle of N vertices (see Figure 9). The corresponding adjacency matrix is very similar to that belonging to [N]polyene,[134]

$$A(C_N) = \begin{array}{c} \\ 1 \\ 2 \\ 3 \\ \cdots \\ \cdots \\ N-1 \\ N \end{array} \begin{array}{cccccccc} 1 & 2 & 3 & \cdots & N-1 & N \\ 0 & 1 & 0 & \cdots & 0 & \boxed{1} \\ 1 & 0 & 1 & \cdots & 0 & 0 \\ 0 & 1 & 0 & \cdots & 0 & 0 \\ \cdot\cdot & \cdot\cdot & & \cdots & \cdot\cdot & \cdot\cdot \\ \cdot\cdot & \cdot\cdot & & \cdots & \cdot\cdot & \cdot\cdot \\ 0 & 0 & 0 & \cdots & 0 & 1 \\ \boxed{1} & 0 & 0 & \cdots & 1 & 0 \end{array} \qquad (85)$$

$A(C_N)$ differs from $A(L_N)$ in two extra entries of *unity* in the top right-hand and bottom left-hand corners of the matrix. This type of matrix is called a *circulant matrix* and its properties and uses have been studied extensively in formal mathematics, quantum chemistry, and mathematical chemistry.[4,135-140]

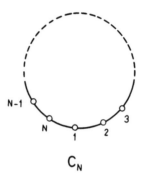

$$C_N$$

FIGURE 9.   A cycle C_N of N vertices depicting [N]annulene.

The zeros of the corresponding characteristic polynomial may be obtained in the closed form as follows.

The general form of this circulant matrix is given by

$$
\begin{bmatrix}
a_1 & a_2 & a_3 & \cdots & a_N \\
a_N & a_1 & a_2 & \cdots & a_{N-1} \\
a_{N-1} & a_N & a_1 & \cdots & a_{N-2} \\
\cdots & \cdots & \cdots & \cdots & \cdots \\
\cdots & \cdots & \cdots & \cdots & \cdots \\
a_2 & a_3 & a_4 & \cdots & a_1
\end{bmatrix}
\tag{86}
$$

The circulant matrix is completely defined when the elements in its first row are specified. The diagonalization of the matrix (Equation 86) produces a set of eigenvectors and eigenvalues,

$$
\begin{bmatrix}
a_1 & a_2 & a_3 & \cdots & a_N \\
a_N & a_1 & a_2 & \cdots & a_{N-1} \\
a_{N-1} & a_N & a_1 & \cdots & a_{N-2} \\
\cdots & \cdots & \cdots & \cdots \\
\cdots & \cdots & \cdots & \cdots \\
a_2 & a_3 & a_4 & \cdots & a_1
\end{bmatrix}
\begin{bmatrix}
1 \\
\rho_i \\
\rho_i^2 \\
\cdots \\
\cdots \\
\rho_i^{N-1}
\end{bmatrix}
= x_i
\begin{bmatrix}
1 \\
\rho_i \\
\rho_i^2 \\
\cdots \\
\cdots \\
\rho_i^{N-1}
\end{bmatrix}
\tag{87}
$$

Thus, the vector

$$V_k = \begin{bmatrix} 1 \\ \rho_i \\ \rho_i^2 \\ \cdot \\ \cdot \\ \cdot \\ \rho_i^{N-1} \end{bmatrix} \tag{88}$$

is an eigenvector of the circulant matrix (Equation 86) with the eigenvalue x_i given by

$$x_i = a_1\rho_i^0 + a_2\rho_i^1 + a_3\rho_i^2 + \ldots + a_N\rho_i^{N-1} \tag{89}$$

where ρ_i is an Nth root of the scalar equation

$$\rho_i^N = 1 \qquad (i = 0,1,2,\ldots, N - 1) \tag{90}$$

The solutions of the above equation are as follows:

$$\rho_i = \cos(2i\pi/N) + i \sin(2i\pi/N) \qquad (i = 0,1,2,\ldots, N - 1) \tag{91}$$

Going back to the circulant matrix of the [N]annulene (Equation 85), it is noted that all elements in the first row vanish, but a_2 and a_N, both being equal to unity. Thus, using this result, Equation 89 reduces to

$$x_i = \rho_i + \rho_i^{N-1} \tag{92}$$

Since the matrix (Equation 85) is a Hermitian matrix

$$\rho_i^{N-1} = \rho_i^* \tag{93}$$

it follows that

$$x_i = \rho_i + \rho_i^* \tag{94}$$

or, by substituting Equation 91 for ρ_i,

$$x_i = 2 \cos(2i\pi/N) \tag{95}$$

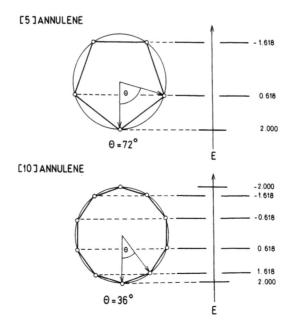

[5]ANNULENE

Θ = 72°

[10]ANNULENE

Θ = 36°

FIGURE 10.   The Frost-Musulin projection diagrams for the eigenvalues of [5]- and [10]annulenes.

Because of (20),

$$E_i = 2 \cos(2i\pi/N) \qquad (i = 0,1,2, ..., N - 1) \qquad (96)$$

This is the equation for calculating the eigenvalues of [N]annulenes.

A mnemonic device of Frost and Musulin[132] may also be used for constructing the energy-level diagrams of [N]annulenes. This is demonstrated for [5]- and [10]annulene in Figure 10.

The following should be pointed out. When N is even, there will be two nondegenerate levels, $+2$ for $i = 0$ and -2 for $i = N/2$. All the others will occur in degenerate pairs. This is yet another demonstration of the applicability of the pairing theorem. When N is odd, there will be an energy level of 2 units for $i = 0$. All the remaining energy levels will appear in degenerate pairs.

VIII. EIGENVALUES OF MÖBIUS ANNULENES

A Möbius [N]annulene is a cyclic structure with at least one phase dislocation, as described in Chapter 3, Section V. The Möbius [N]annulene may be represented graph-theoretically by a Möbius cycle of N vertices denoted

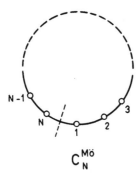

$$C_N^{M\ddot{o}}$$

FIGURE 11. A Möbius cycle $C_N^{M\ddot{o}}$ of N vertices depicting Möbius [N]annulene. A broken line denotes the position of the phase dislocation. At this position the corresponding bond is weighted by -1.

by $C_N^{M\ddot{o}}$ (see Figure 11).[141] The corresponding adjacency matrix is very similar to that of the [N]cycle,

$$
A(C_N^{M\ddot{o}}) = \quad
\begin{array}{c c c c c c}
 & 1 & 2 & 3 & N-1 & N \\
\end{array}
\begin{array}{c}
1 \\
2 \\
3 \\
\cdots \\
\cdot \\
N-1 \\
N
\end{array}
\left[
\begin{array}{c c c c c c}
0 & 1 & 0 & 0 & \circled{-1} \\
1 & 0 & 1 & 0 & 0 \\
0 & 1 & 0 & 0 & 0 \\
\cdot\cdot & \cdot\cdot & \cdots\cdot & \cdot\cdot & \cdot\cdot \\
\cdot\cdot & \cdot\cdot & \cdots\cdot & \cdot\cdot & \cdot\cdot \\
0 & 0 & 0 & 0 & 1 \\
\circled{-1} & 0 & 0 & 1 & 0
\end{array}
\right]
\qquad (97)
$$

$A(C_N^{M\ddot{o}})$ differs from $A(C_N)$ in the sign of the two entries in the top right-hand and bottom left-hand of the matrix.

The eigenvalues of the Möbius [N]annulenes may be obtained in the closed form using similar reasoning as in the case of [N]annulenes because the matrix (Equation 97) is a special case of the circulant matrix. This type of circulant matrix is referred to as a *skew-circulant matrix*.[140] The expression for calculating the eigenvalues of Möbius cycles is given by[142]

$$
E_i = 2 \cos \frac{(2i + 1)\pi}{N} \qquad (i = 0,1,..., N - 1) \qquad (98)
$$

A Möbius mnemonic device may be introduced following the ideas of Frost and Musulin.[132] This was actually done by Zimmerman.[143-145] In order

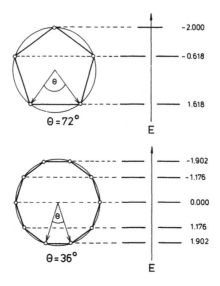

FIGURE 12. The Möbius-Zimmerman projection diagrams for the eigenvalues of Möbius [5]- and [10]-annulenes.

to obtain the Möbius- Zimmerman projection diagrams a vertex at the bottom is replaced by the side of the polygon. Thus, the first two eigenvalues of the Möbius annulene will always be a degenerate pair. In the case of the Möbius annulenes with N = even, all energy levels will appear in the degenerate pairs. An additional feature appears in the Möbius annulenes with $N/2$ = odd, because they will always have a degenerate pair with zero eigenvalue. In the case of Möbius annulene with N = odd, $N − 1$ levels will appear in degenerate pairs. The last energy level will be a single one with the value $−2.00$.

The application of the Möbius-Zimmerman device to Möbius [5]- and [10]annulenes is demonstrated in Figure 12. Inspection of Figure 12 reveals that the pairing theorem holds in the Möbius systems with N = even. This is one of the topological effects that operates in both Möbius and Hückel systems with identical result.

IX. A CLASSIFICATION SCHEME FOR MONOCYCLIC SYSTEMS

In Table 2 the polynomials and spectra of Hückel and Möbius [N]annulenes from $N = 3$ up to $N = 8$ are reported. Inspection of the results in Table 2 shows that the *alternant* Hückel and Möbius annulenes differ only in the value of the a_N coefficient, while the nonalternant systems differ only in the *sign* of the a_N coefficient. This observation is not surprising because the monocyclic systems contain cycles only in the S_N set of Sachs graphs. The construction of Sachs graphs with N vertices for Hückel and Möbius annulenes

TABLE 2
The Characteristic Polynomials and Spectra of Hückel and Möbius [N]Annulenes

Hückel [N]annulenes

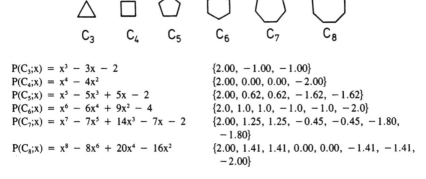

$$C_3 \quad C_4 \quad C_5 \quad C_6 \quad C_7 \quad C_8$$

$P(C_3;x) = x^3 - 3x - 2$ $\{2.00, -1.00, -1.00\}$

$P(C_4;x) = x^4 - 4x^2$ $\{2.00, 0.00, 0.00, -2.00\}$

$P(C_5;x) = x^5 - 5x^3 + 5x - 2$ $\{2.00, 0.62, 0.62, -1.62, -1.62\}$

$P(C_6;x) = x^6 - 6x^4 + 9x^2 - 4$ $\{2.0, 1.0, 1.0, -1.0, -1.0, -2.0\}$

$P(C_7;x) = x^7 - 7x^5 + 14x^3 - 7x - 2$ $\{2.00, 1.25, 1.25, -0.45, -0.45, -1.80, -1.80\}$

$P(C_8;x) = x^8 - 8x^6 + 20x^4 - 16x^2$ $\{2.00, 1.41, 1.41, 0.00, 0.00, -1.41, -1.41, -2.00\}$

Möbius [N]annulenes

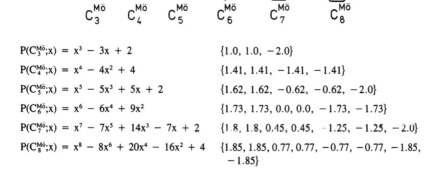

$$C_3^{M\ddot{o}} \quad C_4^{M\ddot{o}} \quad C_5^{M\ddot{o}} \quad C_6^{M\ddot{o}} \quad C_7^{M\ddot{o}} \quad C_8^{M\ddot{o}}$$

$P(C_3^{M\ddot{o}};x) = x^3 - 3x + 2$ $\{1.0, 1.0, -2.0\}$

$P(C_4^{M\ddot{o}};x) = x^4 - 4x^2 + 4$ $\{1.41, 1.41, -1.41, -1.41\}$

$P(C_5^{M\ddot{o}};x) = x^5 - 5x^3 + 5x + 2$ $\{1.62, 1.62, -0.62, -0.62, -2.0\}$

$P(C_6^{M\ddot{o}};x) = x^6 - 6x^4 + 9x^2$ $\{1.73, 1.73, 0.0, 0.0, -1.73, -1.73\}$

$P(C_7^{M\ddot{o}};x) = x^7 - 7x^5 + 14x^3 - 7x + 2$ $\{1.8, 1.8, 0.45, 0.45, -1.25, -1.25, -2.0\}$

$P(C_8^{M\ddot{o}};x) = x^8 - 8x^6 + 20x^4 - 16x^2 + 4$ $\{1.85, 1.85, 0.77, 0.77, -0.77, -0.77, -1.85, -1.85\}$

are given in Table 3. The results in Table 3 lead to the conclusion that the coefficient a_N may be used to establish a simple classification scheme for all monocyclic systems:

1. N = even
    $a_N = 0$ for Hückel [4m]annulenes and Möbius [4m + 2]annulenes
    $$a_N = \begin{cases} -4 \text{ Hückel } [4m+2]\text{annulenes} \\ 4 \text{ Möbius } [4m]\text{annulenes} \end{cases}$$
2. N = odd
    $$a_N = \begin{cases} -2 \text{ Hückel } [4m+1]\text{- or } [4m+3]\text{annulenes} \\ 2 \text{ Möbius } [4m+1]\text{- or } [4m+3]\text{annulenes} \end{cases}$$

TABLE 3
The Sachs Graphs and a_N Coefficients of Hückel and Möbius [N]Annulenes

Hückel [N]annulenes

N = even

$$a_N = 2(-1)^{N/2}\, 2^0 + (-1)^1\, 2^1 = \begin{cases} 0 & \text{for} \quad N = 4m \\ -4 & \text{for} \quad N = 4m + 2 \end{cases}$$

N = odd

$$a_N = (-1)^1\, 2^1 = -2 \quad \text{for} \quad N = 4m + 1 \quad \text{or} \quad N = 4m + 3$$

Möbius [N]annulenes

N = even

$$a_N = 2(-1)^{N/2+0}\, 2^0 + (-1)^{1+1}\, 2^1 = \begin{cases} 4 & \text{for} \quad N = 4m \\ 0 & \text{for} \quad N = 4m + 2 \end{cases}$$

N = odd

$$a_N = (-1)^{1+1}\, 2^1 = 2 \quad \text{for} \quad N = 4m + 1 \quad \text{or} \quad N = 4m + 3$$

This scheme is a graph-theoretical justification of the *generalized* Hückel rule which embraces Hückel and Möbius annulenes and according to which Hückel [4m + 2]- and Möbius [4m]annulenes should exhibit closed-shell stability and aromaticity, while Hückel [4m]- and Möbius [4m + 2]annulenes should exhibit open-shell reactivity and antiaromaticity.[134,142-149] Hückel [4m + 1]- and

TABLE 4
Generalized Hückel Rule Which Embraces Hückel and
Möbius Cyclic Systems

Monocyclic system	Cycles[a]					
	Even-membered		Odd-membered			
	4m	4m + 2	4m + 1		4m + 3	
			Cations	Anions	Cations	Anions
Hückel	−	+	−	+	+	−
Möbius	+	−	+	−	−	+

[a] (+) denotes stable, aromatic species; (−) denotes unstable, antiaromatic species.

[4m + 3]annulenes and Möbius [4m + 1]- and [4m + 3]annulenes are in between the [4m + 2] and [4m] systems. The stabilization or destabilization of these systems may be attained by adding or subtracting π-electrons. Hückel [4m + 1]annulenes become quite *stable* species by *accepting* one π-electron and producing monoanions, i.e., systems with [4m + 2] π-electrons, and quite *unstable* species by *giving away* a π-electron and producing monocations, i.e., monocyclic systems with [4m] π-electrons.[85,150] Hückel [4m + 3]annulenes behave in the opposite way. They stabilize by giving away a π-electron and producing monocations, i.e., [4m + 2] π-electron systems, and destabilize by accepting one π-electron in their half-empty MOs and thus producing monoanions, i.e., monocyclic systems with [4m] π-electrons.[85,150] Similarly, stabilization of Möbius [4m + 1] annulenes may be achieved by giving away a π-electron and producing monocations. Destabilization of such a system is obtained by adding one π-electron and thus producing a monoanion, i.e., a Möbius system with [4m + 2] π-electrons. A Möbius [4m + 3]system stabilizes by receiving a π-electron and producing a monoanion and destabilizes by giving away one π-electron and producing a monocation. All this may be conveniently tabulated as it is done in Table 4. This is also the basis for the use of the Hückel-Möbius concept for qualitatively studying (poly)cyclic molecules and transition states of certain pericyclic reactions (i.e., electrocyclic closures of polyenes) by considering the topology of the corresponding transition states.[11,143-145,151-159]

X. TOTAL π-ELECTRON ENERGY

The total π-electron energy, E_π, is one of the most important pieces of information about the conjugated molecule which may be obtained from the HMO calculations since it can, in a proper way, be related to the thermodynamic stability of conjugated structures. Therefore, E_π is always positive and larger E_π means larger thermodynamic stability for molecules of the same size (the same N). A parametrization scheme, based on thermodynamic data, by Schaad and Hess,[160] has produced agreement with experiment of the same degree of quantitative accuracy as the much more sophisticated SCF π-MO procedure developed by Dewar.[161] In addition, Schaad and Hess[160] have shown that in many instances E_π follows linearly the total (thermodynamically measurable) energy of the conjugated compound. The physical reasons why it is possible that a model as simple as the Hückel model can give not only qualitative, but sometimes also fair quantitative, agreement with the experimental findings are not well understood at present. The "empirical" assumption that the Hückel theory works and that it is a rather useful "pencil and paper" method for everyday use in the field of the chemistry of conjugated structures is accepted.

In this section it will be shown how E_π depends on the structure of a conjugated molecule. The early work on this problem arose from the necessity to obtain numerical values of E_π in the times before the availability of electronic computers for quantum chemical problems. The foremost work in this precomputer era was carried out by Coulson and co-workers.[100,101,162-166] Later when computer facilities were readily available, the emphasis was to uncover the regularities relating E_π to select topological invariants of the molecule.[167-187] The explicit or implicit use of chemical graph theory played a significant role in many of these investigations.[21,22,43,50,188,189]

A. THE FUNDAMENTAL IDENTITY FOR E_π

Let two quantities be defined: E_+ as the sum of the positive elements (bonding energy levels) N_+ of the Hückel spectrum (the Hückel MO energies) and E_- as the sum of the negative elements N_- of the Hückel spectrum, respectively,

$$E_+ = \sum_{i=1}^{N_+} x_i \tag{99}$$

$$E_- = \sum_{i=N_+ + N_0 + 1}^{N} x_i \tag{100}$$

The relationship between E_+ and E_- is defined as follows:

$$E_+ + E_- = 0 \tag{101}$$

$$E_+ - E_- = \sum_{i=1}^{N} |x_i| \tag{102}$$

where $|x_i|$ is the absolute value of x_i. Since all the N_+ (bonding) levels in the molecule are ordinarily doubly occupied and all N_- (antibonding) levels are unoccupied, the total π-electron energy may be also given as

$$E_\pi = 2E_+ \tag{103}$$

By means of Equation 101, Equation 103 becomes

$$E_\pi = E_+ - E_-$$
(104)

Coupling of Equations 104 and 102 produces the identity

$$E_\pi = \sum_{i=1}^{N} |x_i|$$
(105)

This identity is valid for AHs because of the pairing theorem. In fact, the identity (Equation 105) holds if

$$N_+ = N_-$$
(106)

However, if

$$N_+ \neq N_-$$
(107)

the identity is fulfilled approximately. This may be shown in the following way. If $N_+ \neq N_-$ and if there is a filled antibonding level $x_{N/2} < 0$, then

$$E_\pi = \sum_{i=1}^{N} |x_i| + 2x_{N/2}$$
(108)

If,

$$\sum_{i=1}^{N} |x_i| \gg 2x_{N/2}$$
(109)

relation (Equation 105) holds as a good approximation. Similarly, if there is present an empty bonding energy level

$$x_{(N/2)+1} > 0$$
(110)

then,

$$E_\pi = \sum_{i=1}^{N} |x_i| - 2x_{(N/2)+1}$$
(111)

Again if,

$$\sum_{i=1}^{N} |x_i| \gg 2x_{(N/2)+1}$$
(112)

relation (Equation 105) holds approximately. Thus, the relation (Equation 105) is fulfilled either exactly or as a very good approximation for all conjugated molecules. This result (Equation 105) is referred to as the *fundamental identity* for E_π.[190]

B. RELATIONS BETWEEN E_π, THE ADJACENCY MATRIX, AND THE CHARGE DENSITY-BOND ORDER MATRIX

Diagonalization of the adjacency matrix

$$\mathbf{C\,A} = \mathbf{X\,C} \tag{113}$$

produces the diagonal matrix X:

$$\mathbf{X} = \mathrm{diag}(x_1, x_2, \ldots, x_N) \tag{114}$$

Let the function $f(x)$ be defined for all x_i, $i = 1,2, \ldots ,N$. Let also

$$f(x) = \mathrm{diag}[f(x_1),f(x_2),\ldots,f(x_N)] \tag{115}$$

Then, by definition, the matrix function is given by

$$\mathbf{C}\,f(\mathbf{A}) = f(\mathbf{X})\,\mathbf{C} \tag{116}$$

or

$$f(\mathbf{A}) = \mathbf{C}^\dagger\,f(\mathbf{X})\,\mathbf{C} \tag{117}$$

In the scalar form Equation 117 becomes

$$[f(\mathbf{A})]_{rs} = \sum_{i=1}^{N} f(x_i)c_{ir}c_{is} \tag{118}$$

The above relation enables one to find relations between E_π, $A(G)$, and $\mathbf{P}^C$. If a formal function $g = g(x)$ is introduced such that $g(x_i) = g_i$, the following relationship is obtained from Equation 118,

$$\mathbf{P}^C = g(\mathbf{A}) \tag{119}$$

In the case of filled bonding and empty antibonding molecular orbitals,

$$x_i\,g(x_i) = x_i + |x_i| \tag{120}$$

from which it immediately follows,

$$\mathbf{A}\,\mathbf{P}^C = \mathbf{A} + |\mathbf{A}| \tag{121}$$

$|\mathbf{A}|$ denotes the absolute value of the adjacency matrix $\mathbf{A}$ defined as

$$|\mathbf{A}| = \mathbf{C}^\dagger \, \text{diag}(|x_1|, |x_2|, \ldots, |x_N|) \, \mathbf{C} \tag{122}$$

Since the trace of the matrix is not changed by unitary transformation

$$\text{Tr}|\mathbf{A}| = \text{Tr}[\text{diag}(|x_1|, |x_2|, \ldots, |x_N|)] = \sum_{i=1}^{N} |x_i| \tag{123}$$

Use of the fundamental identity (Equation 105) leads to the relation between E_π and $\mathbf{A}$,

$$E_\pi = \text{Tr} \, \mathbf{A} \tag{124}$$

Taking into account that $\text{Tr} \, \mathbf{A} = 0$, and combining Equations 124 and 122, the relation between E_π, and $\mathbf{P}^C$ is obtained:

$$E_\pi = \text{Tr} \, \mathbf{A} \, \mathbf{P}^C \tag{125}$$

Both relations (124) and (125) were first obtained by Ruedenberg[190,191] although they were known to Hall[103] for alternants.

C. THE McCLELLAND FORMULA FOR E_π

The total π-electron energy is a bounded quantity. The bounds may be given, by using the Frobenius theorem, as follows:

$$0 \leq E_\pi \leq N \, D_{max} \tag{126}$$

McClelland[167] was the first to show that much better bounds for E_π can be deduced. Let

$$F = 2M - N(\det \mathbf{A})^{2/N} \tag{127}$$

Note that $F > 0$. Then the McClelland inequalities are

$$0 \leq 2 \, N \, M - E_\pi^2 \leq (N - 1)F \tag{128}$$

The above was later improved:[192]

$$F \leq 2 \, N \, M - E_\pi^2 \leq (N - 1)F \tag{129}$$

Relation (129) holds for all graphs. For bipartite graphs it is slightly altered

$$F \leq 2 \, N \, M - E_\pi^2 \leq (N - 2)F \tag{130}$$

TABLE 5
HMO Energies and the Corresponding McClelland Values of Some Conjugated Molecules

Molecule	HMO energy[a]	McClelland energy[b]
Benzene	8.000	7.806
Fulvene	7.466	7.806
Heptafulvene	9.994	10.409
Styrene	10.424	10.409
o-Xylylene	9.954	10.409
m-Xylylene	9.431	10.409
p-Xylylene	9.925	10.409
Biphenyl	16.383	16.250
Azulene	13.364	13.646
Pentalene	10.456	11.040
Naphthalene	13.683	13.646
Anthracene	19.314	19.472
Phenanthrene	19.448	19.472
Pyrene	22.505	22.685
Naphthacene	24.931	25.296

[a]   Normalized ($\alpha = 0$, $\beta = 1$) Hückel values are taken from Coulson, C. A. and Stretwieser, A., Jr., *Dictionary of π-Electron Calculations*, W.H. Freeman, San Francisco, 1965.
[b]   Normalized McClelland values are obtained from the formula $E_\pi = 0.92 (2MN)^{1/2}$.

The importance of inequalities (Equation 128 to 130) is that they reveal the most important topological factors in determining E_π: (1) the number of atoms N, (2) the number of bonds M, and (3) the determinant of the adjacency matrix *det* **A**.

McClelland[167] has also derived an approximate formula for E_π

$$E_\pi \approx a(2\ M\ N)^{1/2} \qquad (131)$$

where the value of the constant *a* can be obtained by least-squares fitting. The optimal value $a = 0.92$ has reproduced rather closely E_π of various conjugated molecules; the difference between the exact and approximate values being only a few percent (see Table 5). This numerical work clearly shows that the gross part of E_π is determined solely by molecular size. Since molecular size is defined by only two topological parameters — the number of atoms and bonds — all other topological factors play, therefore, a seemingly marginal role. However, in chemical applications one is interested not only in energies, but in energy differences also. Thus, the problem of a few percent of E_π is essential for chemistry. Moreover, the formula (131) fails to differentiate isomeric structures, because isomers have identical values of M and N quantities. Thus, it cannot be used, for example, for stability predictions.[171]

It can be also shown that E_π is proportional to the number of rings in a molecule. The number of rings R in a molecule is given by

$$R = M - N + 1 \tag{132}$$

By substituting $R + N - 1$ for M in Equation 131, one obtains

$$E_\pi \approx a[2N(R + N - 1)]^{1/2} \tag{133}$$

For sufficiently large molecules $N >> R - 1$, E_π should be almost linearly proportional to the number of vertices. Such linear dependences are observed in various homologous series, i.e., annulenes,[193] radialenes,[194] acenes,[195,196] etc. A practical result of Equation 133 is that one cannot compare conjugated molecules with the same N, but different R. For example, naphthalene should not be compared with [10]annulene, because according to Equation 133 naphthalene should necessarily have a greater E_π.

REFERENCES

1. **Hückel, E.**, *Z. Phys.*, 60, 204, 1931.
2. **Hückel, E.**, *Z. Phys.*, 72, 310, 1932.
3. **Hückel, E.**, *Z. Phys.*, 76, 628, 1932.
4. **Salem, L.**, *The Molecular Orbital Theory of Conjugated Systems*, Benjamin, New York, 1966.
5. **Coulson, C. A., O'Leary, B., and Mallion, R. B.**, *Hückel Theory for Organic Chemists*, Academic Press, London, 1978.
6. **Gijzeman, O. L. J. and Sykes, A.**, *Photochem. Photobiol.*, 18, 339, 1973.
7. **Ichikawa, H.**, *J. Am. Chem. Soc.*, 106, 6249, 1984.
8. **Salem, L. and Leforestier, C.**, *J. Am. Chem. Soc.*, 107, 2526, 1985.
9. **Fox, M. A. and Matsen, F. A.**, *J. Chem. Educ.*, 62, 367, 1985; 62, 477, 1985; 62, 551, 1985.
10. **Burdett, J. K. and Lee, S.**, *J. Am. Chem. Soc.*, 107, 3063, 1985.
11. **Norrinder, U., Wennerström, O., and Wennerström, H.**, *Tetrahedron*, 41, 713, 1985.
12. **Haymet, A. D. J.**, *J. Am. Chem. Soc.*, 108, 319, 1986.
13. **Klein, D. J., Schmalz, T. G., Seitz, W. A., and Hite, G. E.**, *Int. J. Quantum Chem.: Quantum Chem. Symp.*, 19, 707, 1986.
14. **Mestechkin, M. N. and Poltavets, V. N.**, *J. Struct. Chem.*, 29, 461, 1988.
15. **Burdett, J. K.**, *Struct. Bonding (Berlin)*, 65, 29, 1987.
16. **Wang, Y., George, T. F., Lindsay, D. M., and Beri, A. C.**, *J. Chem. Phys.*, 86, 3493, 1987.
17. **Dias, J. R.**, *J. Chem. Educ.*, 66, 1012, 1989.
18. **Amić, D. and Trinajstić, N.**, *J. Chem. Soc. Perkin Trans. II*, 1595, 1990.
19. **Brendsdal, E., Cyvin, S. J., Cyvin, B. N., Brunvoll, J., Klein, D. J., and Seitz, W. A.**, in *Quasicrystals, Networks and Molecules with Fivefold Symmetry*, Hargittai, I., Ed., VCH Publishers, New York, 1991, 257.
20. **Marcus, R. A.**, *J. Chem. Phys.*, 43, 2643, 1965.
21. **Gutman, I. and Trinajstić, N.**, *Topics Curr. Chem.*, 42, 49, 1973.

22. **Trinajstić, N.**, in *Semiempirical Methods of Electronic Structure Calculation. Part A. Techniques*, Vol. 7, Segal, G. A., Ed., Plenum Press, New York, 1977, 1.
23. **Gimarc, B. M.**, *J. Am. Chem. Soc.*, 105, 1979, 1983.
24. **Kugler, S. and Laszlo, I.**, *Phys. Rev. B*, 39, 3882, 1989.
25. **Polansky, O. E.**, in *Theoretical Models of Chemical Bonding*, Maksić, Z. B., Ed., Springer-Verlag, Berlin, 1990, 29.
26. **Streitwieser, A., Jr.**, *Molecular Orbital Theory for Organic Chemistry*, John Wiley & Sons, New York, 1961.
27. **Heilbronner, E. and Bock, H.**, *The HMO Model and Its Applications*, John Wiley & Sons, London, 1976.
28. **Hückel, E.**, *Z. Phys.*, 83, 632, 1933.
29. **Hückel, E.**, *Z. Phys. Chem. Abt. B*, 34, 335, 1936.
30. **Hückel, E.**, *Z. Electrochem. Angew. Phys. Chem.*, 43, 752, 1937; 43, 827, 1937.
31. **Murrell, J. N. and Harget, A. J.**, *Semi-Empirical SCF MO Theory of Molecules*, Wiley-Interscience, London, 1972, chap. 1.
32. **Mulliken, R. S.**, *J. Chem. Phys.*, 3, 375, 1935.
33. **Bloch, F.**, *Z. Phys.*, 52, 555, 1929; 61, 206, 1930.
34. **Coulson, C. A. and Streitwieser, A., Jr.**, *Dictionary of π-Electron Calculations*, W.H. Freeman, San Francisco, 1965.
35. **Purcell, W. P. and Singer, J. A.**, *J. Chem. Eng. Data*, 12, 235, 1967.
36. **Ruedenberg, K.**, *J. Chem. Phys.*, 22, 1878, 1954.
37. **Trinajstić, N.**, in *Chemical Graphs Theory: Introduction and Fundamentals*, Bonchev, D. and Rouvray, D. H., Eds., Abacus Press/Gordon & Breach, New York, 1991, 235.
38. **Ruedenberg, K.**, *J. Chem. Phys.*, 34, 1861, 1961.
39. **Günthard, H. H. and Primas, H.**, *Helv. Chim. Acta*, 39, 1645, 1956.
40. **Schmidtke, H.-H.**, *J. Phys. Chem.*, 45, 3920, 1966.
41. **Gutman, I. and Trinajstić, N.**, *Croat. Chem. Acta*, 47, 507, 1975.
42. **Gutman, I. and Trinajstić, M.**, *Math. Chem. (Mülheim/Ruhr)*, 1, 71, 1975.
43. **Graovac, A., Gutman, I., and Trinajstić, N.**, *Topological Approach to the Chemistry of Conjugated Molecules*, Springer-Verlag, Berlin, 1977.
44. **Cvetković, D. M., Doob, M., and Sachs, H.**, *Spectra of Graphs: Theory and Application*, Academic Press, New York, 1980.
45. **Cvetković, D. M., Doob, M., Gutman, I., and Torgašev, A.**, *Recent Results in the Theory of Graph Spectra*, North-Holland, Amsterdam, 1982.
46. **Mallion, R. B. and Rouvray, D. H.**, *Mol. Phys.*, 36, 125, 1978.
47. **Mallion, R. B.**, *Croat. Chem. Acta*, 56, 447, 1983.
48. **Coulson, C. A.**, *Valence*, 2nd ed., University Press, Oxford, 1961, 35.
49. **Cvetković, D., Gutman, I., and Trinajstić, N.**, *Chem. Phys. Lett.*, 29, 65, 1974.
50. **Dias, J. R.**, *Molecular Orbital Calculations Using Chemical Graph Theory*, Springer-Verlag, Berlin, 1992.
51. **Herndon, W. C.**, *Tetrahedron Lett.*, 671, 1974.
52. **Živković, T., Trinajstić, N., and Randić, M.**, *Mol. Phys.*, 30, 517, 1975.
53. **Herndon, W. C. and Ellzey, M. L., Jr.**, *Tetrahedron*, 31, 99, 1975.
54. **Randić, M., Trinajstić, N., and Živković, T.**, *J. Chem. Soc. Faraday Trans. II*, 72, 244, 1976.
55. **D'Amato, S. S., Gimarc, B. M., and Trinajstić, N.**, *Croat. Chem. Acta*, 54, 1, 1981.
56. **Živković, T., Trinajstić, N., and Randić, M.**, *Croat. Chem. Acta*, 49, 89, 1977.
57. **Heilbronner, E.**, *Math. Chem. (Mülheim/Ruhr)*, 5, 105, 1979.
58. **D'Amato, S. S.**, *Mol. Phys.*, 37, 1363, 1979.
59. **D'amato, S. S.**, *Theoret. Chim. Acta*, 53, 319, 1979.
60. **Randić, M.**, *J. Comput. Chem.*, 1, 386, 1980.
61. **Jiang, Y.**, *Sci. Sin.*, 27, 236, 1984.
62. **Knop, J. V., Müller, W. R., Szymanski, K., Trinajstić, N., Kleiner, A. F., and Randić, M.**, *J. Math. Phys.*, 27, 2601, 1986.

63. Lowe, J. P. and Soto, M. R., *Math. Chem. (Mülheim/Ruhr)*, 20, 21, 1986.
64. Herndon, W. C. and Ellzey, M. R., Jr., *Math. Chem. (Mülheim/Ruhr)*, 20, 53, 1986.
65. Lowe, J. P. and Davis, M. W., *J. Math. Chem.*, 5, 275, 1990.
66. Dias, J. R., *J. Mol. Struct. (Theochem.)*, 165, 125, 1988.
67. Dias, J. R., *Theoret. Chim. Acta*, 76, 153, 1989.
68. Longuet-Higgins, H. C., *J. Chem. Phys.*, 18, 265, 1950.
69. Jahn, G. A. and Teller, E., *Proc. R. Soc. London Ser. A*, 161, 220, 1937.
70. Clar, E., Kemp, W., and Stewart, D. C., *Tetrahedron*, 3, 36, 1958.
71. Lin, C. Y. and Krantz, A., *J. Chem. Soc. Chem. Commun.*, 1111, 1972.
72. Živković, T., *Croat. Chem. Acta*, 44, 351, 1972.
73. Lloyd, D., *Carbocyclic Non-Benzenoid Aromatic Compounds*, Elsevier, Amsterdam, 1966.
74. Cvetković, D., Gutman, I., and Trinajstić, N., *Croat. Chem. Acta*, 44, 365, 1972.
75. Cvetković, D., Gutman, I., and Trinajstić, N., *J. Mol. Struct.*, 28, 289, 1975.
76. Gutman, I., Trinajstić, N., and Živković, T., *Tetrahedron*, 29, 3349, 1973.
77. Gutman, I., *Chem. Phys. Lett.*, 26, 85, 1974.
78. Kurosh, A. G., *Higher Algebra*, Mir, Moscow, 1980, 247, 3rd printing.
79. Gutman, I. and Trinajstić, N., *Naturwissenschaften*, 60, 475, 1973.
80. Katz, T. J. and Rosenberg, M., *J. Am. Chem. Soc.*, 84, 865, 1962.
81. Baker, W., in *Perspectives in Organic Chemistry*, Todd, A., Ed., Interscience, New York, 1956, 28.
82. Ginsburg, D., Ed., *Non-Benzenoid Aromatic Compounds*, Interscience, New York, 1959.
83. Clar, E., *Polycyclic Hydrocarbons*, Academic Press, London, 1964.
84. Cava, M. P. and Mitchell, M. J., *Cyclobutadiene and Related Compounds*, Academic Press, New York, 1967.
85. Garratt, P. J. and Sargent, M. V., in *Advances in Organic Chemistry*, Vol. 6, Taylor, E. C. and Wynberg, H., Eds., Interscience, London, 1969, 1.
86. Nozoe, T., Breslow, R., Hafner, K., Itô, S., and Murata, I., Eds., *Topics in Non-Benzenoid Aromatic Chemistry*, Vol. 2, John Wiley & Sons, New York, 1973.
87. Lloyd, D., *Non-Benzenoid Conjugated Carbocyclic Compounds*, Elsevier, Amsterdam, 1984.
88. Dias, J. R., *Handbook of Polycyclic Hydrocarbons. Part A. Benzenoid Hydrocarbons*, Elsevier, Amsterdam, 1987.
89. Dias, J. R., *Handbook of Polycyclic Hydrocarbons. Part B. Polycyclic Isomers and Heteroatom Analogs of Benzenoid Hydrocarbons*, Elsevier, Amsterdam, 1988.
90. Coulson, C. A., *Proc. R. Soc. London Ser. A*, 169, 413, 1939.
91. Mulliken, R. S., *J. Chem. Phys.*, 23, 1841, 1955.
92. Ham, N. S. and Ruedenberg, K., *J. Chem. Phys.*, 29, 1215, 1958.
93. Coulson, C. A. and Golebiewski, A., *Proc. Phys. Soc. (London)*, 78, 1310, 1961.
94. Pauling, L., Brockway, L. O., and Beach, J. Y., *J. Am. Chem. Soc.*, 57, 2705, 1935.
95. Klein, D. J. and Trinajstić, N., Eds., *Valence Bonds Theory and Chemical Structure*, Elsevier, Amsterdam, 1990.
96. Herndon, W. C., *J. Am. Chem. Soc.*, 96, 7605, 1974.
97. Randić, M., *Croat. Chem. Acta*, 47, 71, 1975.
98. Herndon, W. C. and Párkányi, C., *J. Chem. Educ.*, 53, 689, 1976.
99. Ham, N. S., *J. Chem. Phys.*, 29, 1229, 1958.
100. Coulson, C. A. and Longuet-Higgins, H. C., *Proc. R. Soc. London Ser. A*, 192, 16, 1947.
101. Coulson, C. A. and Rushbrooke, G. S., *Proc. Cambridge Philos. Soc.*, 36, 193, 1940.
102. Moffitt, W., *J. Chem. Phys.*, 26, 424, 1957.
103. Hall, G. G., *Proc. R. Soc. London Ser. A*, 229, 251, 1955.
104. McLachlan, A. D., *Mol. Phys.*, 2, 271, 1959.
105. Golebiewski, A., *Acta Phys. Polon.*, 23, 235, 1963.

106. **Koutecký, J.,** *J. Chem. Phys.,* 44, 3702, 1966.
107. **Graovac, A., Gutman, I., Trinajstić, N., and Živković, T.,** *Theoret. Chim. Acta,* 26, 67, 1972.
108. **Mallion, R. B., Schwenk, A. J., and Trinajstić, N.,** in *Recent Advances in Graph Theory,* Fiedler, M., Ed., Academia, Prague, 1975, 345.
109. **Rigby, M. J. and Mallion, R. B.,** *J. Comb. Theory (B),* 27, 122, 1979.
110. **Karadakov, P.,** *Int. J. Quantum Chem.,* 27, 699, 1985.
111. **Meyer, I.,** *Int. J. Quantum Chem.,* 29, 31, 1986.
112. **Vysotskii, Y. B. and Sivyakova, L. N.,** *Theoret. Exp. Chem.,* 25, 253, 1989.
113. **Shen, M.,** *Int. J. Quantum Chem.,* 37, 121, 1990.
114. **Mallion, R. B. and Rouvray, D. H.,** *J. Math. Chem.,* 5, 1, 1990; 8, 399, 1991.
115. **Tyutyulkov, N. and Polansky, O. E.,** *Chem. Phys. Lett.,* 139, 281, 1987.
116. **Karabunarliev, S. and Tyutyulkov, N.,** *Theoret. Chim. Acta,* 76, 65, 1989.
117. **Živković, T.,** *Int. J. Quantum Chem.,* 32, 313, 1987.
118. **McClellan, A. L.,** *Tables of Experimental Dipole Moments,* W.H. Freeman, San Francisco, 1963.
119. **Kirsanov, B. P. and Basilevsky, M. V.,** *Zh. Struct. Khim.,* 5, 99, 1964.
120. **Slee, T. S. and MacDougall, P. J.,** *Can. J. Chem.,* 66, 2961, 1988.
121. **Hall, G. G.,** *Trans. Faraday Soc.,* 53, 573, 1957.
122. **Gimarc, B. M. and Ott, J. J.,** in *Mathematics and Computational Concepts in Chemistry,* Trinajstić, N., Ed., Horwood, Chichester, 1986, 74.
123. **Gimarc, B. M. and Ott, J. J.,** *Angew. Chem. Int. Ed. Engl.,* 23, 506, 1984.
124. **Gimarc, B. M. and Ott, J. J.,** *J. Am. Chem. Soc.,* 108, 4298, 1986.
125. **Ott, J. J. and Gimarc, B. M.,** *J. Am. Chem. Soc.,* 108, 4303, 1986.
126. **Ott, J. J. and Gimarc, B. M.,** *J. Comput. Chem.,* 7, 673, 1986.
127. **Gimarc, B. M. and Ott, J. J.,** *Inorg. Chem.,* 25, 83, 1986; 25, 2708, 1986.
128. **Gimarc, B. M. and Ott, J. J.,** *J. Am. Chem. Soc.,* 109, 1388, 1987.
129. **Aihara, J.-I.,** *Bull. Chem. Soc. Jpn.,* 61, 2309, 1988.
130. **Rutherford, D. E.,** *Proc. R. Soc. Edinburgh Ser. A,* 62, 229, 1947.
131. **Lennard-Jones, J. E.,** *Proc. R. Soc. London Ser. A,* 158, 280, 1937.
132. **Frost, A. A. and Musulin, B.,** *J. Chem. Phys.,* 21, 572, 1953.
133. **Sondheimer, F.,** *Acc. Chem. Soc.,* 5, 81, 1972.
134. **Polansky, O. E.,** *Monatsh. Chem.,* 91, 916, 1960.
135. **Mirsky, L.,** *Introduction to Linear Algebra,* University Press, Oxford, 1955, 36.
136. **Kauzmann, W.,** *Quantum Chemistry: An Introduction,* Academic Press, New York, 1957, 48.
137. **Marcus, H. and Minc, H.,** *A Survey of Matrix Theory and Matrix Inequalities,* Allyn & Bacon, Boston, 1964, 66.
138. **Mallion, R. B.,** *Bull. Chem. Soc. France,* 2799, 1974.
139. **Davis, P. J.,** *Circulant Matrices,* John Wiley & Sons, New York, 1979.
140. **Day, A. C., Mallion, R. B., and Rigby, M. J.,** *Croat. Chem. Acta,* 59, 533, 1986.
141. **Graovac, A. and Trinajstić, N.,** *Croat. Chem. Acta,* 47, 95, 1975.
142. **Heilbronner, E.,** *Tetrahedron Lett.,* 1923, 1964.
143. **Zimmerman, H. E.,** *J. Am. Chem. Soc.,* 88, 1564, 1966; 88, 1566, 1966.
144. **Zimmerman, H. E. and Iwamura, H.,** *J. Am. Chem. Soc.,* 92, 2015, 1970.
145. **Zimmerman, H. E.,** *Acc. Chem. Res.,* 4, 272, 1971.
146. **Mason, S.,** *Nature,* 205, 495, 1965.
147. **Sondheimer, F.,** *Pure Appl. Chem.,* 7, 363, 1963.
148. **Sondheimer, F.,** *Proc. Robert A. Welch Found. Conf. Chem. Res.,* 12, 125, 1968.
149. **Klein, D. J. and Trinajstić, N.,** *J. Am. Chem. Soc.,* 108, 8050, 1984.
150. **Cresp, T. M. and Sargent, M. V.,** in *Essays in Chemistry,* Vol. 4, Bradley, J. N., Gillard, R. D., and Hudson, R. F., Eds., Academic Press, London, 1972.
151. **Shen, K.-W.,** *J. Chem. Educ.,* 50, 238, 1973.
152. **Dewar, M. J. S.,** *Angew. Chem. Int. Ed. Engl.,* 10, 716, 1971.

153. **Zimmerman, H. E. and Little, R. D.**, *J. Am. Chem. Soc.*, 96, 5143, 1974.
154. **Smith, W. B.**, *Molecular Orbital Methods in Organic Chemistry: HMO and PMO*, Marcel Dekker, New York 1974.
155. **Zimmerman, H. E.**, *Tetrahedron*, 38, 753, 1982.
156. **Chiu, Y. N.**, *Theoret. Chim. Acta*, 62, 403, 1983.
157. **Burdett, J. K., Lee, S., and Sha, W. C.**, *Croat. Chem. Acta*, 57, 1193, 1984.
158. **Hua-Ming, Z. and De-Xiang, W.**, *Tetrahedron*, 42, 515, 1986.
159. **Wenneström, O. and Norrinder, U.**, *Acta Chem. Scand. B*, 40, 328, 1986.
160. **Schaad, L. J. and Hess, B. A., Jr.**, *J. Am. Chem. Soc.*, 94, 3068, 1972.
161. **Dewar, M. J. S.**, *The Molecular Orbital Theory of Organic Chemistry*, McGraw-Hill, New York, 1969.
162. **Coulson, C. A.**, *Proc. Cambridge Philos. Soc.*, 36, 201, 1940.
163. **Coulson, C. A. and Longuet-Higgins, H. C.**, *Proc. R. Soc. London Ser. A*, 191, 39, 1947.
164. **Coulson, C. A. and Jacobs, J.**, *J. Chem. Soc.*, 2805, 1969.
165. **Coulson, C. A.**, *Proc. Cambridge Philos. Soc.*, 46, 202, 1950.
166. **Coulson, C. A.**, *J. Chem. Soc.*, 3111, 1954.
167. **McClelland, B. J.**, *J. Chem. Phys.*, 54, 640, 1971.
168. **Gutman, I. and Trinajstić, N.**, *Chem. Phys. Lett.*, 17, 535, 1972.
169. **Gutman, I., Trinajstić, N., and Živković, T.**, *Chem. Phys. Lett.*, 14, 342, 1972.
170. **Hall, G. G.**, *Int. J. Math. Educ. Sci. Technol.*, 4, 233, 273.
171. **Gutman, I., Milun, M., and Trinajstić, N.**, *J. Chem. Phys.*, 59, 2772, 1973.
172. **Gutman, I.**, *Theoret. Chim. Acta*, 35, 355, 1974.
173. **Gutman, I.**, *J. Chem. Phys.*, 66, 1652, 1977.
174. **Hall, G. G.**, *Mol. Phys.*, 33, 551, 1977.
175. **Gutman, I.**, *Theoret. Chim. Acta*, 45, 49, 1977.
176. **Gutman, I. and Petrović, S.**, *Chem. Phys. Lett.*, 97, 292, 1983.
177. **Cioslowski, J.**, *Z. Naturforsch.*, 40a, 1167, 1985.
178. **Cioslowski, J.**, *Z. Naturforsch.*, 40a, 1169, 1985.
179. **Cioslowski, J.**, *Theoret. Chim. Acta*, 68, 315, 1985.
180. **Cioslowski, J. and Gutman, I.**, *Z. Naturforsch.*, 41a, 861, 1986.
181. **Gutman, I. and Polansky, O. E.**, *Mathematical Concepts in Organic Chemistry*, Springer-Verlag, 1986, chap. 12.
182. **Gutman, I.**, *J. Math. Chem.*, 1, 123, 1987.
183. **Dias, J. R.**, *J. Chem. Educ.*, 64, 213, 1987.
184. **Cioslowski, J. and Polansky, O. E.**, *Theoret. Chim. Acta*, 74, 55, 1988.
185. **Cioslowski, J.**, *Int. J. Quantum Chem.*, 34, 417, 1988.
186. **Hosoya, H. and Tsuchiya, A.**, *J. Mol. Struct. (Theochem.)*, 185, 123, 1989.
187. **Cioslowski, J.**, *Math. Chem. (Mülheim/Ruhr)*, 25, 83, 1990.
188. **Gutman, I., Trinajstić, N., and Wilcox, C. F., Jr.**, *Tetrahedron*, 31, 143, 1975; **Wilcox, C. F., Jr., Gutman, I., and Trinajstić, N.**, *Tetrahedron*, 31, 147, 1975.
189. **Tang, A.-C., Kiang, Y.-S., Yan, G.-S., and Tai, S.-S.**, *Graph Theoretical Molecular Orbitals*, Science Press, Beijing, 1986.
190. **Ruedenberg, K.**, *J. Chem. Phys.*, 29, 1232, 1958.
191. **Ruedenberg, K.**, *J. Chem. Phys.*, 34, 1884, 1961.
192. **Gutman, I.**, *Chem. Phys. Lett.*, 24, 283, 1974.
193. **Gutman, I., Milun, M., and Trinajstić, N.**, *Croat. Chem. Acta*, 44, 207, 1972.
194. **Gutman, I., Trinajstić, N., Živković, T.**, *Croat. Chem. Acta*, 44, 501, 1972.
195. **Heilbronner, E.**, *Helv. Chim. Acta*, 37, 921, 1954.
196. **England, W. and Ruedenberg, K.**, *J. Am. Chem. Soc.*, 95, 8769, 1973.

Chapter 7

TOPOLOGICAL RESONANCE ENERGY

I. HÜCKEL RESONANCE ENERGY

The *resonance energy*, RE, is a theoretical quantity which has been used for many years for predicting aromaticity in conjugated structures.[1] The term resonance energy was introduced by Pauling[2] in the analysis of the stability of benzene and the resonance concept was introduced by Heisenberg[3] in connection with the discussion of the quantum states of the helium atom. The concept of *aromaticity* has been developed as a means of characterizing conjugated systems.[4-13] Aromaticity has never been defined unequivocally, because for different people it has different meaning.[13-16] However, a commonly accepted description of aromaticity is as a characteristic of the electronic structure of polycyclic conjugated systems, stabilization of which is due to a cyclic delocalization of electrons.[13,17,18]

The general definition of resonance energy is as follows:

$$RE = E_\pi(\text{conjugated molecule}) - E_\pi(\text{reference structure}) \qquad (1)$$

where $E_\pi(\text{conjugated molecule})$ and $E_\pi(\text{reference structure})$ are the π-electron energies of a given conjugated molecule and the corresponding reference structure, respectively, usually obtained by means of Hückel theory. The main difficulty with the RE concept is the hypothetical nature of the reference structure; its choice being to a great extent arbitrary. Therefore, it is not surprising that there are many proposals in the literature[19] of ways to select the reference structure. The first proposal, the classical *Hückel resonance energy*, HRE,[17] is based on the reference structure containing carbon-carbon double bonds isolated by conjugation barriers:

$$HRE = E_\pi(\text{conjugated molecule}) - 2n_{C=C} \qquad (2)$$

where $E_\pi(\text{conjugated molecule})$ is the Hückel π-electron energy of a conjugated molecule and $n_{C=C}$ is the number of double bonds in one Kekulé structure of the molecule. It should be noted that in Equation 2 use is made of the normalized form of Hückel theory, e.g., the parameter α for the carbon atom is taken as the zero-point energy, $\alpha_C = 0$, and the bond parameter β as the energy unit, $\beta_{CC} = 1$. The HRE is shown to fail in many cases because rather unstable molecules are predicted to be aromatic by virtue of their HREs being large (see Table 1). For example, the HRE/e (HRE per π-electron) values of benzene (0.33) and pentalene (0.31) are comparable, thus, indicating the latter compound to be as aromatic as benzene, in spite of the fact that pentalene

TABLE 1
Hückel Resonance Energies of
Some Conjugated Molecules

Molecule	HRE[a]	HRE/e[b]
Benzene	2.000	0.333
Naphthalene	3.683	0.368
Anthracene	5.314	0.380
Phenanthrene	5.448	0.389
Perylene	8.245	0.412
Coronene	10.572	0.481
Pyrene	6.505	0.407
Pentalene	2.456	0.307
Heptalene	3.618	0.302
Fulvene	1.466	0.244
Heptafulvene	1.994	0.249
Fulvalene	2.799	0.280
Buckminsterfullerene	33.160[c]	0.553

Note: HRE, Hückel resonance energy, and
        HRE/e in β units.

[a] HRE values are calculated using the Hückel
    values taken from Coulson, C. A. and Streit-
    wieser, A., Jr., *Dictionary of π-Electron
    Calculations*, Pergamon Press, Oxford,
    1965.
[b] HRE/e = HRE/N, where N is the number
    of π-electrons in the molecule.
[c] Taken from Randić, M., Nikolić, S., and
    Trinajstić, N., *Croat. Chem. Acta*, 60, 595,
    1987.

is a very unstable molecule.[15,20,21] Similarly, heptalene, fulvene, heptafulvene,
and fulvalene are all predicted to be aromatic by reason of their large HRE
values. Such predictions have been largely disproved by efforts to synthesize
these compounds.[15,20-22] Therefore, the HRE cannot be used as a reliable
criterion for predicting aromatic character of conjugated systems outside a
class of benzenoid hydrocarbons.

The origin of the problems with the HRE can be traced to the fact that
the HRE is grossly proportional to N (the number of π-electrons in the
conjugated molecules).[23] Since

$$n_{C=C} = N/2 \tag{3}$$

it follows,

$$HRE = E_{\pi}(\text{conjugated molecule}) - N \tag{4}$$

Then, by utilizing the McClelland-type formula[24] for approximating the π-energy of a conjugated molecule,

$$E_\pi(\text{conjugated molecule}) \approx a[2N(R + N - 1)]^{1/2} \qquad (5)$$

and using a reasonable assumption $N \gg R - 1$, Equation 4 becomes

$$HRE \approx N(1.4142a - 1) \qquad (6)$$

or for the McClelland value of a (0.92),[24]

$$HRE \approx 0.3N \qquad (7)$$

This relationship reveals that HRE is indeed proportional to N and thus insensitive to other details of molecular structure which may be directly related to the aromatic behavior of the molecule.

II. DEWAR RESONANCE ENERGY

Several attempts have been made to redefine RE in order to obtain better agreement between the theory and experiment.[25] An important improvement was introduced by Dewar,[18] who proposed the use of an *acyclic polyene-like* reference structure in RE calculations. This novel formulation of resonance energy is named *Dewar resonance energy,* DRE.[26] The DRE is defined as

$$DRE = E_\pi(\text{conjugated molecule}) - \sum_{ij} n_{ij}E_{ij} \qquad (8)$$

where n_{ij} and E_{ij} represent, respectively, the number of particular polyene bond types and the corresponding energy parameters. Dewar and co-workers[27-31] have obtained good predictions of aromaticity for many conjugated molecules using this approach (see Table 2).

However, Dewar, in his work, has used an original version[18] of the SCF (self-consistent field) π-MO (molecular orbital) theory,[32] parametrized to reproduce the ground-state properties of conjugated molecules. This led some people to believe that the change of theory from non-SCF to SCF level is responsible for the improvement of the predictions. Subsequent studies have shown that this is not the case. The fundamental step leading to good predictions of aromatic stability was not the use of more sophisticated MO theory, but rather the use of the acyclic polyene-like reference structure.[33,34] This was confirmed completely by Hess and Schaad[33,35] and the Zagreb group,[36] who simply transplanted the Dewar definition of the reference structure from SCF π-MO theory to Hückel theory.

The Zagreb group[36] reported the calculation of DREs in which the acyclic polyene-like reference structure was represented by a two-bond parameter

TABLE 2
Dewar Resonance Energies of Some
Conjugated Molecules

Molecule	Calculated DRE		Status[c]
	DRE[a]	DRE/e[b]	
Cyclobutadiene	−0.78	−0.20	AA
Benzene	0.87	0.15	A
Naphthalene	1.32	0.13	A
Anthracene	1.60	0.11	A
Phenanthrene	1.93	0.14	A
Perylene	2.62	0.13	A
Pyrene	1.82	0.11	A
Pentalene	0.006	0.001	NA
Heptalene	0.09	0.008	NA
Fulvene	0.05	0.008	NA
Cyclooctatetraene	−0.11	−0.01	NA
[18]annulene	0.126	0.007	NA
Acenaphthylene	1.34	0.12	A

Note: DRE, Dewar resonance energy, and DRE/e in eV.

[a] Data from Dewar, M. J. S. and de Llano, C., *J. Am. Chem. Soc.*, 91, 789, 1969; Trinajstić, N., *Rec. Chem. Prog.*, 32, 85, 1971.
[b] DRE/e = DRE/N, where N is the number of π-electrons in the molecule.
[c] A, aromatic; NA, nonaromatic; AA, antiaromatic.

scheme ($E_{C-C} = 0.52$, $E_{C=C} = 2.00$), while Hess and Schaad[35] have used an eight-bond parameter scheme to represent the reference structure. Their parameters are given in Table 3. These values were subsequently corrected by Schmalz et al.[37] They have found a numerical error in the set of Hückel energies of polyenes used by Hess and Schaad and have repeated their work. The corrected values of E_{ij} parameters are only slightly different from the original values. These values are also given in Table 3.

Schaad and Hess[38] have also investigated the origin of the reference structures with the two-bond and eight-bond parameters. They have treated acyclic reference structures as collections of perturbed ethylene molecules. The perturbation treatment through the second order produced the two-bond parameter scheme, while the perturbation treatment through the fourth order produced the eight-bond parameter scheme with the bond energies only slightly different from the original ones (see Table 3). A cluster expansion of the Hückel energy of polyenes[37] produced also a set of parameters close to the original Hess-Schaad set.[35] The cluster expansion values of E_{ij} parameters are also given in Table 3.

TABLE 3
Bond Parameters E_{ij} (in β Units) Determined in Different Ways

Bond type	Original value[a] (corrected value)[b]	Perturbational value[c]	Cluster expansion value[b]
C–C	0.4358 (0.4359)	0.4063	0.4327
HC–C	0.4362 (0.4356)	0.4375	0.4495
HC–CH	0.4660 (0.4649)	0.4688	0.4721
C=C	2.1716 (2.1734)	2.2500	2.1268
HC=C	2.1083 (2.1101)	2.1250	2.0748
H_2C=C	2.0000[d] (2.0000)[d]	2.0000[d]	2.0000[d]
HC=CH	2.0699 (2.0710)	2.0625	2.0437
H_2C=CH	2.0000[d] (2.0000)[d]	2.0000[d]	2.0000[d]

[a] From Hess, B. A., Jr. and Schaad, L. J., *J. Am. Chem. Soc.*, 93, 305, 1971.
[b] From Schmalz, T. G., Živković, T., and Klein, D. J., in *MATH/ CHEM/COMP 1987*, Lacher, R. C., Ed., Elsevier, Amsterdam, 1988, 173.
[c] From Schaad, L. J. and Hess, B. A., Jr., *Isr. J. Chem.*, 20, 281, 1980.
[d] Assigned value.

Additionally, a five-bond parameter scheme (plus a parameter for ethylene) has been proposed by Jiang et al.[39] This is a simpler approach than the Hess-Schaad scheme, but the RE/e values obtained by both schemes are similar. The agreement between the two sets of REs/e (RE[Jiang, Tang, Hoffmann] = RE[JTH] and RE[Hess, Schaad] = RE[HS]) is analyzed by the linear least-squares equation:

$$RE[JTH]/e = a[RE(HS)/e] + b \qquad (9)$$

The following statistical parameters are obtained: $n = 20$, $a = 0.9920$ (± 0.0049), $b = -0.0032$ (± 0.0003), $r = 0.9998$, and $s = 0.002$, where r is the correlation coefficient and s the standard deviation. Consequently, it may be concluded that both schemes have for practical purposes the same predictive power.

For many compounds the two DRE-like methods (the Hess-Schaad method and the Zagreb group method) give very similar results, but there are exceptions due to the fact that a two-bond parameter scheme rather poorly approximates the π-energy of branched polyenes. However, both methods correctly differentiate aromaticity in benzenoid and nonbenzenoid hydrocarbons (see Table 4).

TABLE 4
Resonance Energies (in β Units) of Some Conjugated Molecules

Molecule	Calculated resonance energy				Status[e]
	A_s[a]	A_s/e[b]	RE[c]	RE/e[d]	
Benzene	0.440	0.073	0.39	0.065	A
Naphthalene	0.563	0.056	0.55	0.055	A
Anthracene	0.634	0.045	0.66	0.047	A
Phenanthrene	0.768	0.055	0.77	0.055	A
Chrysene	0.952	0.053	0.96	0.053	A
Perylene	0.965	0.048	0.97	0.049	A
Pyrene	0.785	0.049	0.81	0.051	A
Azulene	0.244	0.024	0.23	0.023	A
Pentalene	−0.144	−0.018	−0.14	−0.018	AA
Heptafulvene	−0.086	−0.011	−0.02	−0.003	NA
Fulvalene	−0.321	−0.032	−0.60	−0.060	AA
Heptalene	−0.022	−0.002	−0.05	−0.004	NA
Fulvene	−0.094	−0.016	−0.01	−0.002	NA
Heptafulvalene	−0.155	−0.011	−0.20	−0.014	AA
Cyclobutadiene	−1.040	−0.260	−1.07	−0.268	AA
Buckminsterfullerene	1.960[f]	0.033	1.870[f]	0.031	A

Note: Resonance energies (RE) calculated using the Zagreb group model and the Hess-Schaad model.

[a]   Data from Milun, M., Sobotka, Ž., and Trinajstić, N., *J. Org. Chem.*, 37, 139, 1972.
[b] $A_s/e = A_s/N$, where N is the number of π-electrons in the molecule.
[c]   Data from Hess, B. A., Jr. and Schaad, L. J., *J. Am. Chem. Soc.*, 93, 305, 1971.
[d]   RE/e = RE/N, where N is the number of π-electrons in the molecule.
[e]   A, aromatic; NA, nonaromatic; AA, antiaromatic.
[f]   Data from Randić, M., Nikolić, S., and Trinajstić, N., *Croat. Chem. Acta,* 60, 595, 1987.

Detailed analysis of DRE and DRE-like methods revealed several limitations of these schemes: results depend on the parametric values chosen for the reference bond energies, the number of the bond parameters increases rapidly when heteroatoms are considered, DRE values can be obtained only for molecules possessing classical structures, thus excluding the number of nonclassical conjugated systems, i.e., conjugated ions and radicals, and homoaromatic structures. (However, recently Hess and Schaad[40] have extended their model to conjugated ions and radicals by introducing a novel reference structure. This extension requires 11 additional parameters.) One would like to have the DRE without these shortcomings. A search for the DRE-like quantity, free of the above limitations, produced an index independent of bond parameters which is directly related to the topology of the molecular π-network. It is named *topological resonance energy,* TRE, and is obtained by translating directly the DRE concept into the formalism of chemical graph theory.[41] Hence, TRE is not really a new aromaticity index, but an optimal nonempirical DRE-like quantity which appears to be applicable to a variety of conjugated systems.

III. THE CONCEPT OF TOPOLOGICAL RESONANCE ENERGY

There have been a number of attempts to produce a topological formula for the resonance energy of π-electron systems.[42] These attempts may be split into two groups. Into the first group go those attempts which aim to produce directly a topological formula for approximating REs of conjugated molecules without worrying about the reference structure. The second group contains the efforts directed toward finding the optimum acyclic reference structure expressed in graph-theoretical terms.

There are several interesting proposals in the first group. Some of these proposals are reviewed below. Green has produced a very simple topological formula for the REs of benzenoid hydrocarbons:[43]

$$RE(Green) = (M/3) + (m/10) \qquad (10)$$

where M is the total number of bonds in a benzenoid hydrocarbon and m is the number of those bonds contained in one benzene ring in the molecule which link two other benzene rings. The Green formula reproduces the HREs and it is useful only in the case of benzenoid hydrocarbons.

Another approximate topological formula for calculating REs of benzenoid hydrocarbons has been proposed by Carter.[44] The Carter formula is given by

$$RE(Carter) = A \, n_{C=C} + B \ln K - 1 \qquad (11)$$

where $n_{C=C}$ is the number of double bonds in one Kekulé structure, K the number of Kekulé structures of a molecule, while A (0.6) and B (1.5) are empirical parameters. The Carter formula could easily be modified to show the logarithmic dependence of RE on the number of Kekulé structures only:

$$RE = A \ln K \qquad (12)$$

This was actually done by Herndon and co-workers[45] although these authors were at first unfamiliar with the work of Carter. They have parametrized the above equation against the Dewar-deLlano SCF π-MO REs[27] and have obtained for A = 1.185,

$$RE(Herndon) = 1.185 \ln K \qquad (13)$$

Equation 12 may also be generalized to include alternant systems with four-membered rings. In this case K should be substituted by ACS, algebraic structure count (see discussion about the ACS in the next chapter). In addition, a correction term for the presence of the four-membered rings should also be added to the equation. This was in fact done by Wilcox:[42]

$$RE(Wilcox) = A \ln (ASC) + B \; n_4 \qquad (14)$$

where n_4 is the number of four-membered rings in the molecule and A (0.445) and B (-0.17) are least-squares parameters obtained by parametrization against the Hess-Schaad REs.[35] McClelland[24] has also derived a topological formula which approximates HREs.

$$RE(McClelland) = A(2NM)^{1/2} - N + r \qquad (15)$$

where $A = 0.30$ and $r = 0$ or 1, depending on whether N is even or odd.

There are also several interesting proposals in the second group. Especially interesting are those of Hosoya et al.[46] and of Aihara.[47] The approximate π-electron energy in their work is given by

$$E_\pi = A \log Z \qquad (16)$$

where Z is the Hosoya index.[48] The Hosoya index is defined as

$$Z = \sum_{n=0}^{[N/2]} p(G;n) \qquad (17)$$

where $p(G;n)$ is the number of ways in which n disconnected bonds are chosen from a molecule. At this point two modified Hosoya indices $\widetilde{Z}$ and Z^* need to be introduced. The $\widetilde{Z}$ index is defined[46] as

$$\widetilde{Z} = \sum_{n=0}^{[N/2]} (-1)^n \; a_{2n} \qquad (18)$$

where a_{2n} ($n = 0,1, \ldots ,[N/2]$) are the coefficients of the characteristic polynomial of a molecular graph G. The Z^* index is defined[47] as

$$Z^* = \prod_{i=1}^{N} (1 + x_i^2)^{1/2} \qquad (19)$$

where $x_i(i = 1,2, \ldots ,N)$ are the zeros of the characteristic polynomial of G.

For acyclic structures

$$\widetilde{Z} = Z \tag{20}$$

$$Z^* = Z \tag{21}$$

For cyclic structures Z is different from $\widetilde{Z}$ or Z^* because Z contains contributions from the cycles present in a molecule. The aromatic stability of conjugated systems can then be predicted using either of the following formulae:

$$RE(\text{Hosoya}) = \widetilde{Z} - Z \tag{22}$$

or

$$RE(\text{Aihara}) = A \log(Z^*/Z) \tag{23}$$

where $A = 6.0846$.

A. TOPOLOGICAL RESONANCE ENERGY

In this section, the theory behind the TRE concept will be outlined. Attention will be paid to the reference structure. The aim is to obtain an acyclic polyene-like reference structure for which construction of all substructural details of a conjugated molecule, except the cycles, should be taken into account. The target is such a reference structure because of the belief that the aromatic stability in a conjugated molecule is related to the contributions from cycles to the total π-electron energy of a system.[49] This was achieved by a convenient adaptation[41] of the Sachs formula[50] for the coefficients of the characteristic polynomial: the cycles are left out of the formula (Equation 7) in Chapter 5,

$$a_n^{ac}(G) = \sum_{s \in S_n} (-1)^{p(s)} \tag{24}$$

This relation considers only the complete set of acyclic Sachs graphs, S_n^{ac}, of a graph G corresponding to a conjugated molecule, or in other words, for the construction of a_n^{ac} coefficients one needs to take into account only K_2 components of G. A Sachs graph s is called acyclic[51] if $c(s) = 0$ and thus,

$$S_n^{ac} = \{s \in S_n | c(s) = 0\} \tag{25}$$

Therefore, the polynomial constructed by means of Equation 24 contains

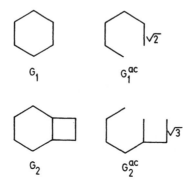

FIGURE 1.   Edge-weighted acyclic graphs (G_1^{ac} and G_2^{ac}) corresponding to the benzene graph (G_1) and benzocyclobutadiene graph (G_2).

coefficients free of any cyclic contributions and represents the optimal approximation to the acyclic reference structure

$$P^{ac}(G;x) = \sum_{n=0}^{N} a_n^{ac}(G)x^{N-n} \qquad (26)$$

The polynomial (Equation 26) is a crucial concept in the topological theory of aromaticity and is named the *acyclic polynomial*.[41] The acyclic polynomial is also referred to as the *reference polynomial*[52] or the *matching polynomial*.[53]

Generally, it is not possible to associate a graph with the acyclic polynomial. However, it is possible for some simple π-electronic structures to give the edge-weighted acyclic graphs.[54,55] An edge-weighted acyclic graph G_{EW}^{ac} corresponding to a graph G, is defined as the graph whose characteristic polynomial $P(G_{EW}^{ac};x)$ equals the acyclic polynomial $P^{ac}(G;x)$ of G. Edge-weighted acyclic graphs corresponding to graphs representing benzene and benzocyclobutadiene are given in Figure 1.

The zeros of the acyclic polynomial x_i^{ac} ($i = 1,2, \ldots ,N$) represent a complete set of Hückel energies of the acyclic reference structure,

$$E_{\pi}(\text{acyclic reference structure}) = \sum_{i=1}^{N} h_i x_i^{ac} \qquad (27)$$

where h_i is the occupancy number of energy levels associated with the reference structure.

Introducing the expression for the total π-electron energy of a conjugated molecule,

$$E_{\pi}(\text{conjugated molecule}) = \sum_{i=1}^{N} g_i x_i \qquad (28)$$

and Equation 27 into Equation 1, the relation for the topological resonance energy is obtained,

$$TRE = \sum_{i=1}^{N} (g_i x_i - h_i x_i^{ac})$$ (29)

which for the case when $g_i = h_i$ reduces to the elegant and condensed expression,[41,56]

$$TRE = \sum_{i=1}^{N} g_i (x_i - x_i^{ac})$$ (30)

The application of Equation 30 is straightforward. What one needs are Hückel energies and the zeros of the acyclic polynomial for the molecule under investigation.

In order to ensure that the TRE values are always real, the zeros of the acyclic polynomial must be real. Since the acyclic polynomial is a combinatorial structure not related, in general, to any symmetric real matrix there is no *a priori* reason to expect all its zeros to be real. However, the proof that the zeros of the acyclic polynomial are real exists in the literature.[57] In addition, the alternative proofs (or partial proofs) have been discussed by several authors.[58-62]

The TRE values can be calculated using Equation 29 or 30. These values can be used for comparing the aromatic stabilities of isomeric structures. However, for molecules of various sizes the TRE values cannot be used meaningfully. For example, the TRE values of benzene (0.276) and naphthalene (0.390) cannot be compared because these two molecules have different sizes (i.e., a different number of atoms and bonds). Therefore, in order to avoid the *size effect*,[63] the TRE must be normalized.

The TRE may be normalized in several ways to produce the intrinsic topological resonance quantity.[64 66] Here is reported the normalization procedure based on the total number of π-electrons in the molecule,[67] N_e, which leads to the TRE per π-electron, TRE/e, value:[56]

$$TRE/e = TRE/N_e$$ (31)

TRE/e formally represents the amount of aromatic stabilization or destabilization that one π-electron contributes to the π-network of the molecule.

The following threshold values of TRE/e are used for classifying conjugated molecules: (1) molecules having TRE/e greater than or equal to 0.01 are considered to be of prevailing aromatic character; (2) molecules having TRE/e in the interval $+0.01$, -0.01 are either ambivalent or nonaromatic species; and (3) molecules having TRE/e equal or less than -0.01 are viewed as antiaromatic.

TABLE 5
Calculation of the TRE and TRE/e Indices of Naphthalene

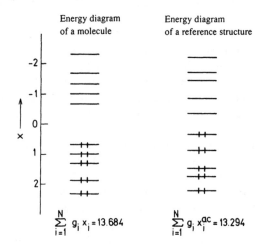

G

(1) Hückel polynomial and spectrum
$P(G;_x) = x^{10} - 11x^8 + 41x^6 - 65x^4 + 43x^2 - 9$
{2.303, 1.618, 1.303, 1.000, 0.618, −0.618, −1.000, −1.303, −1.618, −2.303}

(2) Acyclic polynomial and spectrum
$P^{ac}(G;x) = x^{10} - 11x^8 + 41x^6 - 61x^4 + 31x^2 - 3$
{2.232, 1.732, 1.464, 0.865, 0.354, −0.354, −0.865, −1.464, −1.732, −2.232}

(3) Energy diagrams

Energy diagram of a molecule

Energy diagram of a reference structure

$\sum_{i=1}^{N} g_i x_i = 13.684$

$\sum_{i=1}^{N} g_i x_i^{ac} = 13.294$

(4) TRE and TRE/e
$$TRE = \sum_{i=1}^{N} g_i x_i - \sum_{i=1}^{N} g_i x_i^{ac} = 0.390$$
$TRE/e = 0.039$

This classification should serve as a rough indicator for experimental chemists concerning the difficulties involved in the preparation of a given conjugated molecule, because a molecule labeled aromatic is usually more easily made than one labeled antiaromatic.[1,4-18]

An example of the calculation of TRE and TRE/e values for naphthalene is shown in Table 5.

The TRE model has several favorable features in comparison with DRE-like models: (1) TRE values contain exactly all cyclic and no acyclic contributions to E_π(conjugated molecule); (2) from the form of the TRE formula (Equation 30) it is clear that no parameters are required outside the parameters

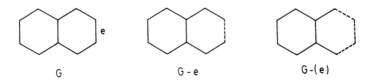

FIGURE 2.   Examples of a graph G and the corresponding subgraphs G−e and G−(e).

for the heteroatoms in the framework of Hückel theory; and (3) TRE can be used for studying all kinds of conjugated systems.

Since acyclic polyenes do not contain cycles, the corresponding characteristic and acyclic polynomials are identical,

$$P(\text{polyene};x) = P^{ac}(\text{polyene};x) \tag{32}$$

The consequence of the identity (Equation 32) is the relation

$$TRE(\text{polyene}) = 0 \tag{33}$$

The TRE model was developed independently by the Zagreb group[41,56] and by Aihara,[52] using different ways of reasoning.

IV. COMPUTATION OF THE ACYCLIC POLYNOMIAL

It appears that the computation of the acyclic polynomial via the truncated Sachs formula (Equation 24), which has shown a great conceptual value, has practical difficulty in the form of an enormous increase of combinatorial possibilities of K_2 components with the increase of the size of the molecules. This problem of generating the coefficients via the Sachs procedure cannot be avoided even by the computer construction of the acyclic polynomial.[68-70] However, the computation of the acyclic polynomial can be simplified considerably by using the recurrence relation,[56,71]

$$P^{ac}(G;x) = P^{ac}(G - e;x) - P^{ac}(G - (e);x) \tag{34}$$

where G−e denotes the subgraph of a graph G obtained by removing an edge *e* from G and G−(e) is the subgraph obtained by removal of an edge *e* and incident vertices from G. Examples of G, G−e, and G−(e) are given in Figure 2.

Equation 34 is a generalization of the Heilbronner formula:[72]

$$P(G;x) = P(G - e;x) - P(G - (e);x) \tag{35}$$

which is valid for polyenes only.

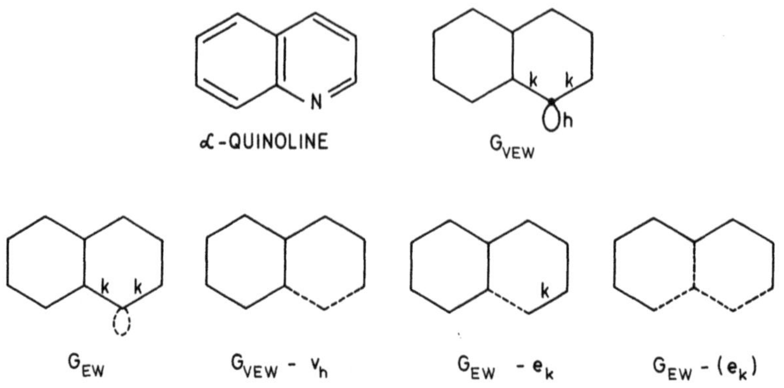

FIGURE 3. Examples of a vertex- and edge-weighted graph G_{VEW} and the corresponding subgraphs G_{EW}, G_{VEW-v_h}, G_{EW-e_k} and $G_{EW}(-e_k)$.

The computation of the acyclic polynomial for heterocyclic structures is also based on Equation 34, which is modified[56,71] in this case to include the Hückel parameters h and k for heteroatoms and heterobonds, respectively,

$$P^{ac}(G_{VEW};x) = P^{ac}(G_{EW};x) - h\,P^{ac}(G_{VEW} - v_h;x) \qquad (36)$$

$$P^{ac}(G_{EW};x) = P^{ac}(G_{EW} - e_k;x) - k^2 P^{ac}(G_{EW} - (e_k);x) \qquad (37)$$

where G_{VEW} is a vertex- and edge-weighted graph with a loop of a weight h attached to the vertex v_h; G_{EW} is an edge-weighted subgraph obtained by deleting the loop at vertex v_h from G_{VEW}; $G_{VEW} - v_h$ is a subgraph obtained by deleting the weighted vertex v_h and incident edges from G_{VEW}; $G_{EW} - e_k$ is a subgraph obtained by removal of the weighted edge e_k from G_{EW}; and $G_{EW} - (e_k)$ is a subgraph obtained by omitting the weighted edge e_k and incident vertices from G_{EW}. Examples of G_{VEW} and the corresponding subgraphs G_{EW}, $G_{VEW} - v_h$, $G_{EW} - e_k$, and $G_{EW} - (e_k)$ are given in Figure 3.

The recurrence formulae (Equations 34, 36, and 37) are applied stepwise until the structure is reduced in a minimum number of steps into linear fragments whose characteristic polynomials are readily available (see Table 1, Chapter 5). In Figure 4 the computation of the acyclic polynomial for naphthalene is given as an example.

Similarly, in Figure 5 the computation of the acyclic polynomial for α-quinoline is given. The acyclic polynomial of α-quinoline, of course, reduces to one belonging to naphthalene for $k = 1$ and $h = 0$.

There is available in the literature[70] a computer program for the computation of the acyclic polynomial, based on the recurrence relations (Equations 34, 36, and 37).

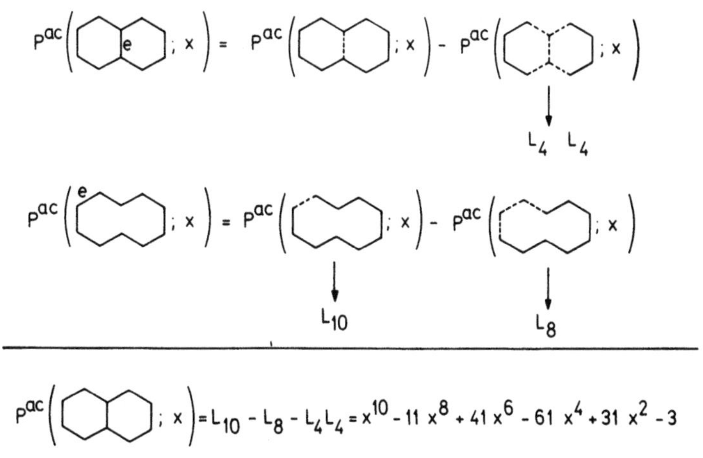

$$P^{ac}\left(\begin{array}{c}\text{[naphthalene with } e\text{]}\end{array}; x\right) = P^{ac}\left(\begin{array}{c}\text{[structure]}\end{array}; x\right) - P^{ac}\left(\begin{array}{c}\text{[structure]}\end{array}; x\right)$$

$$\downarrow$$
$$L_4 \quad L_4$$

$$P^{ac}\left(\begin{array}{c}\text{[10-membered ring with } e\text{]}\end{array}; x\right) = P^{ac}\left(\begin{array}{c}\text{[structure]}\end{array}; x\right) - P^{ac}\left(\begin{array}{c}\text{[structure]}\end{array}; x\right)$$

$$\downarrow \qquad\qquad\qquad\qquad \downarrow$$
$$L_{10} \qquad\qquad\qquad\qquad L_8$$

$$P^{ac}\left(\begin{array}{c}\text{[naphthalene]}\end{array}; x\right) = L_{10} - L_8 - L_4 L_4 = x^{10} - 11\,x^8 + 41\,x^6 - 61\,x^4 + 31\,x^2 - 3$$

FIGURE 4. The computation of the acyclic polynomial for naphthalene. The symbol L_N denotes the characteristic polynomial of [N]polyene.

A. CONNECTION BETWEEN THE CHARACTERISTIC POLYNOMIAL AND THE ACYCLIC POLYNOMIAL

The characteristic polynomial and the acyclic polynomial of G are related in a simple way. For polyenes these two polynomials are identical (see Equation 32). In the case of [N]annulenes the difference between $P(C_N;x)$ and $P^{ac}(C_N;x)$ is rather simple:

$$P(C_N;x) - P^{ac}(C_N;x) = -2 \qquad (38)$$

where C_N is a shorthand notation for the N-cycle representing [N]annulene. If one gives explicitly $P(C_N;x)$ and $P^{ac}(C_N;x)$, the above result immediately arises:

$$P(C_N;x) = \sum_{k=0}^{[N/2]} (-1)^k\, p(C_N;k)x^{N-2k} - 2 \qquad (39)$$

$$P^{ac}(C_N;x) = \sum_{k=0}^{[N/2]} (-1)^k\, p(C_N;k)x^{N-2k} \qquad (40)$$

where $p(C_N;k)$ is the number of distinct k-matchings in a cycle C_N. A k-matching in a graph G is a selection of 2k vertices which are joined pairwise by k edges.[73]

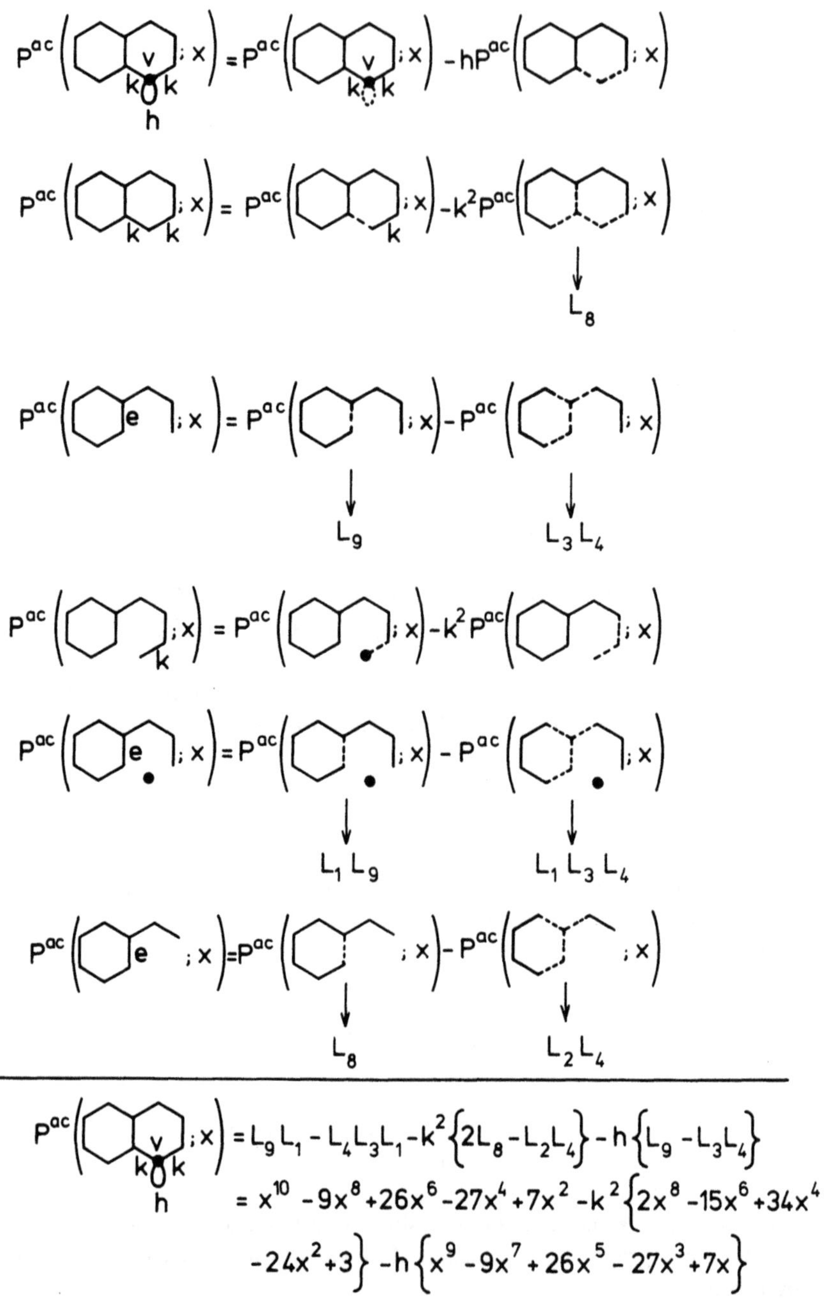

$$P^{ac}\left(\begin{array}{c}\underset{k\overset{\mathbf{v}}{\underset{h}{\bigcirc}}k}{}\end{array};x\right)=P^{ac}\left(\begin{array}{c}\underset{k:k}{\mathbf{v}}\end{array};x\right)-hP^{ac}\left(\begin{array}{c}\end{array};x\right)$$

$$P^{ac}\left(\begin{array}{c}\underset{k\quad k}{}\end{array};x\right)=P^{ac}\left(\begin{array}{c}\underset{k}{}\end{array};x\right)-k^2P^{ac}\left(\begin{array}{c}\end{array};x\right)$$
$$\downarrow$$
$$L_8$$

$$P^{ac}\left(\begin{array}{c}\mathbf{e}\end{array};x\right)=P^{ac}\left(\begin{array}{c}\end{array};x\right)-P^{ac}\left(\begin{array}{c}\end{array};x\right)$$
$$\qquad\qquad\downarrow\qquad\qquad\qquad\downarrow$$
$$\qquad\qquad L_9\qquad\qquad\qquad L_3L_4$$

$$P^{ac}\left(\begin{array}{c}\underset{k}{}\end{array};x\right)=P^{ac}\left(\begin{array}{c}\bullet\end{array};x\right)-k^2P^{ac}\left(\begin{array}{c}\end{array};x\right)$$

$$P^{ac}\left(\begin{array}{c}\mathbf{e}\\\bullet\end{array};x\right)=P^{ac}\left(\begin{array}{c}\bullet\end{array};x\right)-P^{ac}\left(\begin{array}{c}\bullet\end{array};x\right)$$
$$\qquad\qquad\downarrow\qquad\qquad\qquad\downarrow$$
$$\qquad\qquad L_1L_9\qquad\qquad\qquad L_1L_3L_4$$

$$P^{ac}\left(\begin{array}{c}\mathbf{e}\end{array};x\right)=P^{ac}\left(\begin{array}{c}\end{array};x\right)-P^{ac}\left(\begin{array}{c}\end{array};x\right)$$
$$\qquad\qquad\downarrow\qquad\qquad\qquad\downarrow$$
$$\qquad\qquad L_8\qquad\qquad\qquad L_2L_4$$

$$P^{ac}\left(\begin{array}{c}\underset{k\overset{\mathbf{v}}{\underset{h}{\bigcirc}}k}{}\end{array};x\right)=L_9L_1-L_4L_3L_1-k^2\left\{2L_8-L_2L_4\right\}-h\left\{L_9-L_3L_4\right\}$$
$$=x^{10}-9x^8+26x^6-27x^4+7x^2-k^2\left\{2x^8-15x^6+34x^4\right.$$
$$\left.-24x^2+3\right\}-h\left\{x^9-9x^7+26x^5-27x^3+7x\right\}$$

FIGURE 5.   The computation of the acyclic polynomial for α-quinoline. The symbol L_N denotes the characteristic polynomial of [N]polyene.

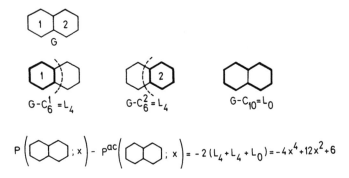

FIGURE 6. The difference between the characteristic polynomial and the acyclic polynomial for the naphthalene graph.

The connection between $P(G;x)$ and $P^{ac}(G;x)$ for polycyclic systems is more complicated:[74]

$$P(G;x) - P^{ac}(G;x) = -2 \sum_m P^{ac}(G-C_m;x) + 4 \sum_{m<n} P^{ac}(G-C_m-C_n;x)$$

$$-8 \sum_{m<n<p} P^{ac}(G-C_m-C_n-C_p;x) + \ldots \tag{41}$$

where the summations go over all pairs, triplets, . . . , etc. of mutually disconnected cycles which are contained in G. $G-C_m$, $G-C_m-C_n$, $G-C_m-C_n-C_p$, . . . , etc. are subgraphs obtained by removing successively C_m; C_m and C_n; C_m, C_n, and C_p; . . . ; and adjacent bonds from G. In Figure 6 the difference between $P(G;x)$ and $P^{ac}(G;x)$ for the naphthalene graph is given.

The relation (Equation 41) may be used for studying the effect of individual cycles on TRE and E_π.[74,75] It has been shown that in polycyclic conjugated systems the $(4m)$-membered cycles always *destabilize* conjugated molecules. On the other hand, the $(4m+2)$-membered cycles *stabilize* conjugated structures, though examples have been found in which $(4m+2)$ rings have a destabilizing effect on the molecular stability.[49] The contribution to E_π(conjugated molecule) which comes from the presence of the C_N (N = 3,4,5, . . .) is called the *cycle energy* (E_{cycle}) and is given as the following π-energy difference:[49]

$$E_{cycle} = E_\pi(\text{conjugated molecule}) - E_\pi(G-C_N) \tag{42}$$

where $E_\pi(G-C_N)$ stands for the π-electron energy of the structure obtained when the cycle C_N is removed from the molecule.[76] Cycle energies of several conjugated molecules are given in Table 6.

TABLE 6
Cycle Energies (in β Units) of
Several Conjugated Molecules

$E_A = 0.1211; A = C_6$
$E_{A+B} = 0.0709; A+B = C_{10}$

$E_A = 0.4252; A = C_6$
$E_B = -0.4287; B = C_4$
$E_{A+B} = -0.2641; A+B = C_8$

$E_A = 0.1050; A = C_6$
$E_C = 0.0114; C = C_5$
$E_{A+B} = 0.0567; A+B = C_{10}$
$E_{A+B} = 0.0049; A+C = C_9$

$E_A = -0.7664; A = C_4$
$E_B = -0.1284; B = C_6$
$E_{A+B} = -0.5746; A+B = C_8$
$E_{A+B+C} = -0.1797; A+B+C = C_{10}$

The results in Table 6 may serve as a quantitative test of the Hückel $(4m+2)$ rule. The Hückel rule was originally proposed for annulenes only.[77] In the original form the rule states that the planar monocyclic systems containing $(4m+2)$ π-electrons are expected to exhibit aromatic stability. Later it was conjectured that it holds for any arbitrary conjugated molecule.[18,78-80] However, these authors, and many others,[1,13,35,82] have been using the Hückel rule as a *qualitative* rule, though it was quantified for the monocycles.[81] Hence, the Hückel rule was used only to indicate whether a certain cycle has a stabilizing or destabilizing effect on the molecular stability, without a knowledge of the magnitude of this effect. The possible effect of electron correlation on the Hückel rule has also been considered.[83] The results of this analysis have provided broader justification for the Hückel rule. The quantitative approach of Gutman and Bosanac,[49] however, has produced an important result (not obtainable by the qualitative studies): in some cases the Hückel $(4m+2)$ rule can be violated.

V. APPLICATIONS OF THE TRE MODEL

The TRE model has been used abundantly for studying the aromatic stability of organic and inorganic molecules.[41,52,56,71,84-112] In this section the application of the TRE model to several classes of conjugated systems will be reviewed.

A. HÜCKEL ANNULENES

Hückel [N]annulenes $(CH)_N$ are regular structures for which it is possible to derive the analytical formulae for TRE. The analytical expressions for E_π of [N]annulenes[113] are given below:

$$E_\pi([N]\text{annulene}) = \begin{cases} 2(\text{ctan}\theta - \tan\theta) & N = 4m \\ 2\text{ctan}\theta \cos\theta & N = 4m + 1 \text{ or } N = 4m + 3 \\ 2(\text{ctan}\theta + \tan\theta) & N = 4m + 2 \end{cases}$$

(43)

with $\theta = \pi/2N$. The analytical expressions for the corresponding reference polynomials can be obtained by starting with Equation 84 given in Chapter 6 for Hückel energies of linear polyenes:

$$E_i = 2 \cos(i\,\pi/N + 1); i = 1,2,\ldots, N \tag{44}$$

The E_π([N]annulene references structure) is given as a sum of occupied E_is,

$$E_\pi([N]\text{annulene reference structure}) = 2 \sum_{i=1}^{L} E_i = 4 \sum_{i=1}^{L} \cos(i\pi/N + 1)$$

(45)

where

$$L = \begin{cases} \dfrac{N}{2} & N = \text{even} \\[2mm] \dfrac{N-1}{2} & N = \text{odd} \end{cases}$$

(46)

Equation 45 can be further simplified using the technique of Polansky:[114]

$$E_\pi([N]\text{annulene reference structure}) = \begin{cases} 2\text{ctg}\theta/\sin\theta & N = \text{even} \\ 2\text{ctg}\theta & N = \text{odd} \end{cases} \tag{47}$$

The TRE expressions for [N]annulenes are obtained by substituting Equations 43 and 47 into Equation 1:

TRE([N]annulene) =

$$
\left\{
\begin{array}{ll}
2\left(\dfrac{\cos\theta - 1}{\sin\theta} - \tan\theta\right) & N = 4m \qquad\qquad (48)\\[3mm]
2\,\dfrac{\cos\theta - 1}{\sin\theta}\cos\theta & N = 4m + 1 \text{ or } N = 4m + 3 \\[3mm]
2\left(\dfrac{\cos\theta - 1}{\sin\theta} + \tan\theta\right) & N = 4m + 2
\end{array}
\right.
$$

The analytical expressions for TRE of annulene ions can also be derived in a similar way.[91,115] Let TRE^{+}, TRE^{++}, TRE^{-}, and TRE^{--}, respectively, stand for topological resonance energies of annulene ions $(CH)_N^{+}$, $(CH)_N^{++}$, $(CH)_N^{-}$, and $(CH)_N^{--}$, respectively. Then

$$
TRE^{+} = TRE - \left\{
\begin{array}{ll}
2\sin\theta & N = 4m \\
-2\sin\theta & N = 4m + 1 \\
2\sin\theta - 2\sin2\theta & N = 4m + 2 \\
2\sin\theta & N = 4m + 3 \quad (49)
\end{array}
\right.
$$

$$
TRE^{++} = TRE - \left\{
\begin{array}{ll}
4\sin\theta & N = 4m \\
2\sin2\theta - 4\sin\theta & N = 4m + 1 \\
4\sin\theta - 4\sin2\theta & N = 4m + 2 \\
2(\sin\theta + \sin2\theta - \sin3\theta) & N = 4m + 3 \quad (50)
\end{array}
\right.
$$

$$
TRE^{-} = TRE - \left\{
\begin{array}{ll}
2\sin\theta & N = 4m \\
2\sin\theta & N = 4m + 1 \\
2\sin\theta - 2\sin2\theta & N = 4m + 2 \\
-2\sin\theta & N = 4m + 3 \quad (51)
\end{array}
\right.
$$

$$
TRE^{--} = TRE - \left\{
\begin{array}{ll}
4\sin\theta & N = 4m \\
2(\sin\theta + \sin2\theta - \sin3\theta) & N = 4m + 1 \\
4\sin\theta - 4\sin2\theta & N = 4m + 2 \\
2\sin2\theta - 4\sin\theta & N = 4m + 3 \quad (52)
\end{array}
\right.
$$

Numerical values for TREs of [N]annulenes and [N]annulene ions are easily obtained from the above formulae. Predictions based on the TRE values agree nicely with the chemistry of annulenes and their ions.

The dependence of TRE/e on the ring size is shown in Figure 7. A plot of TRE/e vs. N = even indicates that both [4m]- and [4m + 2]annulenes become rapidly nonaromatic compounds with increasing value of N and that

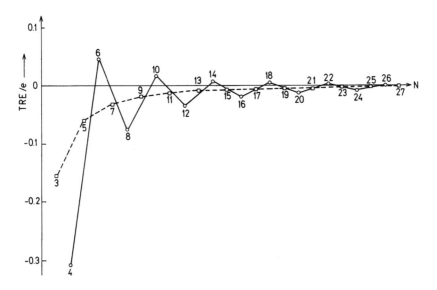

FIGURE 7.   TRE/e vs. the ring size of Hückel [N]annulenes.

the difference between them practically disappears at N = 26. Similarly, the odd-membered annulenes quickly converge to nonaromaticity, though the first few members appear to be antiaromatic species. Annulene positive and negative ions also converge rather rapidly to nonaromaticity (see Table 7).

B. RELATIONSHIP BETWEEN TREs AND RING CURRENTS OF [4m + 2] π-ELECTRON ANNULENES

According to the NMR criterion of aromaticity the diamagnetic ring currents indicate aromaticity while paramagnetic ring currents indicate antiaromaticity.[1,116,117] However, it had not been clear that this magnetic criterion is equivalent to the resonance energy criterion until Aihara[118] proved that the diagmagnetic susceptibility of a polycyclic conjugated system strongly reflects its conjugative stabilization.

An analytical relationship between the Hückel resonance energies and the reduced ring currents, RCs, of [4m + 2] π-electron annulenes is derived by Haddon.[119] In analogy with his work, one can derive the relationship between TREs and RCs of [4m + 2] π-electron annulenes. There are possible five classes of [4m + 2] π-electron annulenes. These are $[4m]^{++}$ annulenes; $[4m]^{--}$ annulenes; $[4m + 1]^{-}$ annulenes; $[4m + 2]$ annulenes,   and $[4m + 3]^{+}$ annulenes. They are all predicted to be aromatic species up to a certain size. TREs of these systems are given analytically in the previous section, but here they will be transformed into more convenient forms for further discussion.

TABLE 7
TRE/e Values of Hückel [N]Annulenes and Their Positive and Negative Ions

N	TRE/e	(TRE/e)$^+$	(TRE/e)$^{++}$	(TRE/e)$^-$	(TRE/e)$^{--}$
3	−0.155	0.268	0.268	−0.366	−0.146
4	−0.307	−0.154	0.152	−0.092	0.051
5	−0.060	−0.230	−0.121	0.053	−0.018
6	0.045	−0.042	−0.173	−0.030	−0.087
7	−0.031	0.038	−0.031	−0.083	−0.027
8	−0.073	−0.029	0.031	−0.023	0.018
9	−0.019	−0.065	−0.026	0.018	−0.013
10	0.016	−0.016	−0.056	−0.013	−0.038
11	−0.013	0.014	−0.014	−0.036	−0.011
12	−0.032	−0.012	0.013	−0.010	0.009
13	−0.009	−0.030	−0.011	0.009	−0.007
14	0.008	−0.008	−0.027	−0.007	−0.021
15	−0.007	0.008	−0.007	−0.020	−0.006
16	−0.019	−0.007	0.007	−0.006	0.005
17	−0.005	−0.017	−0.006	0.005	−0.006
18	0.005	−0.005	−0.016	−0.004	−0.013
19	−0.004	0.005	−0.005	−0.012	−0.004
20	−0.012	−0.004	0.004	−0.004	0.003
21	−0.004	−0.011	−0.004	0.003	−0.003
22	0.003	−0.003	−0.011	−0.003	−0.009
23	−0.003	0.003	−0.003	−0.008	−0.003
24	−0.008	−0.003	0.003	−0.003	0.002

$$TRE = \frac{4}{\sin 2\theta}(1 - \cos\theta) \quad [4m + 2]\text{annulenes} \tag{53}$$

$$TRE = \frac{4}{\tan 2\theta}(1 - \cos\theta) \quad [4m]^+\text{- and } [4m]^{--}\text{annulenes} \tag{54}$$

$$TRE = \frac{2}{\sin\theta}(1 - \cos\theta) \quad [4m + 1]^-\text{- and } [4m + 3]^+\text{annulenes} \tag{55}$$

On the other hand, the reduced ring currents for the same annulene structures are given by[119-122]

$$RC = \frac{2S}{N^2 \sin 2\theta} \quad [4m + 2]\text{annulenes} \tag{56}$$

$$RC = \frac{2S}{N^2 \tan 2\theta} \quad [4m]^{++}\text{- and } [4m]^{--}\text{annulenes} \tag{57}$$

$$RC = \frac{S}{N^2 \sin\theta} \qquad [4m + 1]^- \text{- and } [4m + 3]^+ \text{ annulenes} \qquad (58)$$

where S and N are the area and the size, respectively, of the annulene ring and $\theta = \pi/2N$. When Equations 53 to 55 and 56 to 58 are combined, one obtains the following relationship for all classes of $[4m+2]$ π-electron annulenes:[123]

$$TRE = \frac{2N^2RC}{S}(1 - \cos\theta) \qquad (59)$$

Formula (59) may additionally be simplified. This can be done in the following way. The expansion of the cosine into the corresponding power series gives

$$1 - \cos\theta = 1 - (1 - \frac{\theta^2}{2!} + \frac{\theta^4}{4!} - \frac{\theta^6}{6!} + ...) \qquad (60)$$

After the substitution of $\theta = \pi/2N$, Equation 60 changes into

$$1 - \cos\theta = \frac{\pi^2}{8N^2}(1 - \frac{\pi^2}{48N^2} + \frac{\pi^4}{5760N^4} - ...) \qquad (61)$$

Clearly for large N it is possible to truncate Equation 61 at the first term:

$$1 - \cos\theta = \frac{\pi^2}{8N^2} \qquad (62)$$

Introduction of Equation 62 into Equation 59 produces the elegant formulation of the relationship between TRE and RC:[123]

$$TRE = \frac{\pi^2}{4S} RC \qquad (63)$$

This expression only slightly differs from the Haddon relation given below, which relates DREs and RCs:[119]

$$DRE = \frac{\pi^2}{3S} RC \qquad (64)$$

The difference is due to the fact that DRE is larger than TRE for all aromatic annulenes, both neutral and ionic.[52]

TRE and the diamagnetic susceptibility of an annulene system are related as follows. The diamagnetic susceptibility χ_π of an annulene is given by[124]

$$\chi_\pi = \frac{SI}{cH} \tag{65}$$

where I is the ring current intensity and H the applied magnetic field taken perpendicular to the annulene plane. The ring current intensity is defined by

$$I = [8\pi^2 c(e/hc)^2 H]RC \tag{66}$$

where c, e, and h are fundamental physical constants ($c = 299792458$ ms^{-1}; $e = 1.60217733 \times 10^{-19}$ C; $h = 6.6260755 \times 10^{-34}$ Js).[125] The combination of Equations 63, 65, and 66 leads to the relationship between the topological resonance energy and the diamagnetic susceptibility:

$$TRE = \frac{1}{32S^2}\left(\frac{hc}{e}\right)^2 \chi_\pi \tag{67}$$

Equation 67 indicates that TRE is directly proportional to the diagmetic susceptibility and to the reciprocal value of the area of the annulene ring squared.

There are some results in the literature which support the relationship between the resonance energies and ring currents of [4m + 2] π-electron annulenes. For example, it has been shown that chemical shifts of the inner and outer protons in tetra-*tert*-butyldehydro[N]annulenes correlate well with the Hess-Schaad RE/e values.[126]

C. MÖBIUS ANNULENES

Möbius annulenes are also regular structures which differ from Hückel annulenes in having one phase dislocation resulting from the negative overlap between the adjacent 2p$_z$-orbitals of different sign. The analytical expressions for TRE of Möbius [N]annulenes are, therefore, closely related to those of Hückel [N]annulenes and can be derived following the same line of reasoning. The TRE expressions for Möbius [N]annulenes are given by[91]

TRE(Möbius[N]annulene) =

$$\begin{cases} 2\left(\dfrac{\cos\theta - 1}{\sin\theta} + \tan\theta\right) & N = 4m \qquad\qquad\qquad (68) \\[3mm] 2\,\dfrac{\cos\theta - 1}{\sin\theta}\cos\theta & N = 4m + 1 \text{ or } N = 4m + 3 \\[3mm] 2\left(\dfrac{\cos\theta - 1}{\sin\theta} - \tan\theta\right) & N = 4m + 2 \end{cases}$$

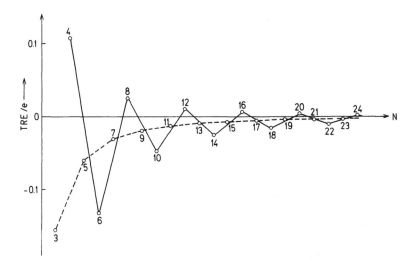

FIGURE 8. TRE/e vs. the ring size of Möbius [N]annulenes.

A plot of TRE/e vs. the ring size of Möbius [N]annulene (see Figure 8) reveals that Möbius [N = even]annulenes follow the reverse, *anti-Hückel rule*. Thus, Möbius [4*m*]annulenes are aromatic while Möbius [4*m* + 2]annulenes are antiaromatic when *m* is low. At large values of *m* both classes of even-membered Möbius annulenes converge to nonaromaticity. The behavior of odd-membered Möbius annulenes parallels the result obtained for odd-membered Hückel annulenes. This must be so because the expressions for TRE of odd-membered Hückel and Möbius annulenes are identical.

The analytical expressions for TRE of Möbius [N]annulene positive and negative ions are also closely related to the corresponding expressions for Hückel [N]annulene positive and negative ions:

$$
TRE^+ = TRE - \begin{cases} -2\sin\theta & N - 4m + 3 \\ 2\sin\theta & N = 4m + 2 \\ 2\sin\theta & N = 4m + 1 \\ 2\sin\theta - 2\sin2\theta & N = 4m \end{cases} \tag{69}
$$

$$
TRE^{++} = TRE - \begin{cases} 2\sin2\theta - 4\sin\theta & N = 4m + 3 \\ 4\sin\theta & N = 4m + 2 \\ 2(\sin\theta + \sin2\theta - \sin3\theta) & N = 4m + 1 \\ 4\sin\theta - 4\sin2\theta & N = 4m \end{cases} \tag{70}
$$

$$
TRE^- = TRE - \begin{cases} 2\sin\theta & N = 4m + 3 \\ 2\sin\theta & N = 4m + 2 \\ -2\sin\theta & N = 4m + 1 \\ 2\sin\theta - 2\sin2\theta & N = 4m \end{cases} \tag{71}
$$

$$\text{TRE}^{--} = \text{TRE} - \begin{cases} 2(\sin\theta + \sin2\theta - \sin3\theta) & N = 4m + 3 \\ 4\sin\theta & N = 4m + 2 \\ 2\sin2\theta - 4\sin\theta & N = 4m + 1 \\ 4\sin\theta - 4\sin2\theta & N = 4m \end{cases} \quad (72)$$

Möbius annulene ions also follow the anti-Hückel rule and converge rather rapidly to nonaromaticity (see Table 8).

D. CONJUGATED HYDROCARBONS AND HETEROCYCLICS

TRE and TRE/e values of conjugated hydrocarbons and heteroconjugated molecules can be obtained straightforwardly by means of the procedure described earlier in this chapter. TRE and TRE/e values of some randomly chosen conjugated molecules are given in Table 9. TRE/e values of conjugated hydrocarbons and heterocyclics parallel in most cases RE/e values Hess and Schaad.[35,127,128]

A plot of TRE/e vs. RE/e for 89 randomly selected conjugated hydrocarbons (45 structures) and heterocyclics (44 structures) is given in Figure 9. A linear least-squares fit produces

$$\text{TRE/e} = 1.016 \, (\pm 0.026) \, \text{RE/e} - 0.006 \, (\pm 0.001) \quad (73)$$

with the correlation coefficient 0.973 and the standard deviation 0.012. Hence, one may conclude that TRE/e and RE/e have practically the same predictive power in the case of conjugated hydrocarbons and heterocyclics. However, this conclusion needs to be slightly modified because of the finding[129] that the difference between TRE/e and RE/e is proportional to the natural logarithm of the number of Kekulé structures K,

$$\text{TRE/e} - \text{RE/e} \approx a \ln K \quad (74)$$

where a (0.69) is the proportionality constant. Therefore, the difference between TRE/e and RE/e is expected to be appreciable for large polycyclic conjugated molecules for which K is usually rather large.

E. CONJUGATED IONS AND RADICALS

The chemistry of aromatic ions and radicals has considerably advanced in the last two decades,[130-132] owing to improvements in preparative techniques.[133-135] For this reason there are many conjugated ions and radicals available and their structures, properties, and reactivities are studied.[136-138] The question about their aromaticity has also been raised.[40] The TRE model can be directly applied to conjugated ions and radicals. Hess and Schaad have also proposed an aromaticity index for studying these species by introducing a novel reference structure into the framework of their aromaticity theory.[40,139]

TABLE 8
TRE/e Values of Möbius [N]Annulenes and Their
Positive and Negative Ions

N	TRE/e	(TRE/e)$^+$	(TRE/e)$^{++}$	(TRE/e)$^-$	(TRE/e)$^{--}$
3	−0.155	−0.732	−0.732	0.134	0.054
4	0.108	−0.072	−0.434	−0.044	−0.145
5	−0.060	0.079	−0.042	−0.153	−0.052
6	−0.133	−0.056	0.059	−0.040	0.030
7	−0.031	−0.111	−0.048	0.028	−0.017
8	0.025	−0.025	−0.092	−0.019	−0.055
9	−0.019	0.022	−0.020	−0.052	−0.017
10	−0.047	−0.018	0.019	−0.015	0.013
11	−0.013	−0.043	−0.016	0.012	−0.010
12	0.011	−0.011	−0.038	−0.010	−0.027
13	−0.009	0.010	−0.010	−0.026	−0.008
14	−0.024	−0.009	0.009	−0.008	0.007
15	−0.007	−0.022	−0.008	0.007	−0.006
16	0.006	−0.006	−0.021	−0.006	−0.016
17	−0.005	0.006	−0.006	−0.015	−0.005
18	−0.015	−0.005	0.005	−0.005	0.004
19	−0.004	−0.014	−0.005	0.004	−0.004
20	0.004	−0.004	−0.013	−0.004	−0.011
21	−0.004	0.004	−0.004	−0.010	−0.003
22	−0.010	−0.003	0.004	−0.003	0.003
23	−0.003	−0.009	−0.003	0.003	−0.003
24	0.003	−0.003	−0.009	−0.003	−0.007

They have normalized their RE index with respect to the number of atoms N_a in the conjugated system and denoted by RE/a:

$$RE/a = RE/N_a \tag{75}$$

The TRE model has been applied to numerous conjugated radicals and ions, achieving in most cases a good agreement between the predictions and experimental findings.[90] The results for the pentalene radicals and ions illustrate nicely the value of the TRE approach to aromatic stability of conjugated ions and radicals. Pentalene and its four ions and radicals together with their TRE and TRE/e values are shown in Figure 10.

TRE (and TRE/e) predicts pentalene cation radical and dication strongly antiaromatic and pentalene anion radical and dianion aromatic. The experimental findings support these theoretical results: the preparation of pentalene dication was attempted unsuccessfully,[140,141] pentalene cation is a very unstable structure,[15,22] substituted pentalene anion has been prepared,[141] while pentalene dianion appears to be stable[142] in contrast to the parent hydrocarbon.[15,20,21]

TABLE 9
Topological Resonance Energies of Some Conjugated Molecules

Molecule	TRE[a]	TRE/e[b]	Status
Benzene	0.276	0.046	i
Naphthalene	0.390	0.039	i
Anthracene	0.476	0.034	i
Phenanthrene	0.546	0.039	i
Pyrene	0.592	0.037	i
Cyclobutadiene	−1.228	−0.307	u
Benzocyclobutadiene	0.392	−0.049	u
Biphenylene	0.120	0.010	i
Fulvene	0.018	0.003	u
Acenaphthylene	0.360	0.030	i
Pyracylene	0.126	0.009	u
Azulene	0.150	0.015	i
Heptalene	−0.144	−0.012	u
Furan	0.042	0.007	i
Pyrrole	0.240	0.040	i
Thiophene	0.198	0.033	i
Oxepin	−0.032	−0.004	u
Azepine	−0.232	−0.029	u
Thiepin	−0.184	−0.023	u
Aza-cyclobutadiene	−0.772	−0.193	—
Pyridine	0.228	0.038	i
Cyclo[3.2.2]azine	0.396	0.033	i

Note: TRE, topological resonance energy, and TRE/e in β units.

[a] Data from Gutman, I., Milun, M., and Trinajstić, N., *J. Am. Chem. Soc.*, 99, 1692, 1977.
[b] TRE/e = TRE/N_e, where N_e is the number of π-electrons in the molecule.
[c] i, isolated; u, unstable; —, unknown.

F. HOMOAROMATIC SYSTEMS

Homoaromatic systems are nonclassical structures created when several double bonds, spatially separated, but favorably oriented, interact through space to close a noose over which π-electrons can be delocalized.[143,144] Hence, the formation of a polygon-like pattern by through-space interactions among disjoint intramolecular fragments is recognized as the essential feature of homoaromatic stabilization or destabilization. A good example to illustrate homoaromatic systems is *cis,cis,cis*-cyclonona-1,4,7-triene, shown in Figure 11 in the "crown" form. The homoconjugation in this structure can be illustrated by the 1,3,5-trishomohexagonal system also shown in Figure 11, which in turn may be depicted by the edge-weighted graph G_{EW} with six vertices, also shown in Figure 11.

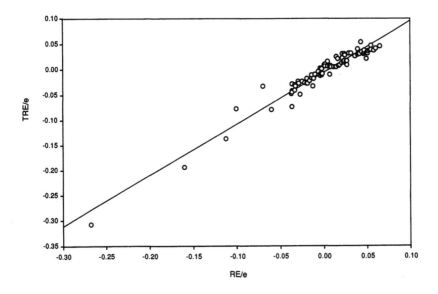

FIGURE 9. A plot of TRE/e vs. RE/e for 89 randomly chosen conjugated hydrocarbons and heteroconjugated molecules. The line represents a linear least-squares fit.

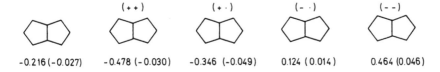

FIGURE 10. Pentalene and its four ions and radicals with their TRE and TRE/e values shown beneath each structure. TRE/e values are given in parentheses.

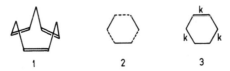

FIGURE 11. Example of a homoaromatic system (1), its planar (2) and graph-theoretical (3) representations. The parameter k signifies through-space conjugative interaction between the neighboring ethylenic fragments.

TREs of homoaromatic systems can also be computed by means of Equation 30. The requisite acyclic polynomial can be obtained by using Equation 37. The crucial quantity in these calculations is the homoaromatic parameter k which signifies the interaction between neighboring, but not connected, $2p_z$ atomic orbitals. The value of this parameter is unknown. However, the numerical values of k have been deduced for some systems from an analysis of their spectral data such as their photoelectron spectra.[145] Thus, for example, the value of k for cyclononatriene is found to be 0.25. This value of k yields the TRE value of 0.011 for *cis,cis,cis*-cyclonona-1,4,7-triene, which places this molecule in the class of nonaromatic species. Therefore, the degree of conjugation in cyclononatriene is rather low (only 4% of that in fully delocalized benzene, TRE of which is 0.276) and it perhaps best explains the nonappearance of any indication of homoaromaticity in the NMR data of this compound.[146,147]

G. AROMATICITY IN THE LOWEST EXCITED STATE OF ANNULENES

The TRE model can be directly used for studying aromaticity of conjugated molecules in the excited states. The pioneering work in this area was carried out by Baird,[148] who extended the DRE model to the lowest π-π^* triplet state of conjugated hydrocarbons by comparing the energy of triplets with the most stable biradical reference structure. Here are reported TREs of Hückel and Möbius annulenes in the lowest excited state, because this is the most interesting state for the majority of photochemical reactions.[149,150]

Topological resonance energy of an annulene in the excited state is denoted by TRE*. The TRE* expressions for both Hückel and Möbius [N]annulenes are given by

		[N]annulene	
		Hückel	Möbius
TRE* = TRE +	$4\ (\sin\theta)$	$N = 4m$	$N = 4m + 2$
	$2\ (\sin2\theta - \sin\theta - \sin3\theta)$	$N = 4m + 1$	$N = 4m + 3$
	$4\ (\sin\theta - \sin2\theta)$	$N = 4m + 2$	$N = 4m$
	$2\ (\sin\theta + \sin2\theta - \sin5\theta)$	$N = 4m + 3$	$N = 4m + 1$

$$(76)$$

where TRE is the topological resonance energy of the ground state, while $\theta = \pi/2N$. Numerical values for TRE* of annulenes are reported in Table 10.

The results in Table 10 reveal several interesting points. For example, the lowest excited state of cyclobutadiene has a large and positive (aromatic) value of TRE*/e (0.076), while benzene has a large and negative (antiaromatic) value of TRE*/e (-0.115). These results are in agreement with the work of Baird[148] and some other studies on the conjugation in annulenes in the lowest excited state.[151] The results in Table 10 thus indicate that the Hückel [4m]annulenes are *aromatic* and the Hückel [4m + 2]annulenes *antiaromatic* up to a certain size in the lowest excited state. The reverse is true for Möbius annulenes: Möbius [4m]annulenes are predicted *antiaromatic* and Möbius [4m + 2]annulenes *aromatic* up to a certain size in the lowest excited state. With the increasing size both Hückel and Möbius [4m]- and [4m + 2]annulenes converge to nonaromaticity. Odd-membered annulenes in

TABLE 10
TRE* and TRE*/e Values of
Hückel and Möbius Annulenes
in the Lowest Excited State

Hückel [N]annulenes				Möbius [N]annulenes			
N	TRE	TRE*	TRE*/e	N	TRE	TRE*	TRE*/e
3	− 0.464	− 1.268	0.423	3	− 0.464	− 1.732	− 0.577
4	− 1.226	0.304	0.076	4	0.431	− 0.867	− 0.217
5	− 0.301	− 1.362	− 0.272	5	− 0.301	− 0.508	− 0.102
6	0.273	− 0.692	− 0.115	6	− 0.799	0.236	0.039
7	− 0.220	− 0.709	− 0.101	7	− 0.220	− 1.044	− 0.149
8	− 0.595	0.186	0.023	8	0.201	− 0.550	− 0.069
9	− 0.172	− 0.836	− 0.093	9	− 0.172	− 0.673	− 0.075
10	0.159	− 0.451	− 0.045	10	− 0.474	0.152	0.015
11	− 0.142	− 0.603	− 0.055	11	− 0.142	− 0.694	− 0.063
12	− 0.394	0.128	0.011	12	0.132	− 0.381	− 0.032
13	− 0.120	− 0.592	− 0.046	13	− 0.120	− 0.537	− 0.041
14	0.113	− 0.329	− 0.024	14	− 0.338	0.110	0.008
15	− 0.104	− 0.479	− 0.032	15	− 0.104	− 0.516	− 0.034
16	− 0.295	0.097	0.006	16	0.099	− 0.290	− 0.018
17	− 0.092	− 0.456	− 0.027	17	− 0.092	− 0.432	− 0.025
18	0.088	− 0.258	− 0.014	18	− 0.262	0.086	0.005
19	− 0.082	− 0.391	− 0.021	19	− 0.082	− 0.409	− 0.022
20	− 0.236	0.078	0.004	20	0.079	− 0.233	− 0.012
21	− 0.075	− 0.371	− 0.018	21	− 0.075	− 0.358	− 0.017
22	0.072	− 0.212	− 0.010	22	− 0.214	0.071	0.003
23	− 0.068	− 0.329	− 0.014	23	− 0.068	− 0.339	− 0.015
24	− 0.197	0.065	0.003	24	0.066	− 0.195	− 0.008

the lowest excited state start with antiaromatic [3]annulene (TRE*/e $= -0.577$) and converge also to nonaromaticity at higher values of m.

The experimental data about the stabilities of Hückel annulenes in the lowest excited state are rather scarce. Perhaps the high antiaromatic character of benzene in the lowest excited state best explains its reactivity in that state. For example, benzene undergoes photocycloaddition with simple olefines to produce 1,2-, 1,3- and 1,4-adducts.[152] Similarly, benzene on irradiation with UV light forms a benzene valence isomer: Dewar benzene.[153]

In the case of Möbius annulenes in the excited state, the experimental data are very limited indeed. However, there are proposals that photochemical reactions with the transition state having the Möbius-type structure are allowed only if the transition state has a $4m + 2$ topology.[149,154] Therefore, if the Dewar-Evans-Zimmerman rules[149,154,155] are valid for predicting qualitatively the allowedness of pericyclic reactions, then the TRE model represents a convenient model for quantitatively treating the aromatic properties of cyclic transition states.

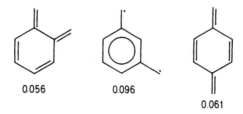

0.056   0.096

0.061

FIGURE 12.   Three isomeric quinodimethanes and their TRE values.

H. FAILURES OF THE TRE MODEL

The increasing use of the TRE model has also led to discovery of cases when the model produces predictions which disagree with the chemical experience.[90,156-160] A typical failure[157] is illustrated in Figure 12.

The *meta*-isomer has the greatest TRE value, although it is a highly reactive biradical. This result warns that the TRE model must be used cautiously in the case of the open-shell systems.[161]

Some other critical remarks, like criticism of the nonphysical reference structure,[162] are really critique of all DRE-like reference structures and not only TRE.

REFERENCES

1. **Lewis, D. and Peters, D.**, *Facts and Theories of Aromaticity*, Macmillan, London, 1975.
2. **Pauling, L.**, *The Nature of the Chemical Bond*, 2nd ed., Cornell University Press, Ithaca, 1958, 12th printing.
3. **Heisenberg, W.**, *Z. Phys.*, 39, 499, 1926.
4. **Craig, D. P.**, in *Non-Benzenoid Aromatic Compounds*, Ginsburg, D., Ed., Wiley-Interscience, New York, 1959, 1.
5. **Snyder, J. P.**, in *Non-Benzenoid Aromatics*, Snyder, J. P., Ed., Academic Press, New York, 1969, 1.
6. **Badger, G. M.**, *Aromatic Character and Aromaticity*, University Press, Cambridge, 1969.
7. **Garratt, P. J.**, *Aromaticity*, McGraw-Hill, New York, 1971.
8. **Lloyd, D. and Marshall, D. R.**, *Angew. Chem. Int. Ed. Engl.*, 11, 404, 1972.
9. **Agranat, I.**, in *Aromatic Compounds*, Zollinger, H., Ed., Butterworths, London, 1973, 139.
10. **Garratt, P. J.**, in *Rodd's Chemistry of Carbon Compounds*, Supplement to the 2nd ed., Vol. VII, Part A, Ansell, M. F., Ed., Elsevier, Amsterdam, 1983, 1.
11. **Garratt, P. J.**, *Endeavour, N. Ser.*, 11, 36, 1987.
12. **Lloyd, D. M. G.**, *The Chemistry of Conjugated Cyclic Compounds (To Be or Not To Be Like Benzene)*, John Wiley & Sons, London, 1989.
13. **Gorelik, M. V.**, *Russ. Chem. Rev.*, 59, 116, 1990.
14. **Balaban, A. T.**, *Pure Appl. Chem.*, 52, 1409, 1980.
15. **Lloyd, D.**, *Non-Benzenoid Conjugated Carbocyclic Compounds*, Elsevier, Amsterdam, 1984.

16. **Párkányi, C. and Boniface, C.**, *Bull. Soc. Chim. Belg.*, 99, 597, 1990.
17. **Streitwieser, A., Jr.**, *Molecular Orbital Theory for Organic Chemists*, John Wiley & Sons, New York, 1961.
18. **Dewar, M. J. S.**, *The Molecular Orbital Theory of Organic Chemistry*, McGraw-Hill, New York, 1969.
19. **Dewar, M. J. S.**, *Pure Appl. Chem.*, 44, 767, 1975.
20. **Bergman, E. D.**, in *Non-Benzenoid Aromatic Compounds*, Ginsburg, D., Ed., Wiley-Interscience, New York, 1959, 141.
21. **Glidewell, C. and Lloyd, D.**, *Tetrahedron*, 40, 4455, 1984.
22. **Lloyd, D.**, *Carbocyclic Non-Benzenoid Aromatic Compounds*, Elsevier, Amsterdam, 1966.
23. **Hall, G. G.**, *Int. J. Math. Educ. Sci. Technol.*, 4, 233, 1973.
24. **McClelland, B. J.**, *J. Chem. Phys.*, 54, 640, 1971.
25. **Trinajstić, N.**, *Croat. Chem. Acta*, 37, 307, 1965.
26. **Baird, N. C.**, *J. Chem. Educ.*, 48, 509, 1971.
27. **Dewar, M. J. S. and de Llano, C.**, *J. Am. Chem. Soc.*, 91, 789, 1969.
28. **Dewar, M. J. S. and Trinajstić, N.**, *J. Chem. Soc. A*, 1754, 1969.
29. **Dewar, M. J. S., Harget, A. J., and Trinajstić, N.**, *J. Am. Chem. Soc.*, 91, 6321, 1969.
30. **Dewar, M. J. S. and Trinajstić, N.**, *J. Am. Chem. Soc.*, 92, 1453, 1970.
31. **Dewar, M. J. S. and Harget, A. J.**, *Proc. R. Soc. (London) Ser. A*, 315, 443, 1970; 315, 457, 1970.
32. **Pople, J. A.**, *Trans. Faraday Soc.*, 49, 1375, 1953.
33. **Hess, B. A., Jr. and Schaad, L. J.**, *J. Org. Chem.*, 37, 4179, 1972.
34. **Gutman, I., Milun, M., and Trinajstić, N.**, *Chem. Phys. Lett.*, 23, 284, 1973.
35. **Hess, B. A., Jr. and Schaad, L. J.**, *J. Am. Chem. Soc.*, 93, 305, 1971.
36. **Milun, M., Sobotka, Ž., and Trinajstić, N.**, *J. Org. Chem.*, 37, 139, 1972.
37. **Schmalz, T. G., Živković, T., and Klein, D. J.**, in *MATH/CHEM/COMP 1987*, Lacher, R. C., Ed., Elsevier, Amsterdam, 1988, 173.
38. **Schaad, L. J. and Hess, B. A., Jr.**, *Isr. J. Chem.*, 20, 281, 1980.
39. **Jiang, Y., Tang, A., and Hoffmann, R.**, *Theoret. Chim. Acta*, 66, 183, 1984.
40. **Hess, B. A., Jr. and Schaad, L. J.**, *Pure Appl. Chem.*, 52, 1471, 1980.
41. **Gutman, I., Milun, M., and Trinajstić, N.**, *Math. Chem. (Mülheim/Ruhr)*, 1, 171, 1975.
42. **Wilcox, C. F., Jr.**, *Croat. Chem. Acta*, 47, 87, 1975.
43. **Green, A. L.**, *J. Chem. Soc.*, 1886, 1956.
44. **Carter, P. C.**, *Trans. Faraday Soc.*, 45, 597, 1949.
45. **Swinborne-Sheldrake, R., Herndon, W. C., and Gutman, I.**, *Tetrahedron Lett.*, 755, 1975.
46. **Hosoya, H., Hosoi, K., and Gutman, I.**, *Theoret. Chim. Acta*, 38, 37, 1975.
47. **Aihara, J.-I.**, *J. Org. Chem.*, 41, 2488, 1976.
48. **Hosoya, H.**, *Bull. Chem. Soc. Jpn.*, 44, 2332, 1971.
49. **Gutman, I. and Bosanac, S.**, *Tetrahedron*, 33, 1809, 1977.
50. **Sachs, H.**, *Publ. Math. (Debrecen)*, 11, 119, 1964.
51. **Graovac, A., Gutman, I., Trinajstić, N., and Živković, T.**, *Theoret. Chim. Acta*, 26, 67, 1972.
52. **Aihara, J.-I.**, *J. Am. Chem. Soc.*, 98, 2750, 1976.
53. **Farrell, E. J.**, *J. Comb. Theory*, 26B, 111, 1979; 27B, 75, 1979.
54. **Herndon, W. C. and Ellzey, M. L., Jr.**, *J. Chem. Inf. Comput. Sci.*, 19, 260, 1979.
55. **Mizoguchi, N.**, *J. Am. Chem. Soc.*, 107, 4419, 1985.
56. **Gutman, I., Milun, M., and Trinajstić, N.**, *J. Am. Chem. Soc.*, 99, 1692, 1977.
57. **Heilmann, O. J. and Leib, E. H.**, *Commun. Math. Phys.*, 25, 190, 1972.
58. **Schaad, L. J., Hess, B. A., Jr., Nation, J. B., Trinajstić, N., and Gutman, I.**, *Croat. Chem. Acta*, 52, 233, 1979.

59. Godsil, C. D. and Gutman, I., Z. *Naturforsch,* 34a, 776, 1979.
60. Gutman, I., *Croat. Chem. Acta,* 54, 75, 1981.
61. Graovac, A., Kasum, D., and Trinajstić, N., *Croat. Chem. Acta,* 54, 91, 1981.
62. Graovac, A., *Chem. Phys. Lett.,* 82, 248, 1981.
63. Krygowski, T. M. and Kruszewski, J., *Quantitative Criteria of Aromaticity,* University of Wroclaw, 1978 (in Polish).
64. Aihara, J.-I., *Bull. Chem. Soc. Jpn.,* 50, 3057, 1977.
65. Ilić, P., Džonova-Jerman-Blažič, B., Mohar, B., and Trinajstić, N., *Croat. Chem. Acta,* 52, 35, 1979.
66. Ilić, P. and Trinajstić, N., *Croat. Chem. Acta,* 56, 203, 1983.
67. Glidewell, C. and Lloyd, D., *J. Chem. Educ.,* 63, 306, 1986.
68. Hess, B. A., Jr., Schaad, L. J., and Agranat, I., *J. Am. Chem. Soc.,* 100, 5268, 1978.
69. Džonova-Jerman-Blažič, B., Mohar, B., and Trinajstić, N., in *Applications of Information and Control Systems,* Lainiotis, D. G. and Tzannes, N. S., Eds., Reidel, Dordrecht, Holland, 1980, 395.
70. Mohar, B. and Trinajstić, N., *J. Comput. Chem.,* 3, 28, 1982.
71. Trinajstić, N., *Int. J. Quantum Chem.,* S 11, 469, 1977.
72. Heilbronner, E., *Helv. Chim. Acta,* 36, 170, 1953.
73. Harary, F., *Graph Theory,* Addison-Wesley, Reading, MA, 1971, 96, 2nd printing.
74. Gutman, I., *Croat. Chem. Acta,* 53, 581, 1980.
75. Gutman, I., *J. Chem. Soc. Faraday Trans. II,* 75, 799, 1979.
76. Bosanac, S. and Gutman, I., *Z. Naturforsch.,* 32a, 10, 1977.
77. Hückel, E., *Z. Phys.,* 76, 628, 1932.
78. Kruszewski, J. and Krygowski, T. M., *Can. J. Chem.,* 53, 945, 1975.
79. Gutman, I. and Trinajstić, N., *Can. J. Chem.,* 53, 1789, 1976.
80. Gutman, I. and Trinajstić, N., *J. Chem. Phys.,* 64, 4921, 1976.
81. Dewar, M. J. S. and Gleicher, G. J., *J. Am. Chem. Soc.,* 87, 689, 1965.
82. Krygowski, T. M. and Kruszewski, J., *Bull. Acad. Pol. Sci.,* 21, 509, 1973.
83. Klein, D. J. and Trinajstić, N., *J. Am. Chem. Soc.,* 106, 8050, 1984.
84. Aihara, J.-I., *J. Am. Chem. Soc.,* 99, 2048, 1977.
85. Sabljić, A. and Trinajstić, N., *J. Mol. Struct.,* 49, 415, 1978.
86. Aihara, J.-I., *J. Am. Chem. Soc.,* 100, 3339, 1978.
87. Aihara, J.-I., *Bull. Chem. Soc. Jpn.,* 51, 1788, 1978.
88. Aihara, J.-I., *Bull. Chem. Soc. Jpn.,* 51, 3540, 1978.
89. Ilić, P. and Trinajstić, N., *Pure Appl. Chem.,* 52, 1495, 1980.
90. Ilić, P. and Trinajstić, N., *J. Org. Chem.,* 45, 1738, 1980.
91. Ilić, P., Sinković, B., and Trinajstić, N., *Isr. J. Chem.,* 20, 258, 1980.
92. Ilić, P., Jurić, A., and Trinajstić, N., *Croat. Chem. Acta,* 53, 587, 1980.
93. Norrinder, U., Tanner, D., Thulin, B., and Wennerström, O., *Acta Chem. Scand. Ser. B,* 35, 403, 1981.
94. El-Basil, S., *Indian J. Chem.,* 20B, 586, 1981.
95. Sabljić, A. and Trinajstić, N., *J. Org. Chem.,* 46, 3457, 1981.
96. Gimarc, B. M. and Trinajstić, N., *Inorg. Chem.,* 21, 21, 1982.
97. Aihara, J.-I., *Pure Appl. Chem.,* 54, 1115, 1982.
98. Ilić, P., Mohar, B., Knop, J. V., Jurić, A., and Trinajstić, N., *J. Heterocycl. Chem.,* 19, 625, 1982.
99. Jurić, A. and Trinajstić, N., *Croat. Chem. Acta,* 56, 215, 1983.
100. Norrinder, U., Tanner, D., and Wennerström, O., *Croat. Chem. Acta,* 56, 269, 1983.
101. Norrinder, U. and Wennerström, O., *Acta Chem. Scand. Ser. A,* 37, 431, 1983.
102. Singh, P. P. and Raudhawa, H. S., *Indian J. Chem.,* 22 B, 901, 1983.
103. Gimarc, B. M., Jurić, A., and Trinajstić, N., *Inorg. Chem. Acta,* 102, 105, 1985.
104. Jurić, A., Sabljić, A., and Trinajstić, N., *J. Heterocycl. Chem.,* 21, 273, 1984.

105. **Jurić, A., Trinajstić, N., and Jashari, G.,** *Croat. Chem. Acta,* 59, 617, 1986.
106. **Mukherjee, A. K.,** *Indian J. Chem.,* 25 A, 574, 1986.
107. **Nikolić, S., Jurić, A., and Trinajstić, N.,** *Heterocycles,* 26, 2025, 1987.
108. **Saničanin, Ž., Jurić, A., Tabaković, I., and Trinajstić, N.,** *J. Org. Chem.,* 52, 4053, 1987.
109. **Náray-Szabó, G., Surján, P. R., and Angyán, J. G.,** *Applied Quantum Chemistry,* Kiadó, Budapest, 1987, sect. 2.6.
110. **Zhou, Z. and Parr, R. G.,** *J. Am. Chem. Soc.,* 111, 7371, 1989.
111. **Jiang, Y. and Zhang, H.,** *Theoret. Chim. Acta,* 75, 279, 1989.
112. **Amić, D., Jurić, A., and Trinajstić, N.,** *Croat. Chem. Acta,* 63, 19, 1990; **Langler, R. F. and Precedo, L.,** *Can. J. Chem.,* 68, 939, 1990.
113. **Gutman, I., Milun, M., and Trinajstić, N.,** *Croat. Chem. Acta,* 44, 207, 1972.
114. **Polansky, O. E.,** *Monatsh. Chem.,* 91, 916, 1960.
115. **Gutman, I., Milun, M., and Trinajstić, M.,** *Croat. Chem. Acta,* 49, 441, 1977.
116. **Haddon, R. C., Haddon, V. R., and Jackman, L. M.,** *Topics Curr. Chem.,* 16, 103, 1971.
117. **Sondheimer, F.,** *Acc. Chem. Res.,* 5, 81, 1972.
118. **Aihara, J.-I.,** *J. Am. Chem. Soc.,* 101, 558, 1979.
119. **Haddon, R. C.,** *J. Am. Chem. Soc.,* 101, 1722, 1979.
120. **London, F.,** *J. Phys. Radium,* 8, 397, 1937.
121. **McWeeny, R.,** *Mol. Phys.,* 1, 311, 1958.
122. **Haigh, C. W. and Mallion, R. B.,** in *Progress in Nuclear Magnetic Spectroscopy,* Vol. 13, Emsley, J. W., Feeney, J., and Sutcliffe, L. H., Eds., Pergamon Press, Oxford, 1979, 303.
123. **Aihara, J.-I.,** *Bull. Chem. Soc. Jpn.,* 53, 1163, 1980.
124. **Salem, L.,** *Molecular Orbital Theory of Conjugated Systems,* Benjamin, New York, 1966, chap. 4.
125. **Mills, I., Cvitaš, T., Homann, K., Kallay, N., and Kuchitsu, K.,** *Quantities, Units and Symbols in Physical Chemistry,* Blackwell Scientific, Oxford, 1988.
126. **Hess, B. A., Jr., Schaad, L. J., and Nakagawa, M.,** *J. Org. Chem.,* 42, 1661, 1977.
127. **Hess, B. A., Jr., Schaad, L. J., and Holyoke, C. W., Jr.,** *Tetrahedron,* 28, 3657, 1972; 31, 295, 1975.
128. **Hess, B. A., Jr. and Schaad, L. J.,** *J. Am. Chem. Soc.,* 95, 3907, 1973.
129. **Gutman, I.,** *Bull. Chem. Soc. (Belgrade),* 43, 191, 1978.
130. **Isaacs, N. S.,** *Reactive Intermediates in Organic Chemistry,* John Wiley & Sons, New York, 1974.
131. **March, J.,** *Advanced Organic Chemistry,* 3rd ed., John Wiley & Sons, New York, 1985.
132. **Isaacs, N. S.,** *Physical Organic Chemistry,* Longman, Burnt Mill, Harlow, Essex, 1987.
133. **Olah, G. A., Staral, J. S., Liang, G., Paquette, L. A., Melega, W. P., and Carmody, M. J.,** *J. Am. Chem. Soc.,* 100, 3349, 1978.
134. **Broxterman, Q. B., Hogeveen, H., and Kingma, R. F.,** *Pure Appl. Chem.,* 58, 89, 1986.
135. **Bock, H.,** *Pure Appl. Chem.,* 62, 383, 1990.
136. **Rabinovitz, M. and Willner, I.,** *Pure Appl. Chem.,* 52, 1575, 1980.
137. **Minsky, A., Meyer, A. Y., Hafner, K., and Rabinovitz, M.,** *J. Am. Chem. Soc.,* 105, 3975, 1983.
138. **Rabinovitz, M.,** *Topics Curr. Chem.,* 146, 99, 1988.
139. **Bates, R. B., Hess, B. A., Jr., Ogle, C. A., and Schaad, L. J.,** *J. Am. Chem. Soc.,* 103, 5062, 1981.
140. **Trost, B. and Kinsom, P. L.,** *J. Am. Chem. Soc.,* 97, 2438, 1975.
141. **Johnson, R. W.,** *J. Am. Chem. Soc.,* 99, 1461, 1977.
142. **Katz, T. J. and Rosenberg, M.,** *J. Am. Chem. Soc.,* 84, 865, 1962.
143. **Winstein, S.,** *Chem. Soc. Special Publ.,* 21, 5, 1967.

144. **Childs, R. F., Mahendran, M., Zweep, S. D., Shaw, G. S., Chadda, S. K., Burke, N. A. D., George, B. E., Faggiani, R., and Lock, C. J. L.,** *Pure Appl. Chem.,* 58, 111, 1986.

145. **Bischof, P., Gleiter, R., and Heilbronner, E.,** *Helv. Chim. Acta,* 53, 1425, 1970.

146. **Untch, K.,** *J. Am. Chem. Soc.,* 85, 345, 1963.

147. **Untch, K. and Kurland, R. J.,** *J. Am. Chem. Soc.,* 85, 346, 1963.

148. **Baird, N. C.,** *J. Am. Chem. Soc.,* 94, 4941, 1972.

149. **Dewar, M. J. S.,** *Angew. Chem. Int. Ed. Engl.,* 10, 761, 1971.

150. **Cowan, D. O. and Drisco, R. L.,** *Elements of Organic Photochemistry,* Plenum Press, New York, 1976.

151. **Ilić, P., Sinković, B., and Trinajstić, N.,** *J. Mol. Struct. (Theochem.),* 136, 155, 1986.

152. **Wilzbach, K. E. and Kaplan, L.,** *J. Am. Chem. Soc.,* 93, 2073, 1971.

153. **Wilzbach, K. E. and Kaplan, L.,** *J. Am. Chem. Soc.,* 87, 4004, 1965.

154. **Zimmerman, H. E.,** in *Perycyclic Reactions,* Vol. 1, March, A. P. and Lehr, R. E., Eds., Academic Press, New York, 1977, 53.

155. **Zimmerman, H. E.,** *Acc. Chem. Res.,* 4, 272, 1971.

156. **Gutman, I.,** *Chem. Phys. Lett.,* 66, 595, 1979.

157. **Gutman, I.,** *Theoret. Chim. Acta,* 56, 89, 1980.

158. **Gutman, I. and Mohar, B.,** *Chem. Phys. Lett.,* 69, 375, 1980; 77, 567, 1981.

159. **Gutman, I. and Mohar, B.,** *Croat. Chem. Acta,* 55, 375, 1982.

160. **Langler, R. F.,** *Austral. J. Chem.,* 44, 297, 1991.

161. **Aihara, J.-I.,** *Chem. Phys. Lett.,* 73, 404, 1980.

162. **Heilbronner, E.,** *Chem. Phys. Lett.,* 73, 377, 1982.

Chapter 8

ENUMERATION OF KEKULÉ VALENCE STRUCTURES

I. THE ROLE OF KEKULÉ VALENCE STRUCTURES IN CHEMISTRY

Kekulé valence structures have been used in organic chemistry[1-4] since 1865, when Kekulé proposed a hexagonal structure (see Figure 1) for benzene.[5-10] The continued interest in Kekulé structures[11] is related to their use in simple classical pictures of delocalized chemical bonding;[12,13] pictures that carry over to modern quantum-chemical theories of benzenoid hydrocarbons.[14,15]

Kekulé valence structures are the basis of several variants of valence bond (VB) resonance-theoretical models[16-18] such as the Pauling-Wheland model,[12,13,19] the Simpson-Herndon model,[20-27] and the conjugated-circuit model[28] (see the next chapter). These models have been shown to be useful for predicting the properties of conjugated molecules[11-15,22,23,29-34] in spite of some rather strong criticisms.[35]

In the computationally simplest form of the resonance theory[19] the ground states of conjugated molecules are described in terms of wave functions built from the Kekulé structures of equal weights. Therefore, the essential step in the application of the resonance theory is the enumeration of Kekulé valence structures. If the number of Kekulé structures K is known, one can, for instance, estimate quite accurately the total π-electron energy, E_π, of a benzenoid hydrocarbon with N = the number of carbon atoms and M = the number of carbon-carbon bonds, by means of the formula[36]

$$E_\pi = 0.442 \text{ N} + 0.788 \text{ M} + 0.34 \text{ K} (0.632)^{M-N} \qquad (1)$$

The resonance energy RE of benzenoid hydrocarbons can also be estimated only from the knowledge of the number of their Kekulé structures,[37-39]

$$RE = A \ln K \qquad (2)$$

where A = 1.185 eV.

Another formula which relates the resonance energy of benzenoid hydrocarbons to the number of their Kekulé structures and rings (R) has been proposed by Hall,[36]

$$RE = 0.4768 (0.6477)^R \text{ K} \qquad (3)$$

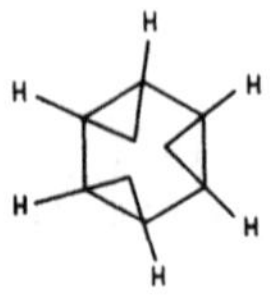

FIGURE 1. Kekulé formula for benzene (1865).

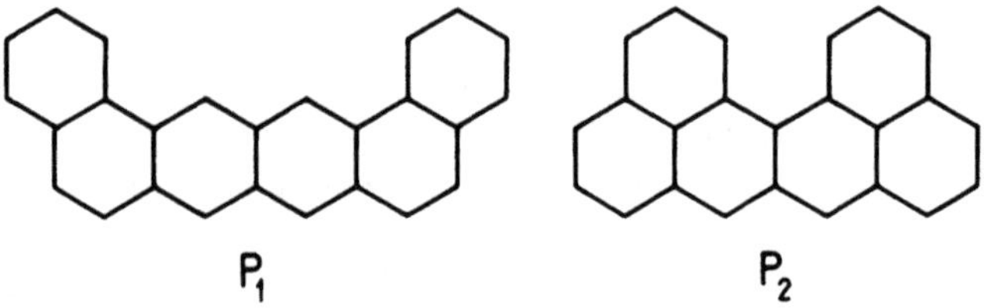

FIGURE 2. A Kekuléan polyhex P_1 ($K = 16$) and a non-Kekuléan polyhex P_2 ($K = 0$).

Bond lengths of benzenoid hydrocarbons can be predicted with a reasonable accuracy by the following formula:[40,41]

$$\ell_{ij} = 1.465 - 0.125 \, p_{ij} \qquad (4)$$

where ℓ_{ij} is the length of the i-j bond with the Pauling bond order p_{ij}.[29,42] The Pauling π-bond order p_{ij} for the i-j bond in a molecule with graph G is given by[40,41,43]

$$p_{ij} = \frac{K_{ij}}{K} \qquad (5)$$

where K_{ij} is the number of Kekulé structures for a subgraph $G - (i\text{-}j)$ obtained from G by removing the i-j bond.

Many additional uses of Kekulé valence structures in the chemistry of benzenoid hydrocarbons and other conjugated systems are described in the literature.[3,4,11-15,22,23,44-47]

It is also important to point to the *Clar postulate*,[48] which states that a polyhex system with *no* Kekulé structures should be an unstable biradical. Planar and nonplanar polyhex hydrocarbons may be partitioned into two classes:[49] Kekuléan polyhex hydrocarbons or benzenoid hydrocarbons ($K > 0$) and non-Kekuléan polyhex hydrocarbons ($K = 0$). Examples of Kekuléan and non-Kekuléan polyhexes are given in Figure 2. In general, every conjugated system may be placed into either a class of Kekuléan systems or a class of non-Kekuléan systems, depending on whether or not it possesses Kekulé structures. The Clar postulate predicts all non-Kekuléan systems to

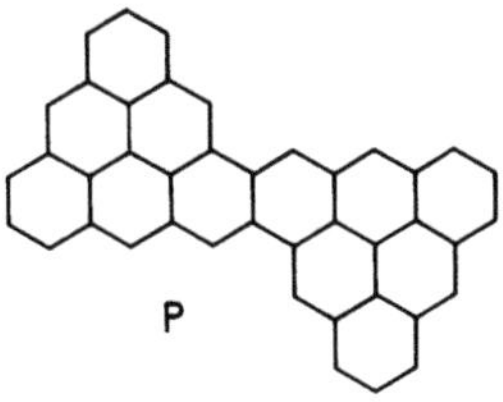

FIGURE 3.   A non-Kekuléan polyhex with A = 0.

be unstable. Since its introduction in 1958, this postulate has remained un-challenged. Therefore, the existence of Kekulé structures is of utmost im-portance for the stability of conjugated systems.

II. THE IDENTIFICATION OF KEKULÉAN SYSTEMS

The identification of Kekuléan systems has been attracting chemists, theoretical chemists, and mathematical chemists for a long time.[3,11] Many necessary or sufficient conditions for a conjugated system to be Kekuléan have been discovered.[50] A simple necessary, but not sufficient criterion may be devised to predict if a benzenoid molecule possibly has Kekulé structures.[51] Polyhexes, representing polyhex hydrocarbons, are necessarily bipartite, that is, their vertices may be separated in a unique way into two groups: starred s and unstarred u, respectively, such that s and u are never connected. Note that $s + u = N$. If a number A is defined as follows:

$$A = s - u \qquad (6)$$

then

$$A = 0 \qquad (7)$$

is a *necessary* condition for the existence of Kekulé structures in a polyhex hydrocarbon. However, this condition is not a *sufficient* condition for the existence of Kekulé structures, because one can find polyhexes with A = 0 which do not possess Kekulé structures. One such example is given in Figure 3. It appears that non-Kekuléan polyhexes with A = 0 require at least 11 hexagons to show this property.[51]

A sufficient condition for K $\neq$ 0 is the existence of a Hamiltonian circuit in a chemical graph. If N_i = 0 (N_i = the number of internal vertices), the boundary of a polyhex is a Hamiltonian circuit and the corresponding poly-hex hydrocarbon must possess Kekulé structures. However, this condition is not necessary, because there are Kekuléan polyhexes without the Hamilto-nian circuits. Two Kekuléan polyhexes, with and without Hamiltonian cir-cuits, respectively, are given in Figure 4. The necessary and sufficient con-

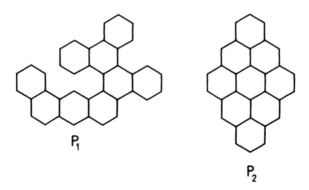

FIGURE 4. A Kekuléan polyhex P_1 (K = 90) with a Hamiltonian circuit and a Kekuléan polyhex P_2 (K = 20) without a Hamiltonian circuit.

ditions for a polyhex system to be Kekuléan have been put forward only recently.[50,52-56] The necessary and sufficient structural requirements for the existence of Kekulé structures in a polyhex hydrocarbon will be given here following arguments by Zhang et al.[54]

Consider a polyhex P whose vertex-set is split into two groups: starred vertices (here denoted by black circles) s and unstarred vertices (here denoted by empty circles) u, respectively. An edge-cut of P is a collection $\{e_1, e_2, \ldots, e_t\}$ of edges of P, such that the following conditions hold:

1. The graph obtained from P by deleting the edges e_1, e_2, $\ldots$,e_t has exactly two components P' and P''
2. For each edge e_i $(i = 1,2, \ldots ,t)$ its s-vertex (black circle) belongs to P'' and its u-vertex (empty circle) to P'
3. Each pair of edges e_i, e_{i+1} $(i = 1,2, \ldots ,t-1)$ belongs to the same hexagon, and e_1 and e_t belong to the boundary of P

Then, a polyhex P has Kekulé structures if, and only if, $s(P) = u(P)$ and if for every edge-cut of P, $s(P') \leq u(P')$. If for at least one edge-cut the above is violated, then P is a non-Kekuléan polyhex. A polyhex P in Figure 5 has no Kekulé structures, because $s(P') > u(P')$.

The application of the above procedure is tedious, because one has to examine every possible edge-cut of a polyhex and they may be numerous for large systems. There are some simplifications of the above procedure possible,[54] but in general there is no simple recipe available to decide whether a conjugated molecule possesses Kekulé structures or not.

III. METHODS FOR THE ENUMERATION OF KEKULÉ STRUCTURES

There are a number of methods available for the enumeration of Kekulé structures.[11,57] Some of these methods will be presented in this section.

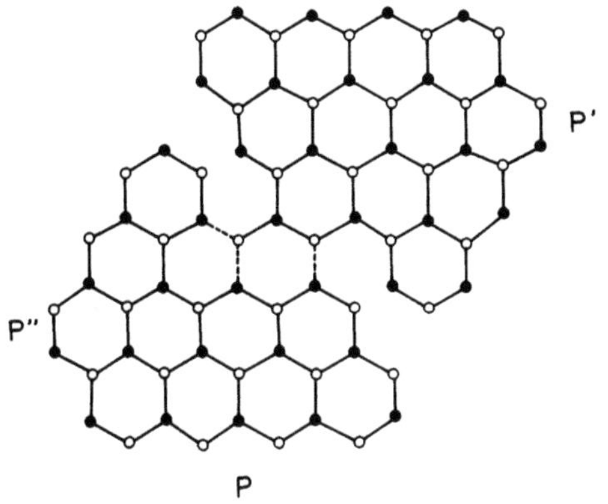

FIGURE 5. A non-Kekuléan polyhex P.

A. THE EMPIRICAL METHOD

The empirical method is based on the systematic drawings of all Kekulé structures of a molecule and counting them at the end. In the words of Pauling:[29] "A few minutes suffice to draw the four unexcited structures for anthracene, the five for phenanthrene or the six for pyrene . . . ; an hour or two might be needed for the 110 structures of tetrabenzoheptacene".

This method is always possible to apply, but it is obviously impractical for large systems. Besides it is prone to errors. However, the empirical method could be speeded up and possible errors eliminated if the systematic drawings of all Kekulé structures are carried out by computer.[58]

B. THE METHOD OF FRAGMENTATION

A variant of the empirical method is the fragmentation method, treated in detail by Randić.[59] However, it appears that this method was already known to Wheland.[60] The fragmentation method is based on the relationship between the number of Kekulé structures of the parent benzenoid hydrocarbon B and of its constituting fragments B′, B″, . . . The method consists of the reduction of a benzenoid hydrocarbon to smaller fragments of a known number of Kekulé structures. For example, the numbers of Kekulé structures for planar benzenoid hydrocarbons with up to nine hexagons are known.[61]

The fragmentation method consists of the following steps:

1. A bond in a benzenoid B is selected and is assumed to be *double*. This bond is then deleted together with all other bonds which can be unambiguously assigned to be single or double. The remaining fragment of B is denoted by B″.
2. The same bond from the above is now assumed to be *single*. This bond is then again deleted and all others (if any) which can be unambiguously assigned. The remaining fragment is denoted by B′.
3. The above process (if necessary) may be repeated on B′ and B″.

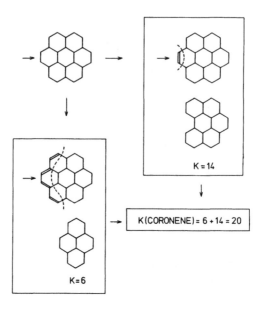

FIGURE 6.   Enumeration of Kekulé structures of coronene using the fragmentation method.

4.   The process ends when structures with known Kekulé numbers appear as fragments B′ and B″. Then

$$K(B) = K(B') + K(B'') \tag{8}$$

The fragmentation method is illustrated on coronene in Figure 6.

The fragmentation method is a very convenient method for enumerating the Kekulé structures of large conjugated systems, because it is rapid and does not require any calculations. The fragmentation may be accelerated by a set of rules.[11,43,59,62-64]

C. METHODS BASED ON GRAPHIC POLYNOMIALS

1. The Characteristic Polynomial

The absolute value of the last coefficient $|a_N (B)|$ of the characteristic polynomial $P(B;x)$ for a benzenoid hydrocarbon B is equal to the square of its Kekulé number,[65]

$$|a_N(B)| = K^2 \tag{9}$$

For example, the characteristic polynomial of the benzenoid hydrocarbon, depicted by a graph B (see Figure 7), is known.[66] The value of the last coefficient is

$$a_{96} = 54,218,191,104 \tag{10}$$

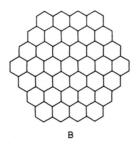

B

FIGURE 7.   A benzenoid graph B with 37 hexagons and 96 vertices.

Consequently, the number of its Kekulé structures is also known by virtue of Equation 9:

$$K = 232,848 \tag{11}$$

2. The Acyclic Polynomial

The absolute value of the last coefficient $\mid a_N^{ac}$ (B) $\mid$ of the acyclic polynomial $P^{ac}(B;x)$ for a benzenoid hydrocarbon B is equal to its Kekulé number,

$$\left| a_N^{ac}(B) \right| = K \tag{12}$$

For example, the acyclic polynomial of coronene is given by

$$P_{ac}(\text{coronene};x) = x^{24} - 30x^{22} + 387x^{20} - 2818x^{18}$$
$$+ 12,783x^{16} - 37,620x^{14} + 72,585x^{12}$$
$$- 90,792x^{10} + 71,256x^8 - 32,968x^6$$
$$+ 8,016x^4 - 816x^2 + 20 \tag{13}$$

The last coefficient is equal to the number of Kekulé structures (K = 20) of coronene.

This approach is also applicable to other alternant systems besides benzenoids and to some nonalternants.

3. The Permanental Polynomial

The value of the last coefficient $a_N(G)$ of the permanent polynomial $\mathcal{P}(G;x)$ for a conjugated structure G is equal to the square of its Kekulé number:

$$a_N(G;x) = K^2 \tag{14}$$

For example, the permanental polynomial of cyclobutadiene is given by

$$\mathscr{P}(\text{cyclobutadiene};x) = x^4 + 4x^2 + 4 \tag{15}$$

It is immediately seen via Equation 15 that the Kekulé number of cyclobutadiene is equal to 2.

D. THE METHOD BASED ON THE COEFFICIENTS OF NONBONDING MOLECULAR ORBITALS

The enumeration of Kekulé valence structures based on the coefficients of nonbonding molecular orbitals, NBMOs, is founded on the recognition of the intimate relationship between the value of K and the coefficients of NBMOs.[67] The number of Kekulé structures for an *odd* alternant is equal to the sum of the *absolute* values of the unnormalized coefficients of the NBMO:

$$\sum_{i=1}^{N} |c_{oi}| = K \tag{16}$$

where the summation is over all the coefficients c_{oi} ($i = 1,2, \ldots ,N$) of the NBMO. It is not difficult to write down the unnormalized coefficients of the NBMO by means of the *zero-sum rule* of Longuet-Higgins.[68] Consider the eigenvalue equation:

$$\mathbf{C}_i \, \mathbf{A} = x_i \, \mathbf{C}_i \tag{17}$$

In the case of $N_0 \neq 0$, $x_i = 0$, and $\mathbf{C}_i = \mathbf{C}_o = $ NBMO,

$$\mathbf{C}_o \, \mathbf{A} = 0 \tag{18}$$

or in scalar form,

$$\sum_{j \to k} c_{oj} \, (\mathbf{A})_{jk} = 0 \qquad (k = 1,2,\ldots, N) \tag{19}$$

where the summation is over all vertices j joined to the vertex k. One set of atoms (unstarred atoms) have zero coefficients in the NBMO, the other set (starred atoms) have simple integers chosen in such a way that the sum of all nonvanishing coefficients around an unstarred atom is zero[68,69] (see example in Figure 8).

The above approach was extended to *even* alternants by Herndon.[70] The first step in the approach is to produce an odd alternant from the even alternant by deleting one carbon atom from the even system. The second step is to generate the coefficients of the corresponding NBMO by one of the available methods.[67-71] Then, the sum of the absolute values of the unnormalized coef-

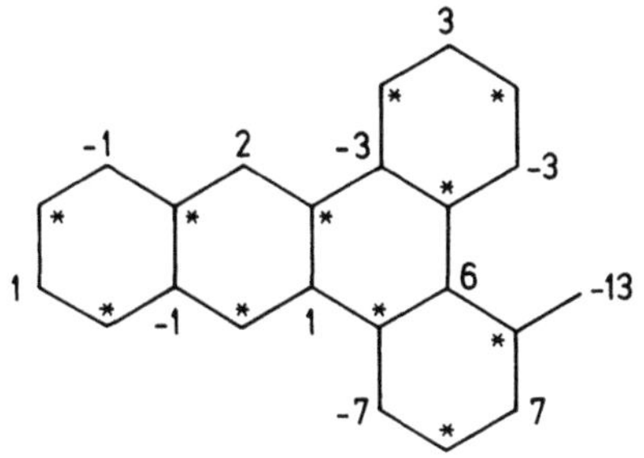

$$K = \sum_{i=1}^{23} |c_{oi}| = 48$$

FIGURE 8. The illustration of the zero-sum rule and Equation 16.

(a) A benzenoid graph B corresponding to anthracene

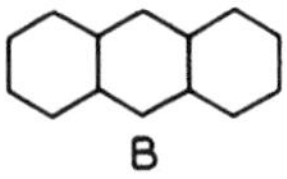

B

(b) Creation of odd-alternant by excising an arbitrary
     atom j from B

(c) Construction of the unnormalized NBMO

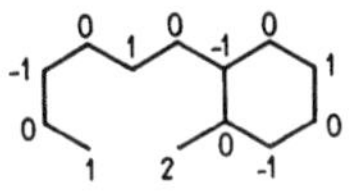

(d) Application of formula (16)

$$K(\text{anthracene}) = |c_{oi}| + |c_{oi}'| + |c_{oi}''| = 4$$

FIGURE 9. The computation of the Kekulé number for anthracene, depicted by a benzenoid graph B, using the Herndon procedure.

ficients c_{oi} at points i adjacent to the deleted atom j is equal to the Kekulé number of the parent alternant,

$$\sum_{i \to j} |c_{oi}| = K \qquad (20)$$

The application of this procedure is illustrated in Figure 9. However, it has been demonstrated that this extension of Herndon[70] may lead to false conclusions.[72]

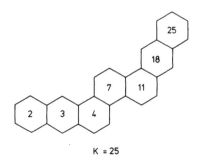

K = 25

FIGURE 10. Enumeration of Kekulé structures of dibenzo[b,n]picene using the Gordon-Davison method.

E. THE METHOD OF GORDON AND DAVISON

The Gordon-Davison method[73] in the original formulation is applicable for the enumeration of Kekulé structures of unbranched catacondensed benzenoid hydrocarbons. This method may be described in the following way. A hexagon at the end of the benzenoid chain is selected and 2 inserted in it (because K = 2 for benzene). 3 is added to the adjacent hexagon (because K = 3 for naphthalene). Subsequent hexagons if fused linearly increase the inscribed value by 1, but if the direction of fusion changes one has to add, instead of 1, the value in the preceding ring. The process continues by adding this number for each subsequent linearly fused hexagon until the next change of fusion direction occurs, when the process is repeated in the same fashion. The process ends when all available hexagons in the benzenoid hydrocarbon are exhausted. An illustrative example is shown in Figure 10.

For a linear chain of h benzene rings called *polyacene*,[47,74] the Gordon-Davison method reduces to a simple counting formula

$$K = h + 1 \tag{21}$$

Similarly, for a zig-zag chain of h benzene rings called *fibonaccene*,[75] the Gordon-Davison method reduces to a *Fibonacci-type* recurrence relation,[76]

$$K_h = K_{h-1} + K_{h-2}; \quad h \geq 2 \tag{22}$$

$$K_0 = 1 \quad \text{(by definition)} \tag{23}$$

$$K_1 = K(\text{benzene}) = 2 \tag{24}$$

Kekulé numbers of fibonaccenes generated by the recurrence relation (Equation 22): 3, 5, 8, 13, 21, 34, 55, 89, 144, 233, etc. belong to the *Fibonacci number* series:[77] 1, 1, 2, 3, 5, 8, 13, 21, 34, 55, 89, 144, 233, 377, 610, 987, etc.

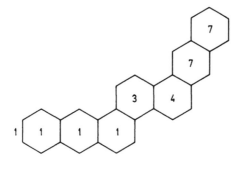

$$K = 1 + 1 + 1 + 1 + 3 + 4 + 7 + 7 = 25$$

FIGURE 11. Enumeration of Kekulé structures of dibenzo[b,n]picene using the numeral-in-hexagon method.

F. THE NUMERAL-IN-HEXAGON METHOD

Cyvin and Gutman[11,78] have introduced a clever and elegant variant of the Gordon-Davison method, named the numeral-in-hexagon method.[3] This method may be described as follows. The process starts by placing 1 outside the first hexagon (because for the trivial case with no rings K = 1 by definition) of an unbranched catabenzenoid hydrocarbon; 1 is also inserted in the first hexagon and is likewise inscribed in all adjacent linearly fused hexagons. If the direction of fusion changes then all numbers in the linearly fused part of a benzenoid preceding the kink are added up and the number obtained is placed in the hexagon at the end of the kink. The same number is placed in the next hexagon if it is linearly fused, otherwise the summing process is repeated. When all hexagons in the benzenoid chain are exhausted, the above process ends. The number of Kekulé structures is then equal to the sum of all numbers. The numeral-in-hexagon method is illustrated in Figure 11, for purposes of comparison, for the same benzenoid hydrocarbon (dibenzo[b,n]picene) as one already used to illustrate the Gordon-Davison method (see Figure 10).

It has been shown[79] that the numeral-in-hexagon method is a simple reformulation of the Gordon-Davison method. However, the numeral-in-hexagon method has some advantages over the Gordon-Davison approach. It can be modified to a form which can be used for counting Kekulé structures of certain classes of peribenzenoids and coronoids. This modification will be presented in Section III.H.

G. THE GENERALIZED GORDON-DAVISON METHOD

The Gordon-Davison method was generalized to include all unbranched catafused alternant systems.[80,81] This was done in the following way. Each nonterminal ring in a chain of catafused even-membered rings can be classified according to the parity of the number of steps along a path between the two

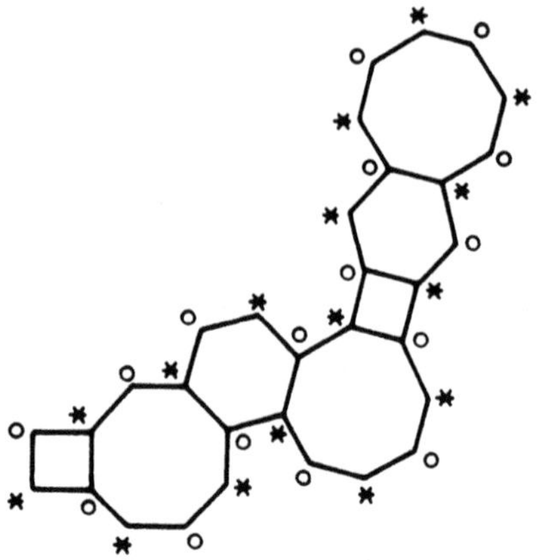

FIGURE 12. A chain of seven even-membered cycles with the same parity sequence and Kekulé count as for benzo[b,n]picene.

points of attachment to adjoining rings. For an even length, the end-points are *alike* if both are starred or unstarred, whereas for an odd length the end-points are *unlike* if one is starred and the other unstarred. Thus, the rings in unbranched catafused alternants are partitioned into like-linked, denoted by ℓ, and unlike-linked, denoted by u, with linking paths of even and odd parity, respectively. For example, dibenzo[b,n]picene (see Figures 10 and 11) has the following sequence of ring types (proceeding from left to right): $\ell u u u \ell$. An example of one possible unbranched catafused alternant hydrocarbon with the same sequence of ring types (and the same Kekulé count $K = 25$) as benzo[b,n]picene is given in Figure 12.

The Kekulé number K_i for the ith ring may be expressed as

$$K_i = \begin{cases} K_{i-1} + K_{i-2}; \text{ ring } i - 1 \text{ is } u \\ K_{i-1} + (K_{i-1} - K_{i-2}); \text{ ring } i - 1 \text{ is } \ell \end{cases} \tag{25}$$

with the initial conditions

$$K_0 = 1 \tag{26}$$
$$K_1 = 2 \tag{27}$$

H. METHODS BASED ON THE LATTICE STRUCTURE OF FUSED BENZENOIDS

1. The Method of Yen

Yen[82] derived elegant equations, based on the work by Gordon and Davison,[73] for the enumeration of the Kekulé structures for several hexagonal lattice systems (also called honeycomb or graphite lattices).[83] Four models for description of the lattice structure of fused benzenoid were considered.

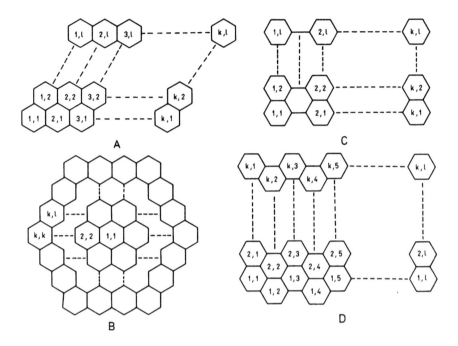

FIGURE 13. Lattice models for benzenoid hydrocarbons: (A) the parallelogram model S[k,ℓ]; (B) the circular model C[k,ℓ]; (C) the rectangular model R[k,ℓ]; (D) the skew strip model Z[k,ℓ].

These are the *parallelogram model* S[k,ℓ], the *symmetric circular model* C[k,ℓ], the *rectangular model* R[k,ℓ], and the *skew strip model* Z[k,ℓ]. They are illustrated in Figure 13.

In the parallelogram model S[k,ℓ], k and ℓ are interchangeable. S[1,1] is benzene, S[2,1] or S[1,2] is naphthalene, S[2,2] is pyrene, etc. The expression for calculating the number of Kekulé structures for the parallelogram model is given by

$$K = \binom{k + \ell}{\ell} \tag{28}$$

The application of this formula is illustrated in Figure 14.

The two-parameter combinatorial formula for computing the number of the Kekulé structures of benzenoid hydrocarbons which can be represented by the circular model as follows:

$$K = \frac{\binom{k + \ell}{\ell}\binom{k + \ell + 1}{\ell} \cdots \binom{k + \ell + (k - 1)}{\ell}}{\binom{\ell + 1}{\ell} \cdots \binom{\ell + (k - 1)}{\ell}} \tag{29}$$

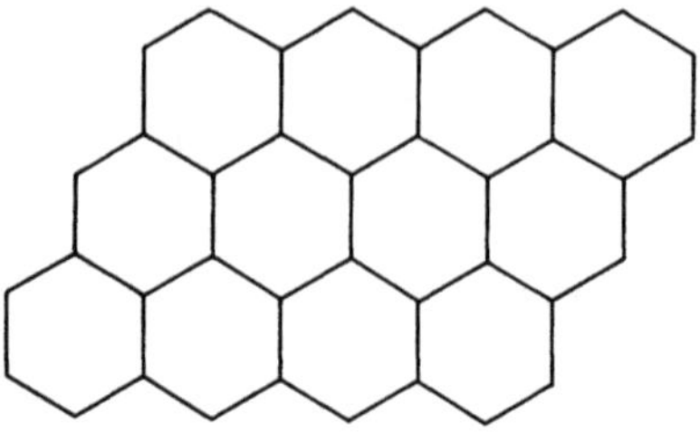

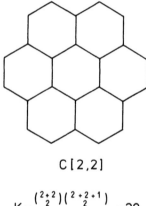

$$S[4,3]$$

$$K = \binom{4 + 3}{3} = 35$$

FIGURE 14.   The enumeration of Kekulé structures for an S[4,3] lattice representing dibenzo[bc,mn]ovalene.

$$C[2,2]$$

$$K = \frac{\binom{2+2}{2}\binom{2+2+1}{2}}{\binom{2+1}{2}} = 20$$

FIGURE 15.   Enumeration of Kekulé structures for a C[2,2] lattice depicting coronene.

This is a reformulation of the corresponding formula given by Gordon and Davison.[73] The application of this formula is shown in Figure 15.

The class of hexagon-shaped benzenoids is a generalization of the circular model, where rows of three different lengths in the three directions are labeled as k, ℓ, and m. The following is the three-parameter combinatorial formula for counting the Kekulé structures of hexagon-shaped benzenoid hydrocarbons:[84]

$$K = \prod_{q=0}^{k-1} \frac{\binom{\ell + m + q}{m}}{\binom{m + q}{m}} \qquad (30)$$

R[2,3]

$$K = (3+1)^2 = 16$$

FIGURE 16.   Enumeration of Kekulé structures for an R[2,3] lattice representing dibenzo[fg,op]anthanthrene.

The formula for computing the Kekulé number of benzenoid hydrocarbons which can be depicted by the rectangular model is given by

$$K = (\ell + 1)^k \qquad (31)$$

This formula is used for the explanatory computation of the number of Kekulé structures of dibenzo[fg,op]anthanthrene (see Figure 16).

The rectangle-shaped benzenoids are also called *prolate rectangles*,[11] where the top and bottom rows (in the IUPAC orientation) contain one more hexagon than their nearest neighbors. A more challenging problem is attached to the *oblate rectangles*,[11] where the top and the bottom rows hold one hexagon less than their nearest neighbors.

Expression (31) reduces for polyphenylenes R[k,1] to $K = 2^k$, for polyacenes R[1,ℓ] to $K = \ell + 1$, for polyrylenes R[k,2] to $K = 3^k$, and for polyanthenes R[k,3] to $K = 4^k$. Expressions (28), (29), and (31) may be collected into a general formula which embraces S[k,ℓ] and R[k,ℓ] lattice models of benzenoid hydrocarbons.

$$K = \left(\frac{\ell + b}{\ell} \right)^c \prod_{q=1}^{k-\ell} \frac{\binom{k + \ell + q}{\ell}}{\binom{\ell + a + q}{\ell}} \qquad (32)$$

where (1) S[k,ℓ]: a = k, b = k, c = 1; (2) C[k,ℓ]: a = 0, b = k, c = 1; and (3) R[k,ℓ]: a = k, b = 1, c = k.

For the skew strip model Z[k,ℓ] with fixed values of ℓ, the explicit formula for counting the Kekulé structures of the corresponding benzenoid hydrocarbons is known only for $\ell = 1$. Since the skew strip Z[k,ℓ] is a zig-zag chain of hexagons, the counting formula is a Fibonacci-type recurrence relation

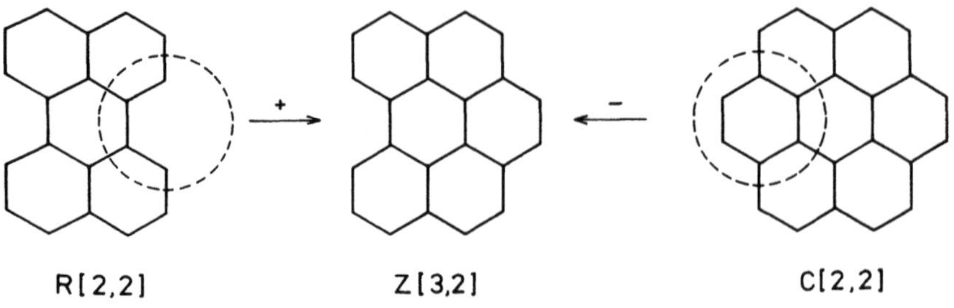

R[2,2]   Z[3,2]   C[2,2]

FIGURE 17.   The generation of the Z[3,2] lattice system from the R[2,2] system by adding one benzene ring to the system and from the C[2,2] system by substracting one benzene ring from the system. Positions at which the addition or the subtraction is performed are denoted by a circle.

(Equation 22). Yen[82] has produced much more complicated formulae for obtaining the same information:

$$K = 2^{-(k+1)} \prod_{q=1}^{(k+2)/2} \binom{k+2}{2q-1} 5^{(q-1)}; \quad k = \text{even} \tag{33}$$

$$K = 2^{-(k+1)} \prod_{q=1}^{(k+3)/2} \binom{k+2}{2q-1} 5^{(q-1)}; \quad k = \text{odd} \tag{34}$$

There have been reported[46,85] a number of recurrence relations and a general formulation for counting Kekulé structures of the skew strip models Z[k,ℓ].

In the case of the skew strip model Z[k,ℓ] with fixed values of k, the K-counting formulae are available for many cases. For the Z[1,ℓ] model the counting formula for K is identical to the one for the polyacene series R[1,ℓ]. The Z[2,ℓ] series contain naphthalene Z[2,1], pyrene Z[2,2], anthanthrene Z[2,3], etc. For this series the number of Kekulé structures is given by

$$K = (\ell + 1)(\ell + 2)/2 = \binom{\ell + 2}{\ell} \tag{35}$$

The next sequence Z[3,ℓ] contain phenanthrene Z[3,1] and 1,12-benzoperylene Z[3,2], etc. The Kekulé number for this series is given by

$$K = (\ell + 1)(\ell + 2)(2\ell + 3)/6 \tag{36}$$

Equation 36 was first given by Gordon and Davison,[73] while Equation 35 was implied by them.

It should be noted that the *addition* of one benzene ring to the R[2,ℓ] series will result in the Z[3,ℓ] series or the substraction of the benzene ring from the C[2,ℓ] series will also yield the Z[3,ℓ] series (see Figure 17).

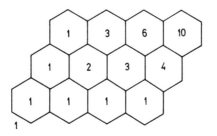

$$K = 7 \cdot 1 + 2 + 2 \cdot 3 + 4 + 6 + 10 = 35$$

FIGURE 18. The enumeration of Kekulé structures of dibenzo[bc,mn] ovalene by the numeral-in-hexagon method.

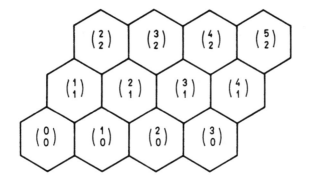

FIGURE 19. Binomial coefficients giving numbers inscribed in hexagons of dibenzo[bc,mn]ovalene in Figure 18.

The counting formulae for Kekulé structures of multiple zig-zag chains $Z[k, \ell]$ have also been derived for the range from $Z[4, \ell]$ to $Z[10, \ell]$.[11,46,84-86]

2. The Numeral-in-Hexagon Method

The numeral-in-hexagon method can be directly applied, for example, to the parallelogram lattice model $S[k, \ell]$ of benzenoids.[78] This is illustrated in Figure 18 for dibenzo[*bc,mn*]ovalene. The digits inside the hexagons in dibenzo[bc,mn]ovalene constitute a part of the Pascal triangle. Hence, they can be expressed in terms of binomial coefficients[76] (see Figure 19).

I. THE TWO-STEP FRAGMENTATION METHOD

The two-step fragmentation method is applicable to unbranched and branched catafused benzenoids (including helicenes), and unbranched and a branched coronoids (corona-condensed rings of benzenes). This method is also a variant of the Randić fragmentation method,[59] which is based on the concept of *partial essential single* and *double bonds* in conjugated systems.[87]

Partial essential single and double bonds are those bonds which remain unchanged in a certain subset of Kekulé structures for the molecule. The enumeration is carried out in the following manner.[11,88] One first deletes partial essential single and double bonds in an arbitrary selected angularly fused or branched benzene ring (this represents the *first step* in the procedure). This will break up the structure into smaller fragments. The Kekulé structures for fragments can be counted by the Gordon-Davison method or by the numeral-in-hexagon method. The resulting number of Kekulé structures is denoted by $K(\alpha)$.

In the second step one deletes partial essential single and double bonds which are now placed in different positions than they were previously in the selected benzene ring. This placement of single and double bonds in a selected benzene ring may affect the distribution of single and double bonds in the adjacent rings. Some of these bonds will remain unchanged in the whole corresponding subset of Kekulé structures and they can also be removed from the molecule. The Kekulé counts for the fragments can again be determined either by the Gordon-Davison method or the numeral-in-hexagon method and the resulting Kekulé number is denoted by $K(\beta)$.

The total number of Kekulé structures depends on whether the considered molecule is a (branched or unbranched) catabenzenoid or a (branched or unbranched) coronoid:

$$K = K(\alpha) + K(\beta) + \text{correction} \tag{37}$$

where the correction is given by

$$\text{correction} = \begin{cases} 2 & \text{for coronoids} \\ 0 & \text{for catabenzenoids} \end{cases} \tag{38}$$

The correction of two for coronoids refers to two Kekulé structures in which the double bonds are placed only on the outer and inner perimeters of the coronoid. These two Kekulé structures, of course, cannot appear in either of the two fragments. For branched coronoids, the value of the correction term depends on the number and size of attached benzenoid chains. This method will be illustrated for a branched catabenzenoid (see Figure 20) and for a coronoid (see Figure 21). Two Kekulé structures of kekulene corresponding to the correction term of 2 in Equation 38 are shown in Figure 22.

J. THE PATH COUNTING METHOD

In the path counting method the number of Kekulé structures of a benzenoid hydrocarbon B is equal to the number of mutually self-avoiding directed peak-to-valley paths,[89]

$$K = \det|\mathbf{P}| \tag{39}$$

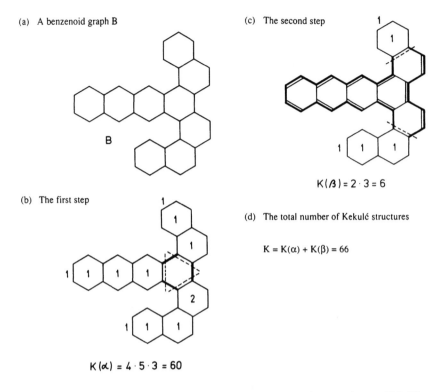

(a) A benzenoid graph B

B

(b) The first step

$K(\alpha) = 4 \cdot 5 \cdot 3 = 60$

(c) The second step

$K(\beta) = 2 \cdot 3 = 6$

(d) The total number of Kekulé structures

$$K = K(\alpha) + K(\beta) = 66$$

FIGURE 20. The application of the two-step fragmentation method to a catabenzenoid B. The enumeration of Kekulé structures of fragments is carried out by the numeral-in-hexagon method. The partial essential single and double bonds are denoted by thicker lines.

where **P** is a matrix with elements $(\mathbf{P})_{ij}$ representing counts of self-avoiding paths in B starting at peak(s) and ending at valley(s). A *peak* is a vertex on the perimeter of B which lies above its adjacent vertices, while a *valley* is a vertex which lies below its nearest neighbors. The identification of peaks and valleys depends on the orientation of a benzenoid hydrocarbon. For convenience, the orientation north-south (the IUPAC orientation)[47] is selected. The number of Kekulé structures is, of course, invariant to the orientation of B. Coronene, for example, has two peaks and two valleys (see Figure 23). The peaks and valleys must match if the Kekulé number is to be non-zero.

The correspondence between the mutually self-avoiding directed peak-to-valley paths and Kekulé structures was already noted by Gordon and Davison[73] and by the Zagreb group,[90-92] but these authors did not discover the counting formula (Equation 39). The efficiency of this formula depends on the enumeration of paths connecting peaks and valleys. One very efficient and elegant way is by means of the Pascal recurrence algorithm based on a part of the Pascal triangle.[55,73,93,94] The use of the path counting method,

(a)   A coronoid graph P representing kekulene

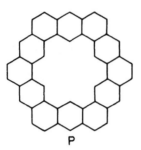

P

(b)   The first step

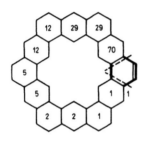

$K(\alpha) = 169$

(c)   The second step

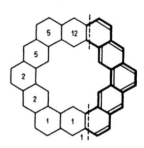

$K(\beta) = 29$

(d)   The total number of Kekulé structures

$$K(\text{kekulene}) = K(\alpha) + K(\beta) + 2 = 200$$

FIGURE 21.   The enumeration of Kekulé structures of kekulene by the two-step fragmentation method. The enumeration of Kekulé structures of fragments is carried out by the numeral-in-hexagon method. The partial essential single and double bonds are denoted by thicker lines.

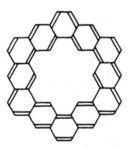

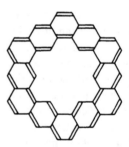

FIGURE 22.   Two Kekulé structures of kekulene with double bonds on the outer and inner perimeters of the molecule. They correspond to the correction term of 2 in Equation 38.

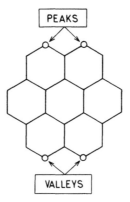

FIGURE 23.   Peaks and valleys in coronene.

together with the Pascal recurrence algorithm, for the enumeration of Kekulé structures is demonstrated in Figure 24.

The path counting method is also applicable in a modified form to coronoid hydrocarbons.[94]

K. THE MATRIX METHOD OF HALL

The matrix method for the enumeration of Kekulé structures of benzenoids has been proposed by Hall.[95,96] Since benzenoids are alternant structures their adjacency matrices may be given in a block form if the numbering of vertices is done in the following way: first starred and then unstarred vertices are numbered:

$$A = \begin{bmatrix} 0 & B \\ B^T & 0 \end{bmatrix} \tag{40}$$

Then,[95]

$$K = \det|B| \tag{41}$$

However, if atoms in a benzenoid hydrocarbon are numbered in a particular way, this will result in a submatrix **B** of a special form whose determinant may be evaluated by a series of matrix multiplications.

The numbering system is designed in the following way. A benzenoid graph may be rotated so that opposite edges in each hexagon become vertical in three different ways. A convenient orientation of a benzenoid is one which has the fewest peaks. These peaks are then numbered consecutively in horizontal lines from left to right, the lines taken from top to bottom. The valleys are then numbered similarly using the same numbers but starred. The remaining vertices are numbered in horizontal lines from left to right at the top line. If there are p peaks then the first of these numbers will be $(p+1)^*$.

(a) A benzenoid graph B depicting dibenzo[hi,st]ovalene

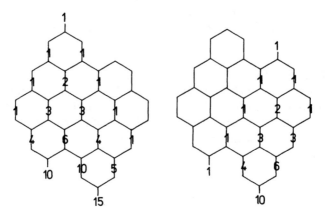

B

(b) Path counting *via* the Pascal triangle (the number of
      copies of B is equal to the number of peaks)

(c) **P**-matrix

$$\mathbf{P} = \begin{bmatrix} 10 & 15 \\ 1 & 10 \end{bmatrix}$$

(d) The number of Kekulé structures of dibenzo[hi,st]ovalene

$$K = \det |\,\mathbf{P}\,| = 10 \cdot 10 - 15 = 85$$

FIGURE 24.   The enumeration of Kekulé structures of dibenzo[hi,sl]-
ovalene using the path counting method.

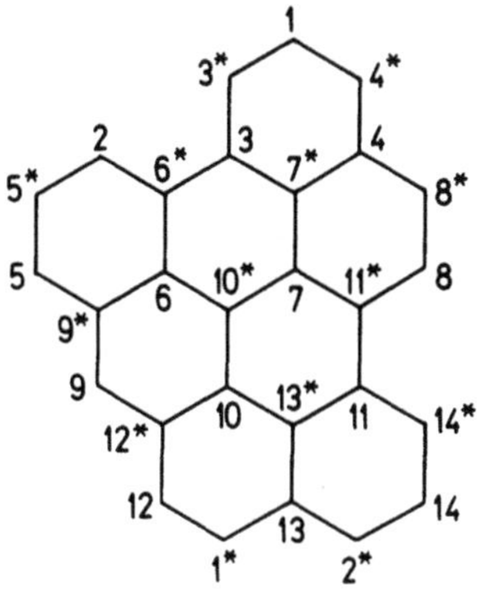

FIGURE 25. A special numbering of vertices in a benzenoid graph B representing naphtho[8,1,2-efg]anthanthrene.

Beneath the first line will be a second line of vertices connected vertically to the first, which will be given the same numbers unstarred. Thus in all the vertical edges the end-points are numbered with the same number, starred above and unstarred below. In Figure 25 vertices of a benzenoid graph B depicting naphtho[8,1,2-efg]anthanthrene are numbered in the above manner.

The results of the above numbering is a **B** matrix of the form,

$$
\mathbf{B} = \begin{bmatrix}
\mathbf{0} & \mathbf{R}_1 & \mathbf{R}_2 \ldots \ldots \ldots \ldots \ldots \ldots \ldots & \mathbf{R}_s \\
\mathbf{C}_1 & \mathbf{I} & \mathbf{0} \ldots \ldots \ldots \ldots \ldots \ldots \ldots & \mathbf{0} \\
\mathbf{C}_2 & \mathbf{A}_1 & \mathbf{I} \ldots \ldots \ldots \ldots \ldots \ldots & \mathbf{0} \\
\mathbf{C}_3 & \mathbf{0} & \mathbf{A}_2 \quad \mathbf{I} \ldots \ldots \ldots \ldots & \mathbf{0} \\
\vdots & \vdots & \vdots \quad \vdots \qquad \vdots & \vdots \\
\mathbf{C}_s & \mathbf{0} & \mathbf{0} \qquad \mathbf{0} \qquad \mathbf{A}_{s-1} \quad \mathbf{I}
\end{bmatrix} \qquad (42)
$$

where **I** and **O** are, respectively, the unit matrix and zero matrix. The $\mathbf{R}_i$ matrices contain information on the connectivities between the unstarred vertices and valleys and the $\mathbf{C}_i$ matrices contain information on the connectivities between the starred vertices and peaks. The $\mathbf{A}_i$ matrices contain information on each row of the slanted bonds between the starred and unstarred vertices in the benzenoid graph. With the help of an auxiliary matrix the development of determinant of the **B** matrix can be given in terms of $\mathbf{A}_i$, $\mathbf{C}_i$, and $\mathbf{R}_i$ matrices.[96] Consequently,

$$\mathbf{K} = |\mathbf{B}| = -\mathbf{R}_1\mathbf{C}_1 + \mathbf{R}_2(\mathbf{A}_1\mathbf{C}_1 - \mathbf{C}_2)$$
$$- \mathbf{R}_3(\mathbf{A}_2\mathbf{A}_1\mathbf{C}_1 - \mathbf{A}_2\mathbf{C}_2 - \mathbf{C}_3) + \dots \qquad (43)$$

For a benzenoid graph B representing naphtho[8,1,2-efg]anthanthrene (see Figure 25) the $\mathbf{A}_i$, $\mathbf{C}_i$, and $\mathbf{R}_i$ matrices are given by

$$
\mathbf{R}_4 = \begin{array}{c} 1^* \\ 2^* \end{array}
\begin{array}{ccc} 12 & 13 & 14 \end{array}
\left[\begin{array}{ccc} 1 & 1 & 0 \\ 0 & 1 & 1 \end{array}\right]
$$

$$
\mathbf{C}_1 = \begin{array}{c} 3^* \\ 4^* \end{array}
\begin{array}{cc} 1 & 2 \end{array}
\left[\begin{array}{cc} 1 & 0 \\ 1 & 0 \end{array}\right]
\qquad
\mathbf{C}_2 = \begin{array}{c} 5^* \\ 6^* \\ 7^* \\ 8^* \end{array}
\begin{array}{cc} 1 & 2 \end{array}
\left[\begin{array}{cc} 0 & 1 \\ 0 & 1 \\ 0 & 0 \\ 0 & 0 \end{array}\right]
$$

$$
\mathbf{A}_1 = \begin{array}{c} 5^* \\ 6^* \\ 7^* \\ 8^* \end{array}
\begin{array}{cc} 3 & 4 \end{array}
\left[\begin{array}{cc} 0 & 0 \\ 1 & 0 \\ 1 & 1 \\ 0 & 1 \end{array}\right]
\qquad
\mathbf{A}_2 = \begin{array}{c} 9^* \\ 10^* \\ 11^* \end{array}
\begin{array}{cccc} 5 & 6 & 7 & 8 \end{array}
\left[\begin{array}{cccc} 1 & 1 & 0 & 0 \\ 0 & 1 & 1 & 0 \\ 0 & 0 & 1 & 1 \end{array}\right]
$$

$$
\mathbf{A}_3 = \begin{array}{c} 12^* \\ 13^* \\ 14^* \end{array}
\begin{array}{ccc} 9 & 10 & 11 \end{array}
\left[\begin{array}{ccc} 1 & 1 & 0 \\ 0 & 1 & 1 \\ 0 & 0 & 1 \end{array}\right]
$$

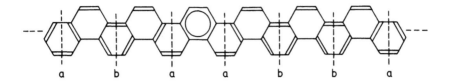

FIGURE 26. A diagram of polyphenanthrene and the corresponding local states.

The expression for computing the K number of B, derived from Equation 43, is given by

$$K = |\mathbf{B}| = \mathbf{R}_4(\mathbf{A}_3\mathbf{A}_2\mathbf{A}_1\mathbf{C}_1 - \mathbf{A}_3\mathbf{A}_2\mathbf{C}_2 - \mathbf{A}_2\mathbf{C}_3 - \mathbf{C}_4)$$

$$= \mathbf{R}_4\mathbf{A}_3\mathbf{A}_2(\mathbf{A}_1\mathbf{C}_1 - \mathbf{C}_2) \tag{44}$$

Introduction of the $\mathbf{A}_1$, $\mathbf{A}_2$, $\mathbf{A}_3$, $\mathbf{C}_1$, $\mathbf{C}_2$, and $\mathbf{R}_4$ matrices from above gives

$$K = \begin{bmatrix} 10 & -4 \\ 9 & -1 \end{bmatrix} = 26 \tag{45}$$

L. THE TRANSFER-MATRIX METHOD

The transfer-matrix method is a powerful tool for enumerating Kekulé structures.[30,31,34,83,97-102] It is very suited for long strips with periodic conditions. In this approach one studies the manner in which a Kekulé structure may be propagated from a position at one side of a unit cell to the other side of the cell. Next the different possible local characters of the Kekulé structures at the boundary of a cell are to be indicated. Then the "local state" (or structure) at each boundary is specified. The last step is to determine the number of ways to propagate from one local state to another at an adjacency boundary. The matrix which counts manners of propagation between possible pairs of adjacent local states is called the *transfer matrix* and is denoted by **T**.

The transfer-matrix method will be illustrated on a polyphenanthrene strip. In Figure 26 this benzenoid polymer[34] is shown, as well as local states at each of the positions marked by a transverse broken line. At these positions only two types of local states, labeled **a** and **b**, are found.

It is readily observed that at a position immediately following a local structure **a** there can occur either two **a** local states (because of K = 2 for the benzene ring separating two **a** local states) or **b** local state. In the case of a local structure **b** there can at a position immediately following it occur either **a** or **b** local states. This can be summarized as

$$\mathbf{a} \rightarrow 2\mathbf{a} + \mathbf{b} \tag{46}$$

$$\mathbf{b} \rightarrow \mathbf{a} + \mathbf{b} \tag{47}$$

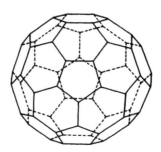

FIGURE 27.   A model of buckminsterfullerene as a truncated icosahedron.

If $K_L(\mathbf{a})$ (or $K_L[\mathbf{b}]$) denotes the number of Kekulé structures for a polyphen-anthrene strip with the length L with local structure $\mathbf{a}$ (or $\mathbf{b}$) at the terminal end, then

$$K_{L+1}(\mathbf{a}) = 2\,K_L(\mathbf{a}) + K_L(\mathbf{b}) \tag{48}$$

$$K_{L+1}(\mathbf{b}) = K_L(\mathbf{a}) + K_L(\mathbf{b}) \tag{49}$$

or upon defining the transfer matrix $\mathbf{T}$,

$$\mathbf{T} = \begin{bmatrix} 2 & 1 \\ 1 & 1 \end{bmatrix} \tag{50}$$

it follows,

$$\begin{bmatrix} K_{L+1}(\mathbf{a}) \\ K_{L+1}(\mathbf{b}) \end{bmatrix} = \mathbf{T} \begin{bmatrix} K_L(\mathbf{a}) \\ K_L(\mathbf{b}) \end{bmatrix} = \mathbf{T}^L \begin{bmatrix} K_1(\mathbf{a}) \\ K_1(\mathbf{b}) \end{bmatrix} \tag{51}$$

The final counting formula is given by

$$K = [2 \quad 1]\mathbf{T}^L \begin{bmatrix} 1 \\ 1 \end{bmatrix} \tag{52}$$

Therefore, the standard matrix multiplications yield the Kekulé structure counts. For example, polyphenanthrene of size $L = 6$ has 987 Kekulé structures.

This transfer-matrix method has been shown to be a very convenient method to compute the number of Kekulé structures of fullerenes.[103] For example, for buckminsterfullerene[104] (see Figure 27), a C_{60} elemental carbon cage, the transfer-matrix method has given 12,500 Kekulé structures.[30,99]

The cage structure of buckminsterfullerene (and other fullerenes) may be viewed as being built up through fusion of several symmetry-equivalent frag-ments or cells. That is, if such a cell is iterated via a symmetry group of operations, the whole surface of a fullerene can be exactly covered. The unit

FIGURE 28.   A unit cell for the C_{60} cage.

TABLE 1
The Number of Kekulé
Structures K for some Fullerenes
C_N of Icosahedral Point Group
Symmetry

Fullerene	K
C_{80}	140,625
C_{140}	2,167,239,697
C_{180}	1,389,029,765,625
C_{240}	21,587,074,966,666,816

cell used in the transfer-matrix computation of the Kekulé number for C_{60} is given in Figure 28.

Buckminsterfullerene can be obtained by gluing together five copies of the unit cell in Figure 28 in a cyclic fashion on the surface of a sphere, so that each cell adjoins on its boundaries to two others. The Galveston group has computed Kekulé numbers of elemental carbon cages C_N of icosahedral point group symmetry using the transfer-matrix method with up to N = 240.[30,99,105] Some of their results are given in Table 1.

Hosoya[106] has calculated the Z-counting polynomial for buckminsterfullerene and thereby deduced, independently of Klein et al.,[30] the K-number for this C_{60} carbon cage. Brendsdal and Cyvin[107] have achieved an analytical derivation (without the aid of the computer) of the same number, emerging from a systematic exploitation of the method of fragmentation.

M. THE METHOD OF DEWAR AND LONGUET-HIGGINS

Dewar and Longuet-Higgins have studied the correspondence between the resonance theory and Hückel molecular orbital (HMO) theory.[108] In this work they have established the relationship between the adjacency matrix A(B) of a benzenoid hydrocarbon B and the corresponding Kekulé number K,

$$\det A(B) = (-1)^{N/2} K^2 \tag{53}$$

This result is also related to the formula (Equation 41) by Hall.[37] The above formula is particularly useful for computer work and a suitable computer program is available in the literature.[109]

N. THE COMPUTATIONAL METHOD BASED ON THE EIGENVALUE SPECTRUM

A convenient computer-oriented method for counting Kekulé structures of benzenoid hydrocarbons is based on the relationship between the eigenvalue spectrum of the adjacency matrix of B and the Kekulé number of a benzenoid hydrocarbon:[95]

$$K = \prod_{i=1}^{N/2} x_i \tag{54}$$

where x_i ($i = 1, 2, \ldots N/2$) are the positive eigenvalues in the eigenvalue spectrum of B. The diagonalization of the adjacency matrix of B can be carried out by a suitable program, e.g., FØAAF of NAG. This procedure has been employed for generating tables of Kekulé numbers of all planar benzenoid hydrocarbons with up to nine hexagons.[61]

IV. THE CONCEPT OF PARITY OF KEKULÉ STRUCTURES

The concept of *parity* of Kekulé valence structures was introduced by Dewar and Longuet-Higgins[108] in their, already quoted (see the proceding section), work on the correspondence principle between the resonance theory and the HMO method. They have found that the Kekulé structures of alternant hydrocarbons (AHs) can be separated into two classes of different parity: "even" (K^+) and "odd" (K^-). The total number of Kekulé structures K is then equal to the sum of positive (even) and negative (odd) Kekulé structures

$$K = K^+ + K^- \tag{55}$$

The difference between the even and odd Kekulé structures has been introduced as the *algebraic structure count*, ASC[110,111] or the *corrected structure count*,[70] CSC.

$$ASC = K^+ - K^- \quad (K^+ > K^-) \tag{56}$$

ASC appears to be an important quantity in the structure-resonance theory when applied to AHs containing 4n rings.[27,112]

The Dewar-Longuet-Higgins definition (definition I) of parity is as follows. Let the N atoms (N = even) of an alternant hydrocarbon G be labeled in such a way that neither two odd-labeled nor two even-labeled atoms are

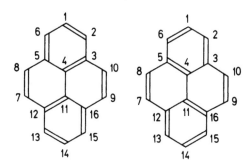

FIGURE 29. Two labeled Kekulé structures of pyrene.

directly linked. The Kekulé structure (1,p[2]), (3,p[4]), (5,p[6]), . . . , (N − 1,p[N]) is then defined to be positive (negative) if atoms 1,3,5, . . . ,N − 1 are paired with an even (odd) permutation of atoms 2,4, . . . ,N, respectively, i.e., if the permutation p is even (odd).

It should be remembered that the permutation p (being even or odd) bears no direct relationship to the number of elements that are permuted. The permutation is even (odd) if, and only if, it can be obtained as a composition of an even (odd) number of transpositions. Transpositions are similar to "basic permutations", in that they keep all except two adjacent elements fixed, but these two are exchanged. For example, the cyclic permutation of an odd number of elements is always even. Thus, for example, in the case of pyrene (see Figure 29) the first Kekulé structure is given by (1,2), (3,4), (5,6), (7,8), (9,10), (11,12), (13,14), (15,16) and the second Kekulé structure is given by (1,6), (3,2), (5,4), (7,8), (9,10), (11,16), (13,12), (15,14), so that the corresponding permutation is $p = (6,2,4,8,10,16,12,14)$. It is composed from two cyclic permutations (6,2,4) and (16,12,14), while elements 8 and 10 are fixed.

p is thus an *even* permutation and both Kekulé structures are of the same parity, as expected, because the Kekulé structures belonging to benzenoid hydrocarbons are all of the same parity.

It should be noted that the assignment of a positive or negative value to one class of parity is arbitrary, because it depends on the initial labeling of a molecule.

For AHs the following proposition holds:[113] The Dewar-Longuet-Higgins relationship to be of the same parity, defined for the set of all Kekulé structures in a manner described above, is an equivalence relationship (i.e., it is reflexive, symmetric, and transitive) and the set of Kekulé structures of an AH splits up into exactly two classes of equivalence. This proposition can be proved in the following way. Reflexivity is trivial, while symmetricity follows from the fact that if p is an even (odd) permutation, so is p^{-1} (inverse permutation). In the case of transitivity, let K_a, K_b, K_c be three Kekulé structures of the AH, with K_a and K_b, and K_b and K_c having the same parity.

As K_a and K_b are of the same parity, K_b can be obtained from K_a by means of an even permutation p, and by the same reasoning K_c can be obtained from K_b by means of an even permutation p'. However, the composition of two even permutations is itself an even permutation, so that K_a and K_c are again of the same parity. The last part of the proposition follows from the fact that each permutation has to be either even or odd (there is no other possibility), and composition of any two odd permutations is also an even permutation.

On the basis of seeming objections against the original Dewar-Longuet-Higgins definition, a new definition (definition II) has been proposed:[114,115] two different Kekulé structures (represented by Kekulé graphs) of an alternant hydrocarbon are of the *same* parity if the number of the $4n$-membered cycles in their superposition graph is *even*.

This definition may be analyzed in the following way. Let the Kekulé structures and the corresponding Kekulé graphs be denoted by K_a, K_b, . . . ,K_n and k_a, k_b, . . . ,k_n, respectively. Let the superposition of k_a and k_b give the graph G_{ab}. Let also the number of cycles of length $4n$ in the graph G_{ab} be $R_{4n}(G_{ab})$. Furthermore, let $p_a = +1$ if K_a is an even and $p_b = -1$ if K_b is an odd Kekulé structure. Then, the relationship

$$p_a p_b = (-1)^{R_{4n}(G_{ab})} \qquad (57)$$

determines the parity of the Kekulé structures. Thus, two Kekulé structures are of the same (opposite) parity if, and only if, $R_{4n}(G_{ab})$ is even (odd). As an example the parity of Kekulé structures of benzocyclobutadiene are shown in Figure 30.

It should be noted that for AHs the above definition of parity and the Dewar-Longuet-Higgins definition are equivalent.[113] This statement can be proved as follows. One should note that if K_b is obtained from K_a by means of just one cyclic permutation then their superposition is just one cycle: this cycle is of length $4n$ if the number of permuted elements is even, i.e., the permutation is odd, and of length $4n + 2$ if the number of permuted elements is odd, i.e., the permutation is even. It should be remembered that any permutation can be obtained as a combination of cyclic permutations. Now let the Kekulé structure K_b be obtained from K_a by means of an even (odd) permutation p. This can be generally decomposed into several cyclic permutations of an odd number of elements (as these permutations are even) and an even (odd) number of cyclic permutations of an even number of elements. This means that the superposition of k_a and k_b gives several $4n + 2$ cycles and an even (odd) number of $4n$ cycles. With this the proof ends. Therefore, there are no problems with the definition of parity of Kekulé structures of alternant hydrocarbons.

The interesting question is whether the concept of parity can be generalized to Kekulé structures of nonalternant hydrocarbons, NAHs. Unfortunately, neither the Dewar-Longuet-Higgins definition nor the definition II can

(i)   Kekulé structures of benzocyclobutadiene

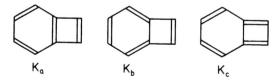

K_a K_b K_c

(ii) Kekulé graphs

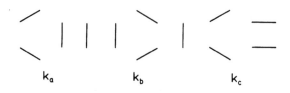

k_a k_b k_c

(iii) Superposition graphs

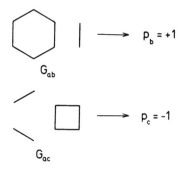

$P_b = +1$

G_{ab}

$P_c = -1$

G_{ac}

(iv) Parity count

$K^+ = 2$

$K^- = 1$

FIGURE 30.   Parity of Kekulé structures of benzocyclobutadiene.

be used as a basis for the required generalization. The Dewar-Longuet-Higgins definition cannot be used at all, because for NAHs the desired labeling does not generally exist. Definition II can be extended to include NAHs, but it does not lead to an equivalence relationship.

It can be shown that the relationship to be of the same parity defined by the definition II is not transitive for a tricyclic system with fused odd-membered cycles. Consider acepentylene which has three Kekulé structures. In

(a) A graph G of acepentylene

G

(b) Kekulé structures

(c) Kekulé graphs

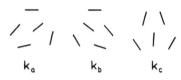

k_a k_b k_c

(d) Superposition graphs

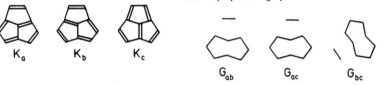

K_a K_b K_c

G_{ab} G_{ac} G_{bc}

(e) Structure count

$$p_a = +1 \quad \text{(by definition)}$$

$$G_{ab} \rightarrow p_b = -1$$

$$G_{ac} \rightarrow p_c = -1$$

$$\left.\begin{matrix} \\ \\ \end{matrix}\right\} \text{ contradiction}$$

$$G_{bc} \rightarrow p_c = +1$$

FIGURE 31.   Parity of Kekulé structures of acepentylene.

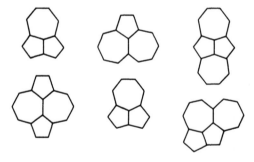

FIGURE 32.   Tricyclic nonalternant hydrocarbons with Kekulé-inseparable structures.

Figure 31 it is shown that the Kekulé structures of acepentylene cannot be separated into two classes. Therefore, acepentylene represents a system with Kekulé-inseparable structures. Several more nonalternant systems with Kekulé-inseparable structures are shown in Figure 32.

Systems with Kekulé-inseparable structures are expected to exhibit low stability and to be rather reactive compounds. For example, the derivative of

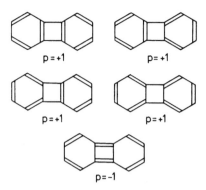

FIGURE 33. The Kekulé structures of biphenylene and their parities.

the acepentylene system is reported[116] as dilithium hexachloroacepentylen-diide, an unstable species detected at a low temperature.

In summary, benzenoid hydrocarbons possess Kekulé structures of the same parity. Similarly, other polycyclic systems with $4n+2$ rings possess all Kekulé structures of the same parity. Nonbenzenoid systems such as ben-zocyclobutadienes possess Kekulé structures of opposite parity. Bicyclic NAHs also possess Kekulé structures of opposite parity and they do not introduce incompatibilities. The breakdown of the parity concept happens in the class of tricyclic NAHs. However, if one of the rings in the tricyclic nonalternant structure is even-membered, such a system does not again generate incom-patibilities. The troublesome systems are tricyclic or polycyclic NAHs with all fused cycles of odd sizes. Their Kekulé structures cannot be separated into two classes.

There are several methods available for computing the ASC values.[117-119] Here three methods will be mentioned. The first method relates the absolute value of the last coefficient $| a_N(G) |$ of the characteristic polynomial of AHs containing 4n cycles and the square of the ASC,[65,120]

$$|a_N(G)| = (ASC)^2 \tag{58}$$

The second method relates the ASC value with the positive eigenvalues of the adjacency matrix of AHs with 4n cycles,[65,120]

$$ASC = \prod_{i=1}^{N/2} x_i \tag{59}$$

If a zero eigenvalue appears in the spectrum, then ASC $= 0$. The consequence of this is identity $K^+ = K^-$. Cyclobutadiene is such a system. The ASC value, for example, of biphenylene, obtained via Equation 59, is 3 and the number of its Kekulé structures, obtained via the acyclic polynomial, is 5.

(a)   A graph G of 1,2-naphthocyclobutadiene

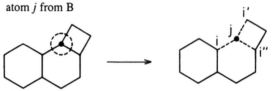

G

(b)   Creation of an odd-alternant by excising an arbitrary
      atom *j* from B

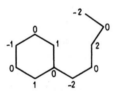

(c)   Construction of the unnormalized NBMO

(d)   Application of eqs. (16) and (60)

$$K = |c_{oi}| + |c_{oi'}| + |c_{oi''}| = 5$$

$$ASC = c_{oi} + c_{oi'} + c_{oi''} = 1$$

FIGURE 34.   The computation of the K and ASC values for 1,2-naphtho-
cyclobutadiene by means of Equations 16 and 60.

These data lead to $K^+ = 4$ and $K^- = 1$. The Kekulé structures of biphenylene
and their parities are shown in Figure 33.

   The third method is based on the coefficients of the NBMO of a molecule.[70] The ASC value is equal to the sum of the unnormalized coefficients
of the nonbonding molecular orbital,

$$\sum_{i=1}^{N} c_{oi} = ASC \tag{60}$$

It should be noted that in the case of benzenoid hydrocarbons Equations 16
and 60 are identical because for this class of alternant hydrocarbons $K =
ASC$. Application of Equation 60 is illustrated in Figure 34.

REFERENCES

1. **Morrison, R. T. and Boyd, R. N.,** *Organic Chemistry,* Allyn & Bacon, Boston, 1973.
2. **Ternay, A. L.,** *Contemporary Organic Chemistry,* W.B. Saunders, Philadelphia, 1979.
3. **Gutman, I. and Cyvin, S. J.,** *Introduction to the Theory of Benzenoid Hydrocarbons,* Springer-Verlag, Berlin, 1989.
4. **Gutman, I. and Cyvin, S. J., Eds.,** *Advances in the Theory of Benzenoid Hydrocarbons,* Springer-Verlag, Berlin, 1990.
5. **Kekulé, A.,** *Bull. Soc. Chim. Fr.,* 3, 98, 1865.
6. **Kekulé, A.,** *Bull. Acad. R. Belg.,* 19, 551, 1865.
7. **Kekulé, A.,** *Justus Liebigs Ann. Chem.,* 137, 129, 1866.
8. **Kekulé, A.,** *Justus Liebigs Ann. Chem.,* 162, 77, 1872.
9. **Wizinger-Anst, R., Gillis, J. B., Helfrich, B., and Wurster, C.,** *Kekulé und seine Benzoformel,* Verlag-Chemie, Weinheim, 1966.
10. **Hafner, K.,** *Angew. Chem. Int. Ed. Engl.,* 18, 641, 1971.
11. **Cyvin, S. J. and Gutman, I.,** *Kekulé Structures in Benzenoid Hydrocarbons,* Springer-Verlag, Berlin, 1988.
12. **Wheland, G. W.,** *The Theory of Resonance and Its Application to Organic Chemistry,* John Wiley & Sons, New York, 1953, 6th printing.
13. **Pauling, L.,** *The Nature of the Chemical Bond,* 2nd ed., Cornell University Press, Ithaca, NY, 1958, 12th printing.
14. **Klein, D. J. and Trinajstić, N., Eds.,** *Valence Bond Theory and Chemical Structure,* Elsevier, Amsterdam, 1990.
15. **Klein, D. J. and Trinajstić, N.,** *J. Chem. Educ.,* 67, 633, 1990.
16. **Klein, D. J.,** *Pure Appl. Chem.,* 55, 299, 1983.
17. **Klein, D. J. and Trinajstić, N.,** *Pure Appl. Chem.,* 61, 2107, 1989.
18. **Klein, D. J.,** *Topics Curr. Chem.,* 153, 57, 1990.
19. **Pauling, L. and Wheland, G. W.,** *J. Chem. Phys.,* 1, 362, 1933.
20. **Simpson, W. T.,** *J. Am. Chem. Soc.,* 75, 597, 1953.
21. **Simpson, W. T. and Looney, C. W.,** *J. Am. Chem. Soc.,* 76, 6285, 1954.
22. **Simpson, W. T.,** *Theories of Electrons in Molecules,* Prentice-Hall, Englewood Cliffs, NJ, 1962.
23. **McGlynn, S. P., Vanquickenborne, L. G., Kinoshita, M., and Carroll, D. G.,** *Introduction to Applied Quantum Chemistry,* Holt, Reinhart & Winston, New York, 1972.
24. **Herndon, W. C.,** *J. Am. Chem. Soc.,* 95, 2404, 1973.
25. **Herndon, W. C.,** *Thermochim. Acta,* 8, 225, 1974.
26. **Herndon, W. C. and Ellzey, M. L., Jr.,** *J. Am. Chem. Soc.,* 96, 663, 1974.
27. **Herndon, W. C.,** *Isr. J. Chem.,* 20, 270, 1980.
28. **Randić, M.,** *Chem. Phys. Lett.,* 38, 68, 1976.
29. **Pauling, L.,** *Acta Crystallogr. Sect. B,* 36, 1898, 1980.
30. **Klein, D. J., Schmalz, T. G., Hite, G. E., and Seitz, W. A.,** *J. Am. Chem. Soc.,* 108, 1701, 1986.
31. **Klein, D. J., Živković, T. P., and Trinajstić, N.,** *J. Math. Chem.,* 1, 309, 1987.
32. **Alexander, S. A. and Schmalz, T. G.,** *J. Am. Chem. Soc.,* 109, 6933, 1987.
33. **Klein, D. J. and Seitz, W. A.,** *J. Mol. Struct. (Theochem.),* 169, 167, 1988.
34. **Seitz, W. A. and Schmalz, T. G.,** in *Valence Bond Theory and Chemical Structure,* Klein, D. J. and Trinajstić, N., Eds., Elsevier, Amsterdam, 1990, 525.
35. **Dewar, M. J. S.,** *Chem. Br.,* 11, 97, 1975.
36. **Hall, G. G.,** *Bull. Inst. Math. Appl.,* 17, 70, 1981; *Int. J. Quantum Chem.,* 39, 605, 1991.
37. **Carter, P. G.,** *Trans. Faraday Soc.,* 45, 597, 1949.
38. **Gutman, I., Trinajstić, N., and Wilcox, C. F., Jr.,** *Tetrahedron,* 31, 143, 1975.
39. **Swinborne-Sheldrake, R., Herndon, W. C., and Gutman, I.,** *Tetrahedron Lett.,* 755, 1975.

40. **Herndon, W. C.,** *J. Am. Chem. Soc.,* 96, 7605, 1974.
41. **Herndon, W. C. and Párkányi, C.,** *J. Chem. Educ.,* 53, 689, 1976.
42. **Pauling, L., Brockway, L. O., and Beach, J. Y.,** *J. Am. Chem. Soc.,* 57, 2705, 1935.
43. **Randić, M.,** *Croat. Chem. Acta,* 47, 71, 1975.
44. **Randić, M.,** *Pure Appl. Chem.,* 55, 347, 1983.
45. **Randić, M. and Klein, D. J.,** in *Mathematics and Computational Concepts in Chemistry,* Trinajstić, N., Ed., Horwood, Chichester, 1986, 274.
46. **Cyvin, S. J. and Gutman, I.,** *Comput. Math. Appl.,* 12 B, 859, 1986.
47. **Dias, J. R.,** *Handbook of Polycyclic Hydrocarbons. Part A. Benzenoid Hydrocarbons,* Elsevier, Amsterdam, 1987.
48. **Clar, E., Kemp, W., and Stewart, D. G.,** *Tetrahedron,* 3, 325, 1958.
49. **Trinajstić, N.,** *J. Math. Chem.,* 5, 171, 1990.
50. **Gutman, I. and Cyvin, S. J.,** *J. Serb. Chem. Soc.,* 53, 319, 1988; **Hall, G. G. and Dias, J. R.,** *J. Math. Chem.,* 3, 233, 1989.
51. **Gutman, I.,** *Croat. Chem. Acta,* 46, 209, 1974.
52. **Zhang, F.-J., Chen, R.-S., and Guo, X.-F.,** *Graphs and Combinatorics,* 1, 383, 1985; **Kostochka, A. V.,** *Proc. 30th. Int. Wiss. Koll. TH Ilmenau,* Heft 5, Vortragsreihe F, 49, 1985.
53. **He, W. J. and He, W. C.,** in *Graph Theory and Topology in Chemistry,* King, R. B. and Rouvray, D. H., Eds., Elsevier, Amsterdam, 1987, 476.
54. **Zhang, F.-J. and Chen, R.-S.,** *Acta Math. Appl. Sin.,* 5, 1, 1989; **Zhang, F.-J., Guo, X.-F. and Chen, R.-S.,** *Topics Curr. Chem.,* 153, 181, 1990.
55. **He, W. C. and He, W. J.,** *Topics Curr. Chem.,* 153, 195, 1990.
56. **Sheng, R.-G.,** *Chem. Phys. Lett.,* 142, 196, 1987; **Sheng, R.-Q., Cyvin, S. J., and Gutman, I.,** *J. Mol. Struct. (Theochem.),* 187, 285, 1989; **Sheng, R.-Q.,** *Topics Curr. Chem.,* 153, 211, 1990.
57. **Trinajstić, N.,** *Rep. Mol. Theory,* 1, 185, 1990.
58. **Harris, F. E., Randić, M., and Stolow, R.,** to be published.
59. **Randić, M.,** *J. Chem. Soc. Faraday Trans. II,* 72, 232, 1976.
60. **Wheland, G. W.,** *J. Chem. Phys.,* 3, 356, 1933.
61. **Knop, J. V., Müller, W. R., Szymanski, K., and Trinajstić, N.,** *Computer Generation of Certain Classes of Molecules,* SKTH, Zagreb, 1985.
62. **Gutman, I. and Trinajstić, N.,** *Croat. Chem. Acta,* 45, 423, 1973.
63. **Gutman, I. and Trinajstić, N.,** *Naturwissenschaften,* 60, 475, 1973.
64. **Gutman, I., Trinajstić, N., and Živković, T.,** *Tetrahedron,* 29, 3449, 1973.
65. **Graovac, A., Gutman, I., Trinajstić, N., and Živković, T.,** *Theoret. Chim. Acta,* 26, 67, 1972.
66. **Herndon, W. C., Radhakrishnan, T. P., and Živković, T. P.,** *Chem. Phys. Lett.,* 152, 233, 1988.
67. **Platt, J. R.,** in *Handbuch der Physik,* Flugge, S., Ed., Springer-Verlag, Berlin, 1961, 173.
68. **Longuet-Higgins, H. C.,** *J. Chem. Phys.,* 18, 265, 1950; 18, 275, 1950; 18, 283, 1950.
69. **Dewar, M. J. S. and Dougherty, R. C.,** *The PMO Theory of Organic Chemistry,* Plenum Press, New York, 1975.
70. **Herndon, W. C.,** *Tetrahedron,* 29, 3, 1973.
71. **Živković, T.,** *Croat. Chem. Acta,* 44, 351, 1972.
72. **Gutman, I.,** *Bull. Chem. Soc. (Belgrade),* 47, 453, 1982.
73. **Gordon, M. and Davison, W. H. T.,** *J. Chem. Phys.,* 20, 428, 1952.
74. **Clar, E.,** *Polycyclic Hydrocarbons,* Academic Press, London, 1964.
75. **Anderson, P. G.,** in *Fibonacci Numbers and Their Applications,* Phillippou, A. N., Bergum, G. E., and Horadam, A. F., Eds., Reidel, Dordrecht, 1986, 2; **Balaban, A. T.,** *Math. Chem. (Mülheim/Ruhr),* 24, 29, 1989.
76. **El-Basil, S. and Klein, D. J.,** *J. Math. Chem.,* 3, 1, 1989.

77. **Cohen, D. I. A.**, *Basic Techniques of Combinatorial Theory*, John Wiley & Sons, New York, 1978, 63.
78. **Cyvin, S. J. and Gutman, I.**, *Math. Chem. (Mülheim/Ruhr)*, 19, 229, 1986.
79. **Živković, T. P. and Trinajstić, N.**, *Chem. Phys. Lett.*, 136, 141, 1987.
80. **Klein, D. J., Schmalz, T. G., El-Basil, S., Randić, M., and Trinajstić, N.**, *J. Mol. Struct. (Theochem.)*, 179, 99, 1988.
81. **Dias, J. R.**, *J. Mol. Struct. (Theochem.)*, 206, 1, 1990.
82. **Yen, T. F.**, *Theoret. Chim. Acta*, 20, 399, 1971.
83. **Klein, D. J., Hite, G. E., Seitz, W. A., and Schmalz, T. G.**, *Theoret. Chim. Acta*, 69, 409, 1986.
84. **Cyvin, S. J.**, *Monatsh. Chem.*, 117, 33, 1986; **Bodroža, O., Gutman, I., Cyvin, S. J., and Tošić, R.**, *J. Math. Chem.*, 2, 287, 1988.
85. **Okhami, N. and Hosoya, H.**, *Theoret. Chim. Acta*, 64, 153, 1983; **Gutman, I. and Cyvin, S. J.**, *Monatsh. Chem.*, 118, 541, 1987; **Cyvin, S. J., Cyvin, B. N., Brunvoll, J., and Gutman, I.**, *Z. Naturforsch.*, 42a, 722, 1987.
86. **Cyvin, S. J., Cyvin, B. N., and Gutman, I.**, *Z. Naturforsch.*, 40a, 1253, 1985.
87. **Dewar, M. J. S.**, *The Molecular Orbital Theory of Organic Chemistry*, McGraw-Hill, New York, 1969, 182.
88. **Bergan, J. L., Cyvin, S. J., and Cyvin, B. N.**, *Chem. Phys. Lett.*, 125, 218, 1986; **Randić, M., Nikolić, S., and Trinajstić, N.**, *Croat. Chem. Acta*, 61, 821, 1988.
89. **Sachs, H.**, *Combinatorica*, 4, 89, 1984; **John, P. and Rempel, J.**, in *Graphs, Hypergraphs and Applications*, Teubner, Leipzig, 1985, 72; **John, P. and Sachs, H.**, in *Graphs, Hypergraphs and Applications*, Teubner, Leipzig, 1985, 80; **John, P. and Sachs, H.**, in *Graphen in Forschung und Unterricht*, Bodendiek, R., Schumacher, H., and Walter, G., Eds., Franzbecker, Bad Salzdetfurth, 1985, 85; **John, P. and Sachs, H.**, *Topics Curr. Chem.*, 153, 145, 1990.
90. **Džonova-Jerman-Blažič, B. and Trinajstić, N.**, *Comput. Chem.*, 6, 121, 1982.
91. **Trinajstić, N. and Křivka, P.**, in *Mathematics and Computational Concepts in Chemistry*, Trinajstić, N., Ed., Horwood, Chichester, 1986, 328.
92. **Křivka, P., Nikolić, S., and Trinajstić, N.**, *Croat. Chem. Acta*, 59, 659, 1986.
93. **He, WJ. and He, WC.**, in *Graph Theory and Topology in Chemistry*, King, R. B. and Rouvray, D. H., Eds., Elsevier, Amsterdam, 1987, 476; see also **He, WJ. and He, WC.**, *Theoret. Chim. Acta*, 75, 389, 1989.
94. **Klein, D. J. and Trinajstić, N.**, *J. Mol. Struct. (Theochem.)*, 206, 135, 1990.
95. **Hall, G. G.**, *Proc. R. Soc. London Ser. A*, 229, 251, 1955; *Int. J. Math. Educ. Sci. Technol.*, 4, 233, 1973.
96. **Hall, G. G.**, *Chem. Phys. Lett.*, 145, 168, 1988.
97. **Klein, D. J., Hite, G. E., and Schmalz, T. G.**, *J. Comput. Chem.*, 7, 443, 1986.
98. **Trinajstić, N., Klein, D. J., and Randić, M.**, *Int. J. Quantum Chem.: Quantum Chem. Symp.*, 20, 699, 1986.
99. **Schmalz, T. G., Seitz, W. A., Klein, D. J., and Hite, G. E.**, *J. Am. Chem. Soc.*, 110, 113, 1988.
100. **Hite, G. E., Živković, T. P., and Klein, D. J.**, *Theoret. Chim. Acta*, 74, 349, 1988.
101. **Chen, R.-S., Cyvin, S. J., Cyvin, B. N., Brunvoll, J., and Klein, D. J.**, *Topics Curr. Chem.*, 153, 227, 1990.
102. **Klein, T. G., Seitz, W. A., and Schmalz, T. G.**, in *Computational Chemical Graph Theory*, Rouvray, D. H., Ed., Nova Science Publishers, New York, 1990, 127.
103. **Kroto, H. W.**, *Nature*, 329, 529, 1987; *Science*, 242, 1139, 1988; *Pure Appl. Chem.* 62, 407, 1990; *Chem. Br.*, 26, 40, 1990.
104. **Kroto, H. W., Heath, J. R., O'Brien, S. C., Curl, R. F., and Smalley, R.**, *Nature*, 368, 6042, 1985.
105. **Schmalz, T. G., Seitz, W. A., Klein, D. J., and Hite, G. E.**, *Chem. Phys. Lett.*, 130, 203, 1985; **Klein, D. J., Seitz, W. A., and Schmalz, T. G.**, *Nature*, 323, 705, 1986.

106. **Hosoya, H.,** *Comput. Math. Appl.,* 12 B, 271, 1986.
107. **Brendsdal, E. and Cyvin, S. J.,** *J. Mol. Struct. (Theochem.),* 188, 55, 1989.
108. **Dewar, M. J. S. and Longuet-Higgins, H. C.,** *Proc. R. Soc. London Ser. A,* 214, 482, 1952.
109. **Brown, R. L.,** *J. Comput. Chem.,* 4, 556, 1983.
110. **Wilcox, C. F., Jr.,** *Tetrahedron Lett.,* 795, 1968.
111. **Wilcox, C. F., Jr.,** *J. Am. Chem. Soc.,* 91, 2732, 1969.
112. **Herndon, W. C.,** *J. Am. Chem. Soc.,* 98, 887, 1976.
113. **Křivka, P. and Trinajstić, N.,** *Coll. Czech. Chem. Commun.,* 50, 291, 1985.
114. **Gutman, I. and Trinajstić, N.,** *Croat. Chem. Acta,* 47, 35, 1975.
115. **Gutman, I., Randić, M., and Trinajstić, N.,** *Rev. Roum. Chim.,* 23, 383, 1978.
116. **Jacobson, I.,** *Chem. Scripta,* 4, 30, 1974.
117. **Randić, M., Ruščić, R., and Trinajstić, N.,** *Croat. Chem. Acta,* 54, 295, 1981.
118. **Eilfeld, P. and Schmidt, W.,** *J. Electron Spectr. Rel. Phenom.,* 24, 101, 1981.
119. **Ruščić, B., Křivka, P., and Trinajstić, N.,** *Theoret. Chim. Acta,* 69, 107, 1986.
120. **Gutman, I., Trinajstić, N., and Wilcox, C. F., Jr.,** *Tetrahedron,* 31, 143, 1975.

Chapter 9

THE CONJUGATED-CIRCUIT MODEL

The *conjugated-circuit model* is a resonance-theoretic model which was introduced by Milan Randić in 1976 for the study of aromaticity and conjugation in polycyclic conjugated systems.[1-3] This model was motivated[4] from an empirical point of view elaborating the Clar aromatic sextet theory.[5] Eric Clar, following the classical chemical ideas[6] of Kekulé[7] and Armit and Robinson,[8] postulated that the aromatic stability correlates with the number of aromatic sextets in benzenoid hydrocarbons. By using the tools of chemical graph theory[9] Randić quantified the above ideas, and those of Fries,[10,11] in his conjugated-circuit model.[1]

The conjugated-circuit model has also a firm quantum-mechanical basis.[12,13] It can be derived rigorously from the Pauling-Wheland resonance theory[14,15] via a Simpson-Herndon model Hamiltonian.[16,17] The conjugated-circuit model also introduces a novel superposition principle not recognized before in quantum chemistry.[18]

I. THE CONCEPT OF CONJUGATED CIRCUITS

A graph-theoretical analysis of Kekulé valence structures produced the concept of conjugated circuits.[1] This concept may be introduced,[19] for example, by using the notion of *cyclic path*.[20] Imagine a Kekulé structure of some polycyclic conjugated hydrocarbon. Start at any carbon atom on the Kekulé structure and traverse any path through bonds, returning to the starting position. If the cyclic path consists of alternating single and double bonds, this path is defined to be a *conjugated circuit;* otherwise it is not. In other words, the conjugated circuits are those circuits within the individual Kekulé structure in which there is a regular alternation of formal CC single and double bonds. Thence, the conjugated circuits are necessarily of *even* length.

The circuit decomposition of individual Kekulé structure of a polycyclic conjugated hydrocarbon gives conjugated circuits of sizes $4n+2$ and/or $4n$ (n = integer). There are possible linearly independent, linearly dependent, and disjoint conjugated circuits. Linearly independent conjugated circuits are those that cannot be represented as a superposition of conjugated circuits of smaller size. Linearly dependent conjugated circuits are those which can be expressed as a linear combination of conjugated circuits of smaller size. Disjoint conjugated circuits are composites of two or more single conjugated circuits, no pair of which share a site.

The conjugated circuits associated only to carbon atoms are called *carbon conjugated circuits* and those of sizes $4n+2$ and $4n$ are denoted by R_n and Q_n. In Figure 1 the Kekulé structures of phenanthrene and their circuit decompositions are given.

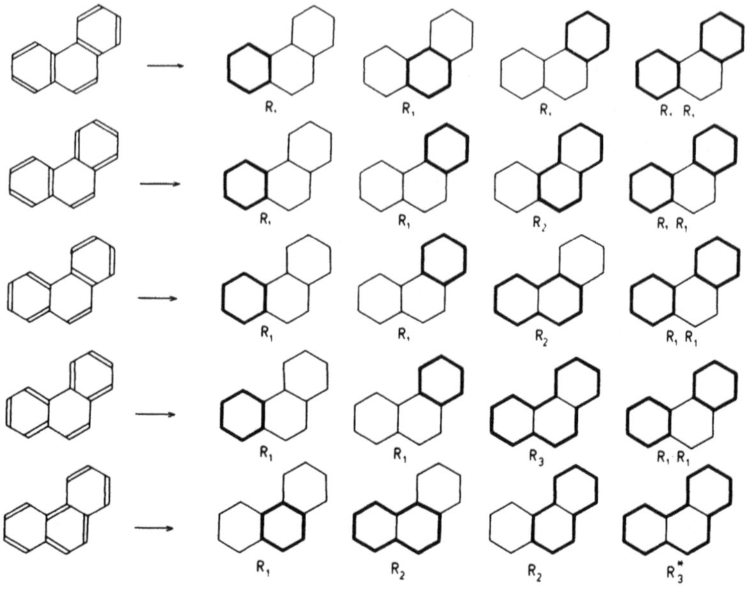

FIGURE 1. The decomposition of Kekulé structures of phenanthrene into the corresponding conjugated circuits. R_3^* is a linearly dependent conjugated circuit which can be obtained by adding twice R_2 and the substracting R_1, i.e., $R_3^* = 2R_2 - R_1$.

The circuit count of phenanthrene is given by

$$10R_1 + 4R_2 + 2R_3 + 4R_1 \cdot R_1 \qquad (1)$$

The total number of conjugated circuits, linearly independent, linearly dependent, and disjoint, within a single Kekulé valence structure is equal to $K - 1$,[21] where $K =$ the number of Kekulé structures for a polycyclic conjugated hydrocarbon. Therefore, by counting conjugated circuits in an arbitrary Kekulé structure, the number of all Kekulé structures of a molecule also becomes known. An illustrative example is shown in Figure 2.

The conjugated-circuit count in one Kekulé structure of benzo[ghi]perylene:

$$3R_1 + 2R_2 + 2R_3 + 3R_1 \cdot R_1 + 2R_1 \cdot R_2 + R_1 \cdot R_1 \cdot R_1 \qquad (2)$$

also gives the Kekulé-structure count ($K = 14$) for the molecule. All $K - 1$ Kekulé structures of benzo[ghi]perylene can be constructed from the above one by considering each conjugated circuit separately and exchanging single and double bonds within the conjugated circuit while leaving all other bonds unchanged.

The concept of conjugated circuits may be extended to heteroatomic systems. If a conjugated circuit contains a heteroatom, it differs from the conjugated circuits involving only carbon atoms and is named a *heteroatomic conju-*

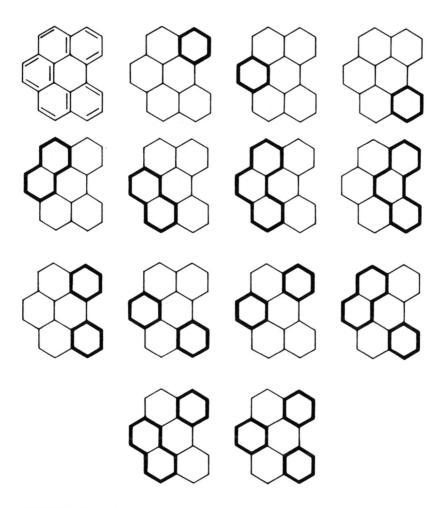

FIGURE 2. The conjugated-circuit count for an arbitrary Kekulé structure of benzo[ghi]perylene.

gated circuit.[22] The $4n + 2$ conjugated circuits containing carbon atoms and a single heteroatom are denoted by H_n, while the $4n$ conjugated circuits containing carbon atoms and a single heteroatom are denoted by H_n''. The conjugated-circuit model is extended so far to include heteroatomic systems[23-28] with the pyridine-like nitrogen, the pyrrole-like nitrogen, the furan-like oxygen, and the thiophene-like sulfur. The conjugated circuits containing these particular heteroatoms are denoted, respectively, by X_n, N_n, O_n, and S_n.

All these heteroatoms, except pyridine-like nitrogen, interrupt the conjugation in the heterocyclic system to some extent, because they contribute, to a somewhat lesser degree their π-electrons to the pool of π-electrons delocalized over the σ-skeleton of a conjugated molecule. The adjacent bonds to the furan-like oxygen or pyrrole-like nitrogen or thiophene-like sulfur may be formally viewed as single bonds. Therefore, one can imagine these het-

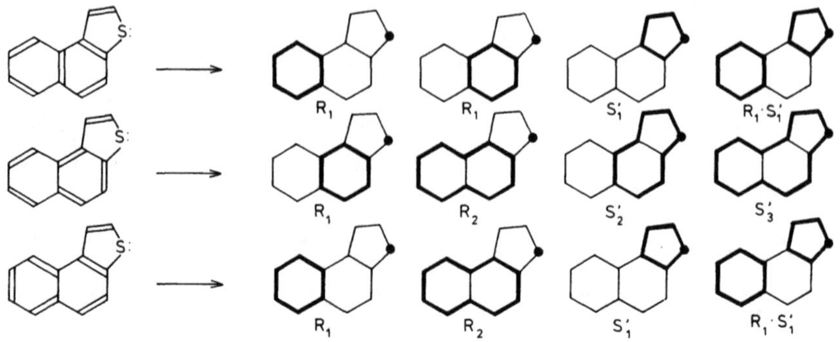

FIGURE 3.   The conjugated circuits belonging to naphtho[1,2-b]thiophene. The black dot denotes the site occupied by the thiophene-like sulfur atom. S'_n stands for $4n+2$ conjugated circuits containing the thiophene-like sulfur. S'_3 is a linearly dependent sulfur-containing conjugated circuit which can be obtained by adding R_2 and S'_2 and substracting R_1, i.e., $S'_3 = R_2 + S'_3 - R_1$.

eroatoms with their two π-electrons as formally equivalent to an isolated double bond "contracted" to a single atomic site. Therefore, the two π-electrons localized on the divalent heteroatom are taken as an internal "double bond" in forming the heteroatomic conjugated circuit. The conjugated circuits containing the pyridine-like nitrogen are, of course, obtained in the same manner as the carbon conjugated circuits. In Figure 3 the conjugated-circuit decomposition of Kekulé structures of naphtho[1,2-b] thiophene is shown. The conjugated-circuit count for naphtho[1,2-b]thiophene is given by

$$4R_1 + 2R_2 + 2S'_1 + S'_2 + S'_3 + 2R_1 \cdot S'_1 \qquad (3)$$

where S'_n ($n = 1,2,3$) stands for $4n+2$ ($n = 1,2,3$) conjugated circuits containing thiophene-like sulfur.

The conjugated-circuit decomposition of Kekulé structures of 1-azaphenanthrene is shown in Figure 4. The conjugated-circuit count for 1-azaphenanthrene is given by

$$6R_1 + 2R_2 + 4X_1 + 2X_2 + 2X_3 + 4R_1 \cdot X_1 \qquad (4)$$

where X_n ($n = 1,2,3$) stands for $4n+2$ ($n = 1,2,3$) conjugated circuits containing pyridine-like nitrogen.

The total number of conjugated circuits within a single Kekulé structure for a heterocyclic system is also equal to $K-1$,[36] where K is the number of Kekulé structures for a parent conjugated hydrocarbon in the case of heterocycles with a divalent heteroatom, or the number of Kekulé structures for a heterocycle with a heteroatom bearing a single π-electron.

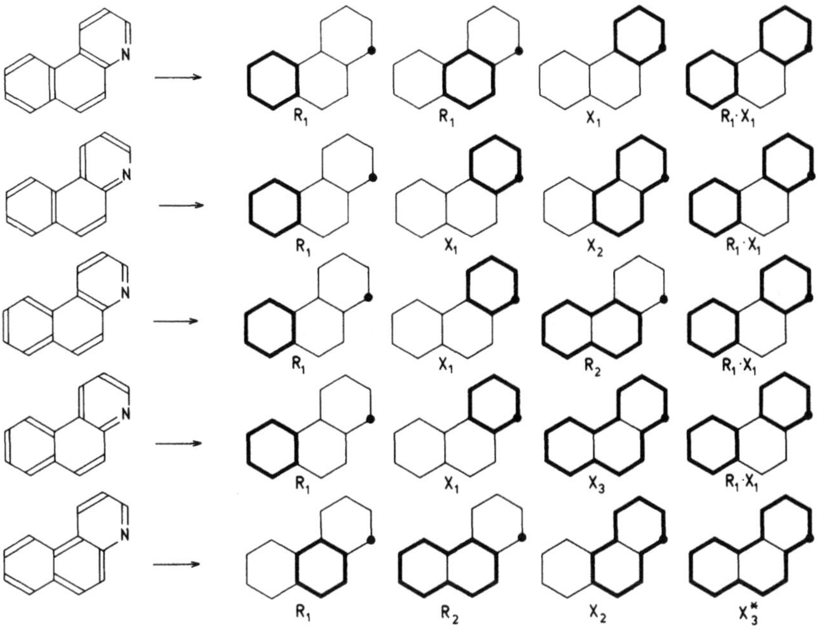

FIGURE 4. The conjugated circuits belonging to 1-azaphenanthrene. The black dot denotes the site occupied by the pyridine-like nitrogen atom. X_n stands for $4n+2$ conjugated circuits containing pyridine-like nitrogen atom. X_3^* is a linearly dependent conjugated circuit which can be obtained by adding R_2 and X_2 and substracting R_1, i.e., $X_3^* = R_2 + X_2 - R_1$.

II. THE π-RESONANCE ENERGY EXPRESSION

The conjugated circuits can be used to generate π-resonance energies, REs, of polycyclic conjugated systems. The RE is given in terms of conjugated circuits as follows:

$$RE = (1/K) \sum_{n \geq 1} (r_n R_n + q_n Q_n + h_n' H_n' + h_n'' H_n'') \qquad (5)$$

where K is the Kekulé-count for the molecule; R_n and Q_n are, respectively, the oppositely signed parameters corresponding to $4n+2$ and $4n$ carbon conjugated circuits; and H_n' and H_n'' are, respectively, the oppositely signed parameters for the heteroatomic conjugated circuits of sizes $4n+2$ and $4n$. The symbols r_n, q_n, h_n', and h_n'' are, respectively, the total counts of R_n, Q_n, H_n', and H_n'' circuits summed over all Kekulé structures of the molecule. In order to simplify the notation the symbols for conjugated circuits (R_n, Q_n, H_n', H_n'') are retained for the corresponding parameters (R_n, Q_n, H_n', H_n'').

The parameters R_n and H_n' (Q_n and H_n'') measure the extent to which a specific conjugated circuit of size $4n+2$ ($4n$) influences the thermodynamic stability of a polycyclic conjugated molecule. The R_n and H_n' circuits contribute

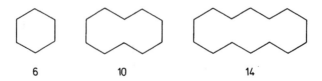

| 6 | 10 | 14 |

FIGURE 5.   Shapes of 6-, 10-, and 14-circuits.

toward the aromatic stability and the Q_n and H_n'' circuits have a destabilizing effect on the aromatic stability of the molecule.

The above RE expression reduces to the following for conjugated hydrocarbons:

$$RE = (1/K) \sum_{n \geq 1} (r_n R_n + q_n Q_n) \qquad (6)$$

Similarly, this expression becomes

$$RE = (1/K) \sum_{n \geq 1} r_n R_n \qquad (7)$$

for benzenoid hydrocarbons, because benzenoids contain only R_n circuits.

Molecules of different sizes are best compared if the RE is normalized. To eliminate the size effect, the RE is usually divided by the number N of π-electrons in the molecule,[29-31]

$$RE/e = RE/N \qquad (8)$$

The RE/e represents formally the aromatic stabilization or destabilization that one π-electron contributes to the π-network of the conjugated molecule.

In the early applications of the conjugated-circuit model[1-3] the set of conjugated circuits was truncated at $n = 4$, but later, especially when the model was extended to heteroconjugated systems,[26] the set of conjugated circuits was truncated at $n = 3$.[19] This truncation, for example, allows Equation 7 to be presented in the explicit form:

$$RE = (1/K)(r_1 R_1 + r_2 R_2 + r_3 R_3) \qquad (9)$$

where R_1, R_2, and R_3 are 6-, 10-, and 14-circuits (see their shapes in Figure 5) and r_1, r_2, and r_3 are the total numbers of these conjugated circuits in the benzenoid hydrocarbon.

Taking into account only the conjugated circuits of the three smallest sizes is in agreement with the empirical finding that only the smallest circuits make appreciable contributions to the RE.[17] In principle, the higher circuits should really be included, but the data needed for establishing the corresponding parameters are rather scarce, especially for heterocyclic systems.

This then limits the extent of parametrization to the dominant conjugated circuits.

Only linearly independent conjugated circuits have been used in most of the RE computations.[1-3,19,32-34] However, it should be pointed out that the application of the conjugated-circuit model to large systems is quite involved because of the necessity to recognize linearly independent conjugated circuits. Therefore, the above requirement was relaxed in some applications and the linearly dependent conjugated circuits have also been used in RE computations.[27,28,35,36]

III. SELECTION OF THE PARAMETERS

A. R_n PARAMETERS

The numerical values of the R_n ($n = 1,2,3$) parameters have usually been obtained from a fitting to the Dewar-de Llano SCF π-MO resonance energies of benzene (B) naphthalene (N), and anthracene (A):[37] RE(B) = 0.869 eV, RE(N) = 1.323 eV, and RE(A) = 1.600 eV. The RE expressions for these three benzenoid hydrocarbons are given by

$$RE = (2R_1)/2 \tag{10}$$

$$RE = (4R_1 + 2R_2)/3 \tag{11}$$

$$RE = (6R_1 + 4R_2 + 2R_3)/4 \tag{12}$$

Introducing the above SCF π-MO REs into Equations 10 to 12, one obtains the numerical values for the R_n ($n = 1,2,3$) parameters (*set I*):

$R_1 = 0.869$ eV
$R_2 = 0.247$ eV
$R_3 = 0.100$ eV

Other selections of parameters are also possible.[38,39] For example, since the Dewar-de Llano REs are known for 32 benzenoid hydrocarbons,[37] a parametrization procedure based on the standard least-squares analysis on all the 32 data points generated the following set (*set II*) of parameters:[40]

$R_1 = 0.827$ eV
$R_2 = 0.317$ eV
$R_3 = 0.111$ eV.

However, in the majority of the conjugated-circuit computations thus far, the set I has been used.

B. A PARAMETRIZATION SCHEME FOR OTHER PARAMETERS

The SCF π-MO REs for alternant systems with $4n$ rings are very limited. A similar situation also occurs with heteroconjugated systems. Therefore, the parametric values for other kinds of conjugated circuits may be approximated in the following way:[17,23,26]

$$(CC)_n = (R_n/R_1)(CC)_1; \; n = 1,2,3 \tag{13}$$

where $(CC)_n$ stands for parameters corresponding to conjugated circuits denoted as Q_n, X_n, N_n, O_n, and S_n. In order to use Formula 13, it is necessary to know only the $(CC)_1$ parameter for each case considered. Formula 13 is based on the assumption that the same relationship that exists between the parameters for $4n + 2$ carbon conjugated circuits also holds for other kinds of circuits.

1. The Q_n Parameters

The numerical values of the Q_n ($n = 1,2,3$) parameters are obtained from the SCF π-MO RE of cyclobutadiene (-0.781 eV)[37] and by using the following approximations:

$$Q_2 = (R_2/R_1)Q_1 \tag{14}$$

$$Q_3 = (R_3/R_1)Q_1 \tag{15}$$

The Q_n parameters thus obtained are

$$Q_1 = -0.781 \text{ eV}$$
$$Q_2 = -0.222 \text{ eV}$$
$$Q_3 = -0.090 \text{ eV}$$

2. The H'_n Parameters

The numerical values of the X'_n, O'_n, N'_n, and S'_n ($n = 1,2,3$) parameters are obtained from the SCF π-MO REs of pyridine (0.908 eV),[41] furan (0.069 eV),[41] pyrrole (0.370 eV),[41] and thiophene (0.282 eV)[42] through the use of the following approximations:

$$H'_2 = (R_2/R_1)H'_1 \tag{16}$$

$$H'_3 = (R_3/R_1)H'_1 \tag{17}$$

The parameters obtained are given in Table 1.

TABLE 1

Parameters (in eV) for Several 4n+2
Heteroatomic Conjugated Circuits

n	X_n'	O_n'	N_n'	S_n'
1	0.908	0.069	0.370	0.282
2	0.258	0.020	0.105	0.080
3	0.105	0.008	0.043	0.032

3. The H_n'' Parameters

The numerical values of the H_n'' ($n = 1,2,3$) parameters may be obtained by means of the following approximations:[27,28]

$$H_n'' = (Q_1/R_1)H_n'; n = 1,2,3 \tag{18}$$

Data necessary to use Equation 18 are available only for some compounds containing divalent sulfur. In this case the above equation transforms into

$$S_n'' = (Q_1/R_1)S_n'; n = 1,2,3 \tag{19}$$

The assumption that the ratios Q_n/R_n and S_n''/S_n' are similar is based on empirical observations for benzene and cyclobutadiene in the case of conjugated hydrocarbons[43-47] and for thiophene and thiirene in the case of heterocyclic compounds containing a single divalent sulfur.[48-52] These molecules represent the extreme cases of aromatic and antiaromatic systems in each class of compounds. Thiophene has been known for a long time to be a stable compound,[48] while thiirene has only recently been observed in an argon matrix at 8 K and is an antiaromatic species.[49-51,53] Some theoretical computations also support the relationship in Equation 19. For example, Hess and Schaad[29,54] produced the following set of RE/e values (in β units) for the above molecules: 0.065 (benzene), -0.268 (cyclobutadiene), 0.032 (thiophene), and -0.114 (thiirene). The ratios cyclobutadiene/benzene (-4.123) and thiirene/thiophene (-3.563) differ by only 14% within the framework of the Hess-Schaad RE model.

Using the parametric values from the above for R_1 (0.869 eV), Q_1 (-0.781 eV), S_1' (0.282 eV), and Equation 19, the following set of S_n'' parameters is obtained:

$$S_1'' = -0.253 \text{ eV}$$
$$S_2'' = -0.072 \text{ eV}$$
$$S_3'' = -0.029 \text{ eV}$$

IV. COMPUTATIONAL PROCEDURE

The computational procedure for the conjugated-circuit model proceeds along the following steps: (1) identification of the polycyclic conjugated systems to be studied; (2) enumeration of Kekulé structures; (3) enumeration of conjugated circuits; (4) setting up the RE expression; (5) selection of parameters and numerical work; and (6) predictions. There is only one combinatorial problem connected with this procedure. This is the enumeration of Kekulé structures and it represents the only computationally involved step in the procedure. However, in the preceding chapter numerous schemes for the enumeration of Kekulé structures have been presented and there are many more available in the literature.[55]

There is seemingly another combinatorial problem in the procedure. This is the enumeration of conjugated circuits. However, since there is a close relationship between the Kekulé structures and conjugated circuits,[1] the known number of the Kekulé structures for the molecule immediately leads to the number of conjugated circuits via the following expression:[56,57]

$$(CC)_n = 2 \sum_{C_n}^{n,G} K(G - C_n) \tag{20}$$

where $G - C_n$ is the structure obtained from G by deletion of a cycle C_n (n $= 4,6,8, \ldots$) and bonds incident to C_n. It should be noted if $G - C_n = 0$ then $K(G - C_n) = 1$ by definition. The enumeration of conjugated circuits is thus reduced to Kekulé structure counts for different fragments of G. The application of the counting formula (Equation 20) is shown in Figure 6.

The linking of the counting formula (Equation 20) with the transfer-matrix method has produced a very efficient and elegant procedure for the computer-assisted enumeration of conjugated circuits.[57-59]

V. APPLICATIONS OF THE CONJUGATED-CIRCUIT MODEL

The conjugated-circuit model has been used in most cases to compute the REs or the REs/e of conjugated systems. This is so because these indices serve well as criteria for aromatic stabilization (destabilization) of the polycyclic π-electron systems. The range of applications of the model to date is considerable and includes all kinds of structures from benzenoid hydrocarbons to infinite strips and high-temperature superconducting materials. These applications are listed in Table 2.

A. BENZENOID HYDROCARBONS

The most studied class of conjugated molecules with the conjugated-circuit model is the class of benzenoid hydrocarbons.[1-3,19,26,32,36,38-40,60-66] The

(a) Benzenoid graph B corresponding to chrysene

B

(b) Consecutive removal of C_6 cycles from B and the Kekulé-structure counts for B-C_6 fragments

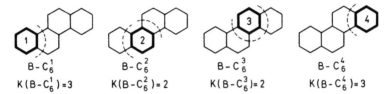

B-C_6^1	B-C_6^2	B-C_6^3	B-C_6^4
$K(B$-$C_6^1) = 3$	$K(B$-$C_6^2) = 2$	$K(B$-$C_6^3) = 2$	$K(B$-$C_6^4) = 3$

(c) Consecutive removal of C_{10} cycles from B and the Kekulé-structure counts for B-C_{10} fragments

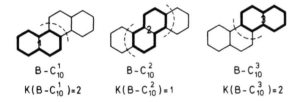

B-C_{10}^1	B-C_{10}^2	B-C_{10}^3
$K(B$-$C_{10}^1) = 2$	$K(B$-$C_{10}^2) = 1$	$K(B$-$C_{10}^3) = 2$

(d) Consecutive removal of C_{14} cycles from B and the Kekulé-structure counts for B-C_{14} fragments

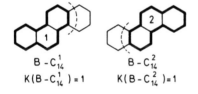

B-C_{14}^1	B-C_{14}^2
$K(B$-$C_{14}^1) = 1$	$K(B$-$C_{14}^2) = 1$

(e) Conjugated-circuits count

$(CC)_6 = R_1 = 20;$ $(CC)_{10} = R_2 = 10;$ $(CC)_{14} = R_3 = 4$

FIGURE 6.   Application of the counting formula (Equation 20) to chrysene.

REs of a number of benzenoid hydrocarbons (see Figure 7) calculated using parameters in the *set I* are given in Table 3.

The REs obtained by the conjugated-circuit model are tested against the REs (SCF) by means of the following linear correlation:

$$RE(SCF) = p \; RE(\text{conjugated circuits}) + q \qquad (21)$$

The statistical parameters: $p = 0.9878(\pm 0.0109)$, $q = 0.0315(\pm 0.0314)$, the correlation coefficient = 0.998, the standard deviation = 0.053, F-ratio = 8223.8, indicate that the REs computed by the conjugated-circuit model reproduce closely the SCF π-MO REs of benzenoid hydrocarbons.

TABLE 2
The List of Molecules Studied by the Conjugated-Circuit Model[a]

Polycyclic conjugated hydrocarbons	Elemental carbon systems	Heterocyclic conjugated compounds	Extended systems	Miscellaneous organic systems	Inorganic systems
Benzenoid hydrocarbons[b-l]	Buckminsterfullerene[v-z]	Azabenzenoids[ac]	Benzenoid polymers[a,o,ai-ap]	Lowest excited states of benzenoids[ar,as]	High-temperature superconductors[ay,az]
Non-benzenoid hydrocarbons[c,d]	Fullerenes[v-z]	Furan and its derivatives[ad,ae]	Conjugated polymers[p,ap]	Benzenoid hydrocarbon radicals[at]	
Macrocyclic benzenoid systems[m]	Graphite[aa]	Pyrrole and its derivatives[ad,ae]		Polycyclic conjugated hydrocarbon (di)cations[au]	
Polyacenes[b-d,n-p]	Toroidal structures[ab]	Thiophene and its derivatives[ad-ah]		Polycyclic conjugated hydrocarbon (di) anions[av,aw]	
Helicenes[a,o]				Möbius systems[ax]	
Corannulenes[r]					
Benzoannelates annulenes[s,t]					
Phenylenes[p]					
Nonalternant systems[c,d]					
Fractal hydrocarbons[a]					

[a] In many reports listed below different sets of parameters for carbon conjugated circuits have been used in the computations. These are as follows: (1) the Herndon set (R_1 = 0.841 eV; R_2 = 0.336 eV; Q_1 = −0.650 eV; Q_2 = −0.260 eV); (2) the original Randić set (R_1 = 0.869 eV; R_3 = 0.246 eV; R_4 = 0.041 eV; Q_1 = −1.600 eV; Q_2 = −0.450 eV; Q_3 = −0.150 eV; Q_4 = −0.060 eV) and (3) the truncated set (R_1 = 0.869 eV; R_2 = 0.246 eV; R_3 = 0.100 eV; Q_1 = −0.781 eV; Q_2 = −0.222 eV; Q_3 = −0.090 eV).

[b] Randić, M., *Chem. Phys. Lett.*, 38, 68, 1976.

[c] Randić, M., *J. Am. Chem. Soc.*, 99, 444, 1977.

d   Randić, M., *Tetrahedron*, 33, 1905, 1977.

e   Randić, M., *Int. J. Quantum Chem.*, 17, 549, 1980.

f   Randić, M., *Chem. Phys. Lett.*, 128, 193, 1986.

g   Randić, M. and Trinajstić, N., *J. Am. Chem. Soc.*, 109, 6923, 1987.

h   Randić, M., Solomon, V., Grossman, S. C., Klein, D. J., and Trinajstić, N., *Int. J. Quantum Chem.*, 32, 35, 1987.

i   Trinajstić, N., Plavšić, D., and Klein, D. J., *Croat. Chem. Acta*, 62, 709, 1989.

j   Nikolić, S., Randić, M., Klein, D. J., Plavšić, D., and Trinajstić, N., *J. Mol. Struct. (Theochem.)*, 198, 223, 1989.

k   Plavšić, D., Nikolić, S., and Trinajstić, N., *Croat. Chem. Acta*, 63, 683, 1990.

l   Plavšić, D., Nikolić, S., and Trinajstić, N., *J. Math. Chem.*, 8, 113, 1991.

m   Randić, M., Henderson, L. L., Stout, R., and Trinajstić, N., *Int. J. Quantum Chem.: Quantum Chem. Symp.*, 22, 127, 1988.

n   Randić, M., Gimarc, B. M., and Trinajstić, N., *Croat. Chem. Acta*, 59, 345, 1986.

o   Randić, M., Gimarc, B. M., Nikolić, S., and Trinajstić, N., *Gazz. Chim. Ital.*, 119, 1, 1989.

p   Trinajstić, N., Schmalz, T. G., Nikolić, S., Klein, D. J., Hite, G. E., Seitz, W. A., and Živković, T. P., *New J. Chem.*, 15, 27, 1991.

r   Randić, M. and Trinajstić, N., *J. Am. Chem. Soc.*, 106, 4428, 1984.

s   Volger, H. and Trinajstić, N., *Theoret. Chim. Acta*, 73, 437, 1988.

t   Vogler, H. and Trinajstić, N., *J. Mol. Struct. (Theochem.)*, 164, 235, 1988.

u   Klein, D. J., Cravey, M. J., and Hite, G. E., *Polycyclic Aromatic Compounds*, 2, 163, 1991.

v   Klein, D. J., Schmalz, T. G., Hite, G. E., and Seitz, W. A., *J. Am. Chem. Soc.*, 108, 1301, 1986.

w   Schmalz, T. G., Seitz, W. A., Klein, D. J., and Hite, G. E., *Chem. Phys. Lett.*, 130, 203, 1986.

x   Klein, D. J., Seitz, W. A., and Schmalz, T. G., *Nature*, 323, 705, 1986.

y   Randić, M., Nikolić, S., and Trinajstić, N., *Croat. Chem. Acta*, 60, 595, 1987.

z   Schmalz, T. G., Seitz, W. A., Klein, D. J., and Hite, G. E., *J. Am. Chem. Soc.*, 110, 1113, 1987.

aa Hite, G. E., Živković, T. P., and Klein, D. J., *Theoret. Chim. Acta*, 74, 349, 1988.

ab Klein, D. J., work in progress.

ac Randić, M., Trinajstić, N., Knop, J. V., and Jeričević, Ž., *J. Am. Chem. Soc.*, 107, 849, 1985.

ad Randić, M., Nikolić, S., and Trinajstić, N., in *Graph Theory and Topology in Chemistry*, King, R. B. and Rouvray, D. H., Eds., Elsevier, Amsterdam, 1987, 429.

ae Nikolić, S., Jurić, A., and Trinajstić, N., *Heterocycles*, 26, 2025, 1987.

af Randić, M., Nikolić, S., and Trinajstić, N., *Coll. Czech. Chem. Commun.*, 53, 2023, 1988.

ag Randić, M. and Trinajstić, N., *Sulfur Rep.*, 6, 379, 1986.

ah Randić, M., Gimarc, B. M., Nikolić, S., and Trinajstić, N., *J. Mol. Struct. (Theochem.)*, 181, 111, 1981.

TABLE 2 (continued)
The List of Molecules Studied by the Conjugated-Circuit Model[a]

ai Seitz, W. A., Klein, D. J., Schmalz, T. G., and Garcia-Bach, M. A., *Chem. Phys. Lett.*, 115, 139, 1985.

aj Klein, D. J., Schmalz, T. G., Hite, G. E., Metropoulos, A., and Seitz, W. A., *Chem. Phys. Lett.*, 120, 367, 1985.

ak Gutman, I. and El-Basil, S., *Chem. Phys. Lett.*, 115, 416, 1985.

al Klein, D. J., Hite, G. E., and Schmalz, T. G., *J. Comput. Chem.*, 7, 443, 1986.

am Hite, G. E., Metropoulos, A., Klein, D. J., Schmalz, T. G., and Seitz, W. A., *Theoret. Chim. Acta*, 69, 369, 1986.

an Klein, D. J., Živković, T. P., and Trinajstić, N., *J. Math. Chem.*, 1, 309, 1987.

ao Seitz, W. A., Hite, G. E., Schmalz, T. G., and Klein, D. J., in *Graph Theory and Topology in Chemistry*, King, R. B. and Rouvray, D. H., Eds., Elsevier, Amsterdam, 1987, 458.

ap Seitz, W. A. and Schmalz, T. G., in *Valence Bond Theory and Chemical Structure*, Klein, D. J. and Trinajstić, N., Eds., Elsevier, Amsterdam, 1990, 525.

ar Randić, M., Plavšić, D., and Trinajstić, N., *J. Mol. Struct. (Theochem.)*, 183, 29, 1989.

as Garcia-Bach, M. A., Valenti, R., and Klein, D. J., *J. Mol. Struct. (Theochem.)*, 185, 287, 1989.

at Plavšić, D., Trinajstić, N., Randić, M., and Venier, C., *Croat. Chem. Acta*, 62, 717, 1989.

au Randić, M. and Nettleton, F. E., *Int. J. Quantum Chem.: Quantum Chem. Symp.*, 20, 203, 1986.

av Randić, M., *J. Phys. Chem.*, 86, 3970, 1982.

aw Randić, M., Plavšić, D., and Trinajstić, N., *J. Mol. Struct. (Theochem.)*, 185, 29, 1989.

ax Randić, M. and Zimmerman, H. E. *Int. J. Quantum Chem.: Quantum Chem. Symp.*, 20, 201, 1986.

ay Rokhsar, D. S. and Kivelson, S. A., *Phys. Rev. Lett.*, 61, 2376, 1988.

az Kivelson, S. A., *Phys. Rev. B*, 39, 259, 1989.

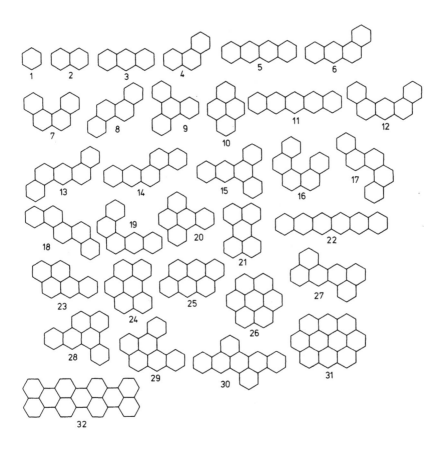

FIGURE 7.   Graphs of 32 benzenoid hydrocarbons.

B. NONALTERNANT HYDROCARBONS

Several novel features emerged from the application of the conjugated-circuit model to nonalternant hydrocarbons. The decomposition of conjugation contained in Kekulé structures of nonalternants reveals that they can be partitioned into two classes of compounds: one class is that in which only conjugated circuits of $4n+2$ type arise and the other class contains structures with both types ($4n+2$ and $4n$) of conjugated circuits. For example the four Kekulé structures of acepleiadylene (see Figure 8) give only R_n circuits ($4R_1 + 2R_2 + 6R_3$), although this molecule is a nonalternant structure.

In contrast four Kekulé structures of another nonalternant azupyrene (see Figure 9) possess the following circuit content: $8R_2 + 2R_3 + 2Q_3$.

Therefore, in analogy with the classification of *alternants* into *benzenoids* and *nonbenzenoids*,[67] the *nonalternant* systems may be classified into *azulenoids* and *nonazulenoids*.[3] The label *azulenoid* is used in view that azulene is one of the simplest and most commonly encountered nonalternant structures.[68] Examples of azulenoids and nonazulenoids are shown in Figure 10 and Figure 11.

TABLE 3
The Resonance Energy Expressions in Terms of Conjugated Circuits and the SCF Resonance Energies (in eV) of Some Benzenoid Hydrocarbons

Benzenoid hydrocarbon	The RE expression[a]	RE	
		Conjugated circuits[b]	SCF π-MO[c]
1	$(2R_1)/2$	0.869	0.869
2	$(4R_1 + 2R_2)/3$	1.323	1.323
3	$(6R_1 + 4R_2 + 2R_3)/4$	1.600	1.600
4	$(10R_1 + 4R_2 + R_3)/5$	1.956	1.933
5	$(8R_1 + 6R_2 + 4R_3)/5$	1.767	1.822
6	$(16R_1 + 8R_2 + 3R_3)/7$	2.311	2.291
7	$(20R_1 + 10R_2 + 2R_3)/8$	2.506	2.478
8	$(20R_1 + 10R_2 + 2R_3)/8$	2.506	2.483
9	$(26R_1 + 6R_2 + 3R_3)/9$	2.708	2.654
10	$(12R_1 + 8R_2 + 4R_3)/6$	2.134	2.098
11	$(10R_1 + 8R_2 + 6R_3)/6$	1.878	2.004
12	$(36R_1 + 16R_2 + 6R_3)/12$	2.986	2.948
13	$(36R_1 + 16R_2 + 6R_3)/12$	2.986	2.948
14	$(30R_1 + 18R_2 + 6R_3)/11$	2.829	2.823
15	$(42R_1 + 14R_2 + 5R_3)/13$	3.112	3.058
16	$(40R_1 + 20R_2 + 5R_3)/13$	3.092	3.072
17	$(40R_1 + 20R_2 + 5R_3)/13$	3.092	3.071
18	$(40R_1 + 20R_2 + 5R_3)/13$	3.092	3.071
19	$(30R_1 + 18R_2 + 6R_3)/11$	2.829	2.823
20	$(32R_1 + 14R_2 + 7R_3)/11$	2.906	2.853
21	$(24R_1 + 12R_2)/9$	2.647	2.619
22	$(12R_1 + 10R_2 + 8R_3)/7$	2.586	2.584
23	$(22R_1 + 14R_2 + 7R_3)/9$	1.957	2.160
24	$(42R_1 + 26R_2 + 12R_3)/14$	3.151	3.128
25	$(24R_1 + 18R_2 + 12R_3)/10$	2.650	2.665
26	$(64R_1 + 48R_2 + 27R_3)/20$	3.509	3.524
27	$(24R_1 + 12R_2)/9$	2.647	2.694
28	$(58R_1 + 26R_2 + 11R_3)/17$	3.407	3.375
29	$(52R_1 + 26R_2 + 13R_3)/16$	3.307	3.283
30	$(100R_1 + 40R_2 + 10R_3)/25$	3.911	3.862
31	$(200R_1 + 160R_2 + 110R_3)/50$	4.486	4.539
32	$(432R_1 + 216R_2)/81$	5.293	5.309

[a]   Only linearly independent conjugated circuits are considered.

[b]   A three-parameter conjugated-circuits model is used with parameters in the *set I*: $R_1 = 0.869$ eV, $R_2 = 0.247$ eV, $R_3 = 0.100$ eV.

[c]   Data are taken from Dewar, M. J. S. and de Llano, C., *J. Am. Chem. Soc.*, 91, 789, 1969.

Another interesting result is that different nonalternant hydrocarbons may have *identical* conjugated-circuit decompositions and consequently, identical REs. Molecules with identical circuit decomposition are called *isoconjugated molecules*[3] and they are expected to exhibit similar aromatic behavior. As an illustrative example, the following pair of isoconjugated azulenoids serve well: cycloheptacenaphthylene and naphthazulene (see Figure 12). They have iden-

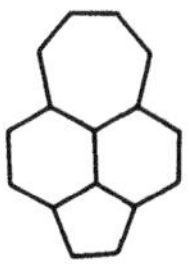

G

FIGURE 8.   A graph G of acepleiadylene.

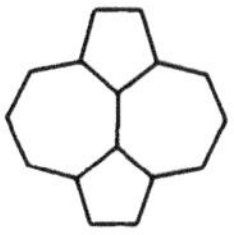

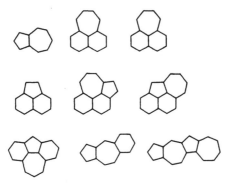

G

FIGURE 9.   A graph G of azapyrene.

FIGURE 10.   Graphs of azulenoids, i.e., nonalternant hydrocarbons containing only 4n + 2 conjugated circuits.

tical circuit decomposition, $4R_1 + 4R_2 + 4R_3$, and identical resonance energy, 1.216 eV. Other theoretical methods also predict cycloheptacenaphthylene and naphthazulene to possess similar aromatic stabilities, because they have produced identical values of the aromaticity indices for them.[69] However, these other approaches could not trace the origin of this identity. There are many such pairs, some of which are shown in Figure 13.

C. AROMATICITY POSTULATE

Randić[2] has proposed the following *aromaticity postulate:* "Systems which possess *only* (4n + 2) conjugated circuits are *aromatic*". This postulate led to the formulation of the *generalized Hückel rule:* "Systems having *only* (4n + 2) conjugated circuits in their Kekulé valence structures are *aromatic* and represent *generalized Hückel* (4n + 2) systems". In Figure 14, graphs of fully

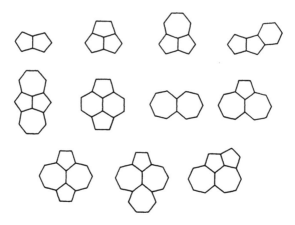

FIGURE 11.   Graphs of nonazulenoids, i.e., nonalternant hydrocarbons containing $(4n+2)$ and $(4n)$ conjugated circuits.

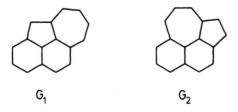

G_1 G_2

FIGURE 12.   Graphs G_1 and G_2 belonging, respectively, to cyclohepta-cenaphthylene and naphthazulene.

aromatic systems (systems containing only $4n+2$ conjugated circuits) are given. The important point here is that the aromaticity postulate does not discriminate between alternant and nonalternant hydrocarbons.

The aromaticity postulate and the Hückel rule may also be extended to structures which contain only $4n$ conjugated circuits: "Systems having *only* $4n$ conjugated circuits are *antiaromatic* and they represent the *generalized* Hückel $4n$ systems". Examples of fully antiaromatic systems containing only $4n$ conjugated circuits are given in Figure 15. Here again the aromaticity postulate makes no difference between alternants and nonalternants.

Molecules which contain both $4n+2$ and $4n$ conjugated circuits represent an intermediate class of compounds. In some cases contributions from the $4n+2$ conjugated circuits may be more important and in other cases those from the $4n$ conjugated circuits. The former systems exhibit features similar to aromatic structures, while the latter systems resemble antiaromatic species. Examples of both these classes of molecules are given in Figure 16 and Figure 17.

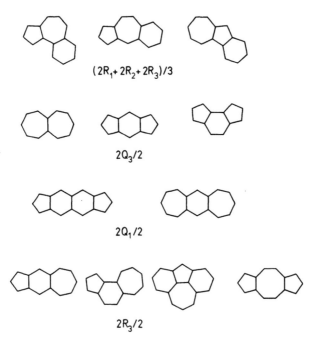

FIGURE 13. Graphs of isoconjugated nonalternants.

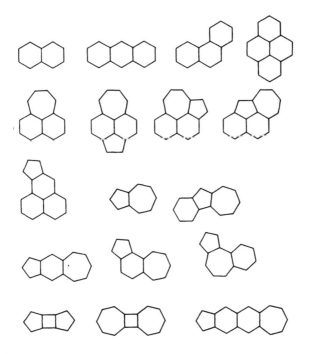

FIGURE 14. Graphs of aromatic molecules which contain only $4n+2$ conjugated circuits in their Kekulé structures.

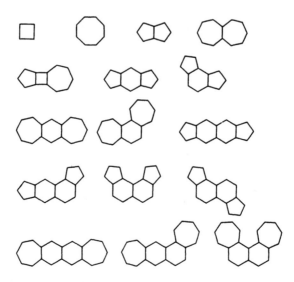

FIGURE 15. Graphs of antiaromatic molecules which contain only 4n conjugated circuits in their Kekulé structures.

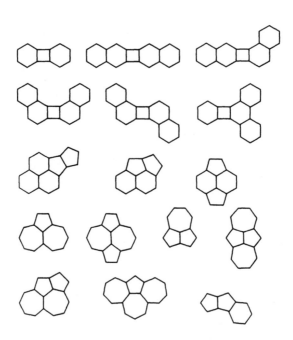

FIGURE 16. Graphs of molecules which contain both 4n + 2 and 4n conjugated circuits, with prevailing contribution from 4n + 2 circuits.

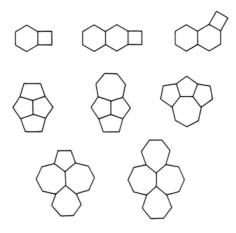

FIGURE 17.   Graphs of polycyclic conjugated molecules which contain
both 4n + 2 and 4n conjugated circuits, with prevailing contribution from
4n circuits.

VI. PARITY OF CONJUGATED CIRCUITS

So far it has been shown that the concept of conjugated circuits is very
useful for computing resonance energies, for defining aromaticity, for classifying polycyclic conjugated systems, etc. In this section the parity of Kekulé
structures will be discussed by means of conjugated circuits. It should be
remembered that the parity is defined for a pair of Kekulé valence structures
and has significance only when the relative effects of such structures are
discussed. It cannot be assigned to an individual Kekulé structure, since it is
not invariant to symmetry operations. Thus, for example, two symmetry-
equivalent Kekulé valence structures of cyclobutadiene have opposite parity.

In the previous chapter it has been seen that the parity concept breaks
down for odd-membered polycyclic conjugated hydrocarbons with three or
more cycles. In addition, it appears that the assumption that the Kekulé
structures of the same parity make the same kind of contribution to the stability
of the molecule is questionable. However, the concept of parity may be
generalized by applying it to conjugated circuits rather than to Kekulé structures.[70] Thus, all $4n + 2$ conjugated circuits are given *positive* parity and
conjugated circuits of $4n$ size are given *negative* parity. This, then permits
assigning to each Kekulé structure an integer number, called ACC (algebraic
circuit count), which is defined as the difference between the numbers of
$4n + 2$ and $4n$ conjugated circuits.

Since a count of conjugated circuits is symmetry invariant, the ACC can
be assigned, in contrast to the parity, to individual Kekulé structures. The
ACC label is introduced in parallel to ASC (algebraic structure count),[71] which
is defined as the difference between the numbers of Kekulé structures of
opposing parities. A closer examination of the ACC shows its full agreement

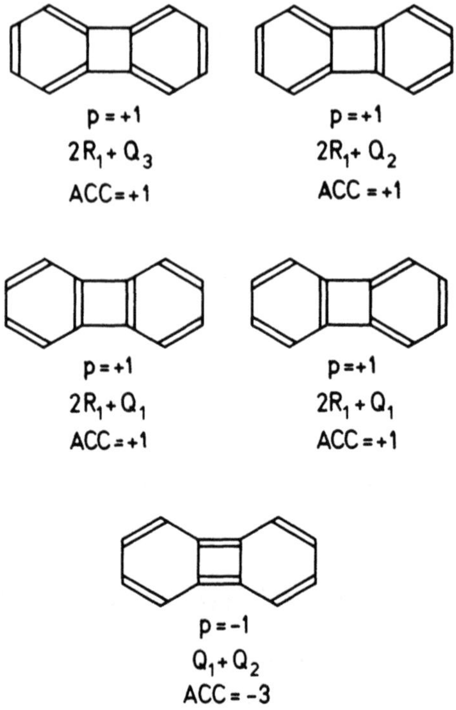

$p = +1$

$2R_1 + Q_3$

$ACC = +1$

$p = +1$

$2R_1 + Q_2$

$ACC = +1$

$p = +1$

$2R_1 + Q_1$

$ACC = +1$

$p = +1$

$2R_1 + Q_1$

$ACC = +1$

$p = -1$

$Q_1 + Q_2$

$ACC = -3$

FIGURE 18.   The parity assignment, the conjugated circuit decomposition, and the ACC values of Kekulé structures of biphenylene.

with the ASC in many structures for which the concept of parity operates, while ambiguities in other instances have been removed. An important conceptual distinction, however, is introduced: the ACC values are *absolute* parameters, not *relative* as before. This is an advantage, because a new label can be associated with individual structural formulae. In the case of the ASC scheme, this concept gives opposite parity to symmetry-equivalent structures and, consequently, they make opposite contributions to the resonance energy, e.g., two Kekulé structures of cyclobutadiene. Such difficulties vanish when the ACC parameter is used, since it can take any positive, zero, or negative value, in contrast to "$+1$" and "-1" values for the relative parities. Thus, for example, in cyclobutadiene both Kekulé structures have ACC $= -1$, they are both negative, and the sum of ACCs is suggestive that the molecule is antiaromatic. In general, Kekulé valence structures with positive ACC will play a stabilizing role, structures with ACC $= 0$ will be nonstabilizing, while those with negative ACC will be destabilizing. The ACC values of biphenylene are given in Figure 18. The Kekulé structures of biphenylene of the same parity also have similar ACC count. The important result is that the two alternative schemes in this case lead to the same conclusion: one of the five Kekulé structures is of a different kind.

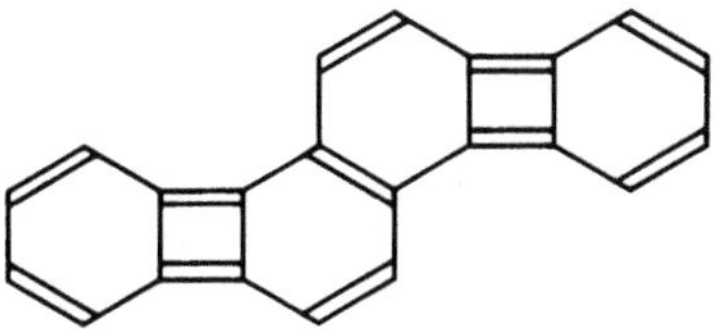

$$p = +1$$
$$2Q_1 + 4Q_2$$
$$ACC = -6$$

FIGURE 19. A Kekulé structure of a biphenylene derivative for which there is a disagreement between the parity parameter and the ACC parameter.

There are cases where the incompatibility between the parity concept and the ACC concept occurs. One such case is shown in Figure 19.

A troublesome case of acepentylene (see Figure 31 in Chapter 8) for which the parity criterion is ambiguous, gives no trouble at all to the ACC concept. The ACC values can be assigned to Kekulé structures of acepentylene, and since these are all equivalent, they all have the same ACC value of -2.

REFERENCES

1. **Randić, M.,** *Chem. Phys. Lett.,* 38, 68, 1976.
2. **Randić, M.,** *J. Am. Chem. Soc.,* 99, 444, 1977.
3. **Randić, M.,** *Tetrahedron,* 33, 1905, 1977.
4. **Graovac, A., Gutman, I., Randić, M., and Trinajstić, N.,** *J. Am. Chem. Soc.,* 95, 6267, 1973.
5. **Clar, E.,** *The Aromatic Sextet,* John Wiley & Sons, London, 1972.
6. **Klein, D. J. and Trinajstić, N.,** *J. Chem. Educ.,* 67, 633, 1990.
7. **Kekulé, A.,** *Bull. Soc. Chim. Fr.,* 3, 98, 1865; *Justus Liebigs Ann. Chem.,* 137, 129, 1866.
8. **Armit, J. W. and Robinson, R.,** *J. Chem. Soc.,* 827, 1922; 1604, 1925.
9. **Gutman, I. and Trinajstić, N.,** *Topics Curr. Chem.,* 42, 49, 1973.
10. **Fries, K.,** *Justus Liebigs Ann. Chem.,* 454, 121, 1927.
11. **Fries, K., Walter, R., and Schilling, K.,** *Justus Liebigs Ann. Chem.,* 516, 245, 1935.
12. **Klein, D. J. and Trinajstić, N.,** *Pure Appl. Chem.,* 61, 2107, 1989.
13. **Klein, D. J.,** *Topics Curr. Chem.,* 153, 57, 1990.
14. **Pauling, L.,** *J. Chem. Phys.,* 1, 280, 1933.
15. **Pauling, L. and Wheland, G. W.,** *J. Chem. Phys.,* 1, 362, 1933; 1, 606, 1933; 2, 482, 1933.
16. **Simpson, W. T.,** *J. Am. Chem. Soc.,* 75, 593, 1953.
17. **Herndon, W. C.,** *J. Am. Chem. Soc.,* 95, 2404, 1973; *Isr. J. Chem.,* 20, 270, 1980.
18. **Gomes, J. A. N. F.,** *Croat. Chem. Acta,* 53, 561, 1980; *Theoret. Chim. Acta,* 59, 333, 1981.

19. Randić, M. and Trinajstić, N., *J. Am. Chem. Soc.*, 109, 6923, 1987.
20. Harary, F., *Graph Theory*, Addison-Wesley, Reading, MA, 1971, 13, 2nd printing.
21. Gutman, I. and Randić, M., *Chem. Phys.*, 41, 265, 1979.
22. Nikolić, S., Trinajstić, N., and Klein, D. J., *Comput. Chem.*, 14, 313, 1990; Trinajstić, N., Nikolić, S., and Klein, D. J., *J. Mol. Struct. (Theochem.)*, 229, 63, 1991.
23. Randić, M., Trinajstić, N., Knop, J. V., and Jeričević, Ž., *J. Am. Chem. Soc.*, 107, 849, 1985.
24. Randić, M. and Trinajstić, N., *Sulfur Rep.*, 6, 379, 1986.
25. Nikolić, S., Jurić, A., and Trinajstić, N., *Heterocycles*, 26, 2025, 1986.
26. Randić, M., Nikolić, S., and Trinajstić, N., in *Graph Theory and Topology in Chemistry*, King, R. B. and Rouvray, D. H., Eds., Elsevier, Amsterdam, 1987, 429.
27. Randić, M., Gimarc, B. M., Nikolić, S., and Trinajstić, N., *J. Mol. Struct. (Theochem.)*, 181, 111, 1988.
28. Randić, M., Nikolić, S., and Trinajstić, N., *Coll. Czech. Chem. Commun.*, 53, 2023, 1988.
29. Hess, B. A., Jr. and Schaad, L. J., *J. Am. Chem. Soc.*, 93, 305, 1971.
30. Ilić, P., Džonova-Jerman-Blažič, B., Mohar, B., and Trinajstić, N., *Croat. Chem. Acta*, 52, 35, 1979.
31. Glidevell, C. and Lloyd, D., *J. Chem. Educ.*, 63, 306, 1986.
32. Randić, M., *Int. J. Quantum Chem.*, 17, 549, 1980.
33. Randić, M. and Trinajstić, N., *J. Am. Chem. Soc.*, 106, 4428, 1984.
34. Randić, M., Plavšić, D., and Trinajstić, N., *J. Mol. Struct. (Theochem.)*, 183, 29, 1989.
35. Randić, M., Henderson, L. L., Stout, R., and Trinajstić, N., *Int. J. Quantum Chem.: Quantum Chem. Symp.*, 22, 127, 1988.
36. Nikolić, S., Randić, M., Klein, D. J., Plavšić, D., and Trinajstić, N., *J. Mol. Struct. (Theochem.)*, 198, 223, 1989.
37. Dewar, M. J. S. and de Llano, C., *J. Am. Chem. Soc.*, 91, 789, 1969.
38. Trinajstić, N., Plavšić, D., and Klein, D. J., *Croat. Chem. Acta*, 62, 711, 1989.
39. Plavšić, D., Nikolić, S., and Trinajstić, N., *Croat. Chem. Acta*, 63, 683, 1990.
40. Plavšić, D., Nikolić, S., and Trinajstić, N., *J. Math. Chem.*, 8, 113, 1991.
41. Dewar, M. J. S., Harget, A. J., and Trinajstić, N., *J. Am. Chem. Soc.*, 91, 6321, 1969.
42. Dewar, M. J. S. and Trinajstić, N., *J. Am. Chem. Soc.*, 92, 1453, 1970.
43. Clar, E., *Polycyclic Hydrocarbons*, Academic Press, London, 1964.
44. Watts, L., Fitzpatrick, J. D., and Pettit, R., *J. Am. Chem. Soc.*, 87, 3253, 1965.
45. Cava, M. P. and Mitchell, M. J., *Cyclobutadiene and Related Compounds*, Academic Press, New York, 1967.
46. Sondheimer, F., *Chimia*, 28, 163, 1974.
47. Dias, J. R., *Handbook of Polycyclic Hydrocarbons. Part A. Benzenoid Hydrocarbons*, Elsevier, Amsterdam, 1987.
48. Harthough, H. D., *Thiophene and Its Derivatives*, Interscience, New York, 1952.
49. Torres, M., Lown, E. M., Gunning, H. E., and Strausz, O. P., *Pure Appl. Chem.*, 52, 1623, 1980.
50. Krantz, A. and Laureni, J., *J. Am. Chem. Soc.*, 103, 486, 1981.
51. Dittmar, D. C., in *Comprehensive Heterocyclic Chemistry*, Vol. 7, Katritzky, A. R. and Rees, C. W., Eds., Pergamon Press, Oxford, 1984, 131.
52. Gronowitz, S., Ed., *Thiophene and Its Derivatives*, Parts I–III, John Wiley & Sons, New York, 1985–1986.
53. Csizmadia, I. G., Lucchini, V., and Modena, G., *Gazz. Chim. Ital.*, 108, 543, 1978.
54. Hess, B. A., Jr. and Schaad, L. J., *J. Am. Chem. Soc.*, 95, 3907, 1973.
55. Cyvin, S. J. and Gutman, I., *Kekulé Structures in Benzenoid Hydrocarbons*, Springer-Verlag, Berlin, 1988.
56. Seibert, J. and Trinajstić, N., *Int. J. Quantum Chem.*, 23, 1829, 1983.

57. **Klein, D. J., Seitz, W. A., and Schmalz, T. G.,** in *Computational Chemical Graph Theory,* Rouvray, D. H., Ed., Nova Science Publishers, New York, 1990, 127.
58. **Klein, D. J., Hite, G. E., and Schmalz, T. G.,** *J. Comput. Chem.,* 7, 443, 1986.
59. **Hite, G. E., Živković, T. P., and Klein, D. J.,** *Theoret. Chim. Acta,* 74, 349, 1988.
60. **Randić, M.,** *Chem. Phys. Lett.,* 128, 193, 1986.
61. **Randić, M., Gimarc, B. M., and Trinajstić, N.,** *Croat. Chem. Acta,* 59, 345, 1986.
62. **Randić, M., Solomon, V., Grossman, S. C., Klein, D. J., and Trinajstić, N.,** *Int. J. Quantum Chem.,* 32, 35, 1987.
63. **Randić, M., Nikolić, S., and Trinajstić, N.,** *Gazz. Chim. Ital.,* 117, 69, 1987.
64. **Randić, M., Plavšić, D., and Trinajstić, N.,** *Gazz. Chim. Ital.,* 118, 441, 1988.
65. **Randić, M., Gimarc, B. M., Nikolić, S., and Trinajstić, N.,** *Gazz. Chim. Ital.,* 119, 1, 1989.
66. **Nikolić, S. and Trinajstić, N.,** *Gazz. Chim. Ital.,* 120, 685, 1990.
67. **Coulson, C. A. and Longuet-Higgins, H. C.,** *Proc. R. Soc. London Ser. A,* 192, 16, 1947.
68. **Heilbronner, E.,** in *Non-Benzenoid Aromatic Compounds,* Ginsburg, D., Ed., Wiley-Interscience, New York, 1959, 171.
69. **Hess, B. A., Jr. and Schaad, L. J.,** *J. Am. Chem. Soc.,* 94, 3068, 1972.
70. **Randić, M.,** *Mol. Phys.,* 34, 849, 1977.
71. **Wilcox, C. F., Jr.,** *Tetrahedron Lett.,* 795, 1968.

Chapter 10

TOPOLOGICAL INDICES

A single number that can be used to characterize the graph of a molecule is called a *topological index*.[1] (The term *graph-theoretical index* would be more accurate than topological index, but the latter is more common in the chemical literature.) A topological index, thus, appears to be a convenient device for converting chemical constitution into a number. Evidently, this number must have the same value for a given molecule regardless of ways in which the corresponding graph is drawn or labeled. Such a number is referred to by graph theorists as a *graph invariant*[2] (see Chapter 2, Section II). For example, one of the simplest graph invariants (topological indices) is the number of vertices in the graph (the number of atoms in the molecule). Hence, it could be simply said that topological indices are graph invariants. It should also be pointed out that topological indices do not generally allow the reconstruction of the molecular graph, implying that a certain loss of information has occurred during their creation.

The interest[3,4] in topological indices is in the main related to their use in *nonempirical*[5] quantitative structure-property relationships (QSPR) and quantitative structure-activity relationships (QSAR). The latter use in such areas as pharmacology, toxicology, environmental chemistry, and drug design is intensively studied by many authors.[6-16]

I. DEFINITIONS OF TOPOLOGICAL INDICES

Topological indices were introduced (albeit unknowingly) 150 years ago,[17,18] and the very fact that they are still in use today is demonstration of their durability and versatility. There are more than 120 topological indices (including information-theoretic indices[19]) available to date in the literature,[4] with no sign that their proliferation will stop in the near future. This large (and every increasing[20-22]) number of topological indices indicates that perhaps a clear and unambiguous criterion for their selection and verification is still missing,[23] although some attempts along these lines have been reported.[24] Moreover, a large number of topological indices also lead to a question to what extent are they orthogonal? In other words, is it possible that some topological indices express predominantly the same type of constitutional information: the difference residing in the scaling factor? Several analyses on the example of alkane trees with up to 12 vertices indicate that a number of topological indices are strongly intercorrelated,[25-27] i.e., that many of them contain to a great extent the same type of structural information.

Most of the proposed topological indices are related to either a vertex adjacency relationship (connectivity) in the molecular graph G or to graph-theoretical (topological) distances in G. Therefore, the origin of topological indices can be traced either to the adjacency matrix of a molecular graph or to the distance matrix of a molecular graph. Furthermore, since the distance matrix can be generated from the adjacency matrix,[28] most of the topological indices are really related to the latter matrix.[29]

A. TOPOLOGICAL INDICES BASED ON CONNECTIVITY

Topological indices in this class are based either on the total sum of some combination of valencies of the adjacent vertices or on the graph spectrum. Several of these indices will be detailed below.

1. The Zagreb Group Indices

In the early work of the Zagreb group on the topological basis of π-electron energy,[30] two terms appeared in the approximate formula for the total π-energy of a conjugated system which may be used separately as topological indices,[31]

$$M_1 = \sum_i D^2(i) \tag{1}$$

$$M_2 = \sum_{\{i,j\}} D(i)D(j) \tag{2}$$

The symbol $D(i)$ stands for the valency of a vertex i. The sum in Equation 1 is over all vertices of G, while the sum in Equation 2 is over all edges in G.

These two indices can also be defined by means of the adjacency matrix A of G:[29]

$$M_1 = \sum_i (A^2)_{ii}(A^2)_{ii} \tag{3}$$

$$M_2 = \sum_{\{i,j\}} (A^2)_{ii}(A^2)_{jj} \tag{4}$$

It should also be noted that

$$D(i) = (A^2)_{ii} \tag{5}$$

2. The Connectivity Index

The connectivity index $\chi(G)$ of a graph G was introduced by Randić[32] and is similar to the Zagreb group indices. It was obtained after a search for the specific index to quantify the notion of molecular branching, unlike the indices $M_1(G)$ and $M_2(G)$, which appeared in the topological formula for the π-electron energy of conjugated systems. Nevertheless, the connectivity index has found considerable application in both QSPR and QSAR.[5,6,9-16,20,22,33-47] Actually $\chi(G)$ is the most used topological index in QSPR and QSAR to date.[11]

TABLE 1
Connectivity Weights for
Edge Types in the
Graphs Corresponding to
Carbon Compounds

Edge type $D(i),D(j)$	Connectivity weight $[D(i)\ D(j)]^{-1/2}$
1,1	1
1,2	0.7071
1,3	0.5773
1,4	0.5
2,2	0.5
2,3	0.4082
2,4	0.3536
3,3	0.3333
3,4	0.2887
4,4	0.25

The *connectivity index* $\chi = \chi(G)$ of a graph G was defined by Randić as a bond additive quantity[32]

$$\chi = \sum_{\{i,j\}} [D(i)D(j)]^{-1/2} \qquad (6)$$

Therefore, the connectivity index may be viewed as a sum of bond weights given in terms of $[D(i)\ D(j)]^{-1/2}$.

This index may also be redefined by means of the adjacency matrix $\mathbf{A}$ of a graph G in an obvious way:[29]

$$\chi = \sum_{\{i,j\}} [(\mathbf{A}^2)_{ii}(\mathbf{A}^2)_{jj}]^{-1/2} \qquad (7)$$

For carbon compounds, Equation 6 can be given in a closed form. In the graphs representing the skeletons of the carbon compounds there are possible only *four* types of vertices according to their valencies. These give rise to 10 edge types (bond types) whose connectivity weights are given in Table 1.

If the number of each edge type is denoted by b_{ij} ($i = 1, \ldots ,4; j = i, \ldots ,4$) and if the corresponding connectivity weights (given in Table 1) are used, then Equation 6 becomes

$$\chi = b_{11} + 0.7071\ b_{12} + 0.5773\ b_{13} + 0.5\ (b_{14} + b_{22})$$
$$+ 0.4882\ b_{23} + 0.3536\ b_{24} + 0.3339\ b_{33} \qquad (8)$$
$$+ 0.2887\ b_{34} + 0.25\ b_{44}$$

In this way, the computation of the connectivity index for carbon compounds transforms into the enumeration of their bond types.

In the case of polycyclic graphs representing the carbon skeletons of polycyclic conjugated hydrocarbons there are possible only vertices of *two* types, according to their valencies. They give rise to three edge types (bond types): b_{22}, b_{23}, and b_{33}. If only these three edge types are taken into account, Equation 8 for polycyclic conjugated hydrocarbons reduces to a rather simple expression:

$$\chi = 0.5\, b_{22} + 0.4082\, b_{23} + 0.3333\, b_{33} \tag{9}$$

The connectivity index was generalized in two ways. The first generalization was to include *weighted paths* p_ℓ of the length ℓ instead of only weighted edges (weighted paths p_ℓ of the length $\ell = 1$) in the graph:[48]

$$^\ell\chi = \sum_{\text{paths}} [D(i)D(j) \ldots D(\ell + 1)]^{-1/2} \tag{10}$$

where $D(i)$, $D(j)$, . . . ,$D(\ell+1)$ are valencies of vertices $i,j,$ . . . ,$\ell+1$ in the considered path of the length ℓ. From Equation 10 naturally follows the three connectivity indices most often used:[6,11]

(1) The zero-order connectivity index $^0\chi$

$$^0\chi = \sum_{i=1}^{N} [D(i)]^{-1/2} \tag{11}$$

where N is the total number of vertices in G or the total number of weighted paths p_0 of the length $\ell = 0$.

(2) The first-order connectivity index $^1\chi$

$$^1\chi = \sum_{i,j=1}^{M} [D(i)D(j)]^{-1/2} \tag{12}$$

where M is the total number of edges in G or the total number of weighted paths p_1 of the length $\ell = 1$. The first-order connectivity index is, of course, identical to the original connectivity index of Randić.

(3) The second-order connectivity index $^2\chi$

$$^2\chi = \sum_{i,j,k=1}^{P_2} [D(i)D(j)D(k)]^{-1/2} \tag{13}$$

(i)   A tree T corresponding to isopentane

T

(ii)   The zero-order connectivity index

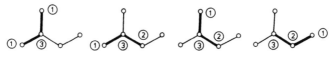

$^0\chi = 3 \cdot 1 + 0.5773 + 0.7071 = 4.2844$

(iii)   The first-order connectivity index

$^1\chi = 0.7071 + 2 \cdot 0.5773 + 0.4082 = 2.2699$

(iv)   The second-order connectivity index

$^2\chi = 3 \cdot 0.4082 + 0.5773 = 1.8019$

FIGURE 1.   The calculation of $^0\chi$, $^1\chi$, and $^2\chi$ indices for isopentane represented by a tree T. The encircled numbers stand for the valencies of the vertices in T.

where p_2 is the number of all paths of the length two in G. Higher-order connectivity indices may also be obtained in the same way from Equation 10 if needed.

The calculation of $^0\chi$, $^1\chi$, $^2\chi$ will be illustrated for isopentane in Figure 1.

The second generalization of the connectivity index was to include atoms other than carbon,[49] which represents an essential step if one wishes to consider the heteroatomic systems. The connectivity index, when it includes heteroatoms, is called the *valence connectivity index* and is denoted by χ^v. This index is defined as[49]

$$\chi^v = \sum_{edges} [\delta(i)\delta(j)]^{-1/2} \tag{14}$$

TABLE 2
Valence Delta Values (δ) for
Carbon, Nitrogen, and Oxygen
in Various Bonding
Arrangements

Atom i	δ_i
-CH$_3$	1
>CH$_2$	2
>CH	3
=CH-	3
=CH	3
>C-	4
=C<	4
=C=	4
=C-	4
-NH$_3^+$	2
-NH$_2$	3
>NH$_2^+$	3
>NH	4
=NH	4
>N-	5
=N-	5
=N	5
-OH$_2^+$	4
-OH	5
=OH$^+$	5
-O-	6
=O	6

where $\delta(i)$ and $\delta(j)$ are weights (valence delta values) of vertices (atoms) i and j making up the i-j edge (bond) in a vertex-weighted graph (heteroatomic system) G_{vw}. Valence delta values are given by

$$\delta(i) = (Z_i^v - H_i)/Z_i - Z_i^v - 1) \tag{15}$$

where Z_i^v stands for the number of valence electrons in the atom i, Z is its atomic number, and H_i is the number of hydrogen atoms attached to i. Valence delta values[11,49] for carbon, nitrogen, and oxygen in a given chemical environment are presented in Table 2.

The zero-order and higher-order valence connectivity indices can be calculated by means of Equations 10 to 13, if valencies of the vertices are replaced by their weights in terms of valence delta values. The first-order valence connectivity index $^1\chi^v$ is also given by Equation 14. As an example, the calculation of $^1\chi^v$ for ethylisopropylether is given in Figure 2.

(i)   A vertex-weighted tree T_{VW} corresponding to the molecular skeleton of ethylisopropylether

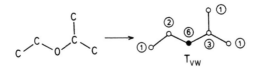

(ii)   The first-order valence connectivity index

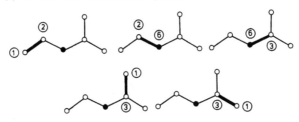

$$^1\chi^v = 0.7071 + 0.2887 + 0.2357 + 2 \cdot 0.5773 = 2.3861$$

FIGURE 2.   The calculation of $^1\chi^v$ for ethylisopropylether depicted by a vertex-weighted tree T_{VW}. The black dot denotes the position of oxygen. The vertex weights are given in terms of valence delta values of the corresponding atoms. In the case of carbon these numbers are identical to vertex valencies. The vertex-weights are encircled.

a. The Relationship Between the π-Electronic Energy and the Connectivity Index

As was already pointed out, the connectivity index $^1\chi = \chi$ represents the summation of bond weights $[D(i) D(j)]^{-1/2}$. These bond weights may also be viewed as *primitive bond orders*.[50] Since E_π in the normalized form of the Hückel molecular orbital (HMO) theory ($\alpha = 0$, $\beta = 1$) can be expressed in terms of bond orders p_{ij} as[51]

$$E_\pi = A' \sum_{ij} p_{ij} + B' \tag{16}$$

(where $A' = 2$ and $B' = 0$), one may replace $\sum_{ij} p_{ij}$ by $\sum_{ij} [D(i) D(j)]^{-1/2}$ and study the following relationship:

$$E_\pi = A \sum_{ij} [D(i)D(j)]^{-1/2} + B \tag{17}$$

or

$$E_\pi = A\chi + B \tag{18}$$

This assumption was tested[52] for 30 arbitrarily selected benzenoid hydrocarbons. The collection of these 30 benzenoids included all catafused benzenoids with five or less rings, hexacene, and selected pericondensed benzenoids from four to ten fused benzene rings. Their connectivity indices were computed using Equation 9. The relationship between the π-electronic energy and the connectivity index is obtained by regression analysis.

$$E_\pi(\text{benzenoids}) = 2.944(\pm 0.022)\chi - 1.080(\pm 0.223) \qquad (19)$$

with the following statistical parameters: $r = 0.999$, $s = 0.282$, and F = 18378 (r = correlation coefficient, s = standard deviation, and F = F-ratio).

Since the above statistical parameters exhibit a high level of significance of correlation, a similar analysis is also performed for 30 arbitrarily chosen nonbenzenoid hydrocarbons. This collection included alternant and nonalternant nonbenzenoids with and without exocyclic bonds. The following linear relationship between the π-electronic energy and the connectivity index is obtained by regression analysis:

$$E_\pi(\text{nonbenzenoids}) = 2.900(\pm 0.048)\chi - 1.151(\pm 0.293) \qquad (20)$$

with the statistical parameters ($r = 0.996$, $s = 0.477$, F = 3611) again of a high level of significance of the correlation.

Analogous analyses have also been carried out by several groups,[50,53] who used in their studies the valence connectivity index. Their results are much the same as the ones reported above: the π-electronic energy and the (valence) connectivity index are highly correlative quantities for polycyclic conjugated hydrocarbons.

b. The Degeneracy of the Connectivity Index

A major drawback of most topological indices is their *degeneracy*, i.e., two or more isomers possess the same topological index. The connectivity index is no exception. Two pairs of alkane trees (corresponding to 3-methylheptane and 4-methylheptane and to 2-ethyl-3-methylpentane and 3-ethyl-2-methylpentane) occur in the octane family with identical connectivity indices.[54] These structures are given in Figure 3.

The above finding did not harm the connectivity index. In spite of being a rather low-discriminatory index, it has been extensively used in QSPR and QSAR, because of its many attractive features, some of which have been presented earlier in this section.

3. The Connectivity ID Number

The identification (ID) number of a graph G was introduced by Randić[55] with an aim to design a topological index with a high discriminatory power.

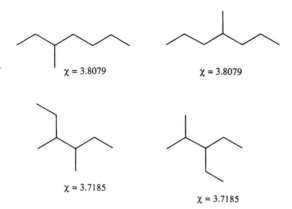

FIGURE 3. Two pairs of octane trees with identical values of the connectivity indices.

This number is based on the connectivity weights (given in Table 1) and path counts.

The connectivity weight $w(e)$ of an edge e in G is given by[32]

$$w(e) = w(i,j) = [D(i)D(j)]^{-1/2} \qquad (21)$$

Instead of taking into account weighted edges, one may consider weighted paths,

$$w(p_\ell) = \prod_{i=1}^{\ell} w(e_i) \qquad (22)$$

where p_ℓ is a weighted path of length ℓ and e_i $(i = 1,2, \ldots ,\ell)$ represents a set of weighted edges making up the path. The ID number is then given by[56]

$$ID = N + (1/2) \sum_{p_\ell} w(p_\ell) \qquad (23)$$

where N is the number of vertices in G and the summation is taken over all weighted paths in G. The factor 1/2 signifies that the ID number contains only *different* weighted paths. Because this number is obtained by means of connectivity weights, it is named the *connectivity ID number*.

In Table 3 as an example the calculation of the connectivity ID number for a graph G representing, for example, the π-system of methylenecyclopropene (see Figure 4) is given.

The connectivity ID number is a highly selective topological index, but is not a unique descriptor of a molecular graph. It has been found[56] that the

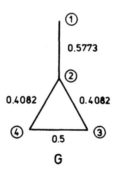

FIGURE 4.   A graph G depicting the π-system of methylenecyclopropene. The vertex labels are encircled. Numbers along the edges represent
connectivity weights.

TABLE 3
Calculation of the
Connectivity ID Number for a
Graph G in Figure 4

Vertex	Path p_ℓ	Length ℓ	Weight $W(p_\ell)$
1	1-2	1	0.5773
	1-2-3	2	0.2357
	1-2-4	2	0.2357
	1-2-3-4	3	0.1178
	1-2-4-3	3	0.1178
2	2-1	1	0.5773
	2-3	1	0.4082
	2-4	1	0.4082
	2-3-4	2	0.2041
	2-4-3	2	0.2041
3	3-2	1	0.4082
	3-4	1	0.5
	3-2-1	2	0.2357
	3-2-4	2	0.1666
	3-4-2	2	0.2041
	3-4-2-1	3	0.1178
4	4-2	1	0.4082
	4-3	1	0.5
	4-2-1	2	0.2357
	4-2-3	2	0.1666
	4-3-2	2	0.2041
	4-3-2-1	3	0.1178

Note: ID $= 4 + (1/2)\,6.3514 = 7.1757$

ID = 27.99646260308284997235

FIGURE 5. The smallest pair of alkane trees with the same connectivity ID number.

smallest pair of nonisomorphic trees (see Figure 5) that possess the same connectivity ID number appears in the family of pentadecanes. This family contains 4347 different trees.[57] In higher families of alkane trees with N >15, pairs of trees with the same connectivity ID number have also been found. For example, in the alkane family with 20 vertices (366,319 members)[57] there are 88 such pairs. There have also been found cases with three nonisomorphic alkane trees with the same connectivity ID number. The first such case is registered in the family of alkane trees with 18 vertices (60,523 members).[57]

There is one problem with the connectivity ID number. This is the enumeration of paths which becomes an obstacle, even if the enumeration is done on computer, when larger acyclic or polycyclic systems are considered.[58] This is so because the number of paths proliferates very fast with the size of a molecular graph. Therefore, the range of applicability of the connectivity ID number is rather limited because it cannot be easily computed for an arbitrary structure.

The connectivity ID number has been used successfully in several QSAR studies such as for classifying anticholinergic compounds,[59] in the search for optimum antitumor compounds,[60] for modeling the antitumoric activity of mitindomines,[61] for modeling the antiemetic activity of phenothiazines,[62] in the study of the antitumoric activity of phenyltriazines,[14,63,64] etc.

4. The Prime ID Number

The prime ID number was also introduced by Randić[65] with an aim to design an ID number with even greater selectivity than one that was exhibited by the connectivity ID number. The mathematical formulation of the prime ID number is exactly the same as that of the connectivity ID number (see Equation 23). The only difference is that now the connectivity weights are computed by means of the number 1 and the first nine prime numbers (2,3,5,7,11,13,17,19,23). The prime number weights are given in Table 4. Since this ID number is based on the prime number weights it is named the *prime ID number*. For example, the prime ID number for a graph G from Figure 4, with its prime weights shown in Figure 6, is given by

$$ID = 4 + (1/2)4.714 = 6.357 \qquad (24)$$

TABLE 4
Prime Number Weights

Edge type	Prime number weight
1,1	$(1)^{-0.5} = 1$
1,2	$(2)^{-0.5} = 0.7071$
1,3	$(3)^{-0.5} = 0.5773$
1,4	$(5)^{-0.5} = 0.4472$
2,2	$(7)^{-0.5} = 0.3780$
2,3	$(11)^{-0.5} = 0.3015$
2,4	$(13)^{-0.5} = 0.2773$
3,3	$(17)^{-0.5} = 0.2425$
3,4	$(19)^{-0.5} = 0.2294$
4,4	$(23)^{-0.5} = 0.2085$

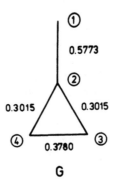

FIGURE 6.   A graph G from Figure 4. Numbers along the edges represent prime number weights.

The prime ID number appears to be a very selective topological index indeed, because in the field of all alkane trees with up to 20 vertices (618,050 trees)[57] only *one* case of degeneracy was found:[66] the smallest and only degenerate pair of alkane trees in this field is detected in the case of isomeric alkane trees with 20 vertices. This pair of alkane trees is shown in Figure 7.

The above result shows that the prime ID number is a highly discriminative topological index, much more discriminative than the connectivity ID number, but it is still not unique. However, a high discriminatory power of the prime ID number was expected[66] as the following discussion will show. For graphs depicting carbon compounds, the connectivity ID number is an element of the following field:

$$F = Q(\sqrt{2}, \sqrt{3}) \qquad (25)$$

while the prime ID number is an element of the field

$$F' = Q(\sqrt{2}, \sqrt{3}, \sqrt{5}, \sqrt{7}, \sqrt{11}, \sqrt{13}, \sqrt{17}, \sqrt{19}, \sqrt{23}) \qquad (26)$$

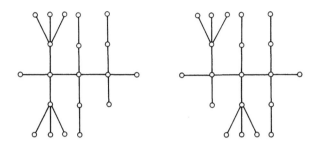

ID = 33.46070577034835559819

FIGURE 7. The smallest pair of alkane trees with the same prime ID number.

The basis of F as a vector space over the rational numbers Q has the dimension 4 ($= 2^2$), while the dimension of F' over Q is 512 ($= 2^9$).[67] Therefore, the representation of a molecular graph in F' should have a discriminatory power higher than in F. The numerical analysis in the field of alkane trees confirmed this expectation.

The prime ID number was not so far used in QSPR or QSAR.

5. The Largest Eigenvalue as a Topological Index

The largest eigenvalue, x_1, in the graph spectrum may be used as a topological index.[31,54,68] For example, it has been found[69] that x_1 can be employed as a measure of branching and that (alkane) trees can be well ordered according to x_1. In Figure 8 as an example, the ordering of alkane trees with seven vertices is shown. The smallest value of x_1 belongs to C_7 chain and the largest value of x_1 to the most branched C_7 alkane tree. The largest eigenvalue is not a very discriminative index, because in many cases the same x_1 value belongs to two (or more) nonisomorphic molecular graphs. One such degenerate pair appears also in the alkane trees shown in Figure 8.

The basis of the finding that x_1 appears to be a measure of branching is the following: x_1 can be interpreted as an approximate mean vertex valency

$$x_1 \approx (1/N) \sum_i D(i) \tag{27}$$

However, in the case of the regular graphs,

$$x_1 = D_{max} \tag{28}$$

Since the number of walks, $w(\ell)$, of the length ℓ in the regular graph are related exactly to the maximum vertex valency,[68]

$$w(\ell) = N(D_{max})^\ell \tag{29}$$

ALKANE TREE x_1

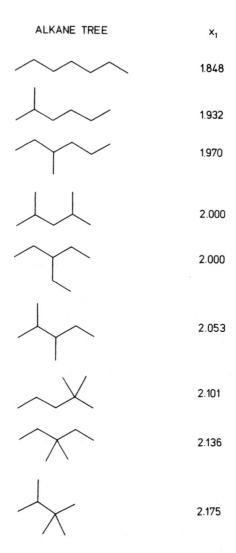

1.848

1.932

1.970

2.000

2.000

2.053

2.101

2.136

2.175

FIGURE 8.   The ordering of alkane trees with seven vertices according to the increasing value of x_1. This order follows the intuitive notion of branching.

the largest eigenvalue and $w(\ell)$ are also exactly related for regular graphs by virtue of Equation 27,

$$w(\ell) = N(x_1)^\ell \qquad (30)$$

For nonregular graphs this equation holds only approximately.

The above result is important because it reveals that there is a close relationship between spectral and combinatorial properties of a graph.

$$p(G;0) = 1$$
$$p(G;1) = 6$$
$$p(G;2) = 9$$
$$\underline{p(G;3) = 2}$$
$$Z(G) = 18$$

FIGURE 9. The computation of the Z-index for a graph G representing the π-system of benzene.

6. The Z-Index

The Z-index, $Z = Z(G)$, of a graph G was introduced by Hosoya.[1] This index is defined as

$$Z = \sum_{k=0}^{[N/2]} p(G;k) \tag{31}$$

where $p(G;k)$ is a selection of k mutually independent edges in G and $[N/2]$ is the maximum k number for G. It should be noted that $p(G;0) = 1$ by definition, $p(G;1) =$ the number of edges in G and $p(G;N/2) =$ the number of 1-factors of G. An example of computing the Z-index is given in Figure 9.

The $p(G;k)$ values appear as coefficients in the Z-counting polynomial $Q(G;x)$ of a graph G,[1]

$$Q(G;x) = \sum_{k=0}^{[N/2]} p(G;k)x^N \tag{32}$$

The Z-index may be obtained from this polynomial for $x = 1$,

$$Z = Q(G;x = 1) \tag{33}$$

For example, the Z-counting polynomial of naphthalene is

$$Q(\text{naphthalene};x) = 1 + 11x + 41x^2 + 61x^3 + 31x^4 + 3x^5 \tag{34}$$

and the corresponding Z-index,

$$Z = Q(\text{naphthalene};x = 1) = 148 \tag{35}$$

The acyclic polynomial $P^{ac}(G;x)$ is related to the Z-counting polynomial through the relationship

$$P^{ac}(G;x) = x^N Q(G;x = -x^{-2}) \tag{36}$$

For example, the above Z-counting polynomial of naphthalene converts directly into the corresponding acyclic polynomial via Equation 36,

$$P^{ac}(\text{naphthalene};x) = x^{10} - 11x^8 + 41x^6 - 61x^4 + 31x^2 - 3 \quad (37)$$

The Z-index may also be, of course, obtained from the acyclic polynomial:

$$Z = P^{ac}(G;|x| = 1) \quad (38)$$

The connection between the Z-counting polynomial and the adjacency matrix may be seen through its relationship with the characteristic polynomial.[70] For polyenes, represented by chains L_N, because of the identity between the characteristic polynomial and the acyclic polynomial (see Chapter 7), this relationship is rather simple,

$$P(L_N;x) = x^N Q(L_N;x = -x^{-2}) \quad (39)$$

and identical to Equation 36 when $G = L_N$. For annulenes, represented by cycles C_N, this relationship is also simple,

$$P(C_N;x) = x^N Q(C_N;x = -x^{-2}) - 2 \quad (40)$$

For polycyclic systems, represented by graphs G, it is somewhat more complicated,

$$P(G;x) = x^N Q(G;x = -x^{-2}) - 2x^{N-m} \sum_m Q(G - C_m;x = -x^{-2})$$

$$+ 4x^{N-(m+n)} \sum_{m>n} Q(G - C_m - C_n;x = -x^{-2}) + ... \quad (41)$$

where $G - C_m$, $G - C_m - C_n$, etc. are subgraphs obtained by the removal of cycles C_m, C_m and C_n, etc. from G. For example, the Z-counting polynomial of naphthalene (see Equation 34) converts via Equation 41 immediately into the characteristic polynomial of naphthalene:

$$P(\text{naphthalene};x) = x^{10} - 11x^8 + 41x^6 - 65x^4 + 43x^2 - 9 \quad (42)$$

The Z-indices have been extensively tabulated[71-73] and used in QSPR studies.[1,23,70,74,75]

B. TOPOLOGICAL INDICES BASED ON DISTANCES

The distance matrix $D = D(G)$ of the graph G contains information on distances in G. Early topological indices, which were used for relating the structure to properties of alkanes, were based on the distance matrix, though their authors were not aware of this fact.[76-80] Several indices based on distances will be described in this section.

1. The Wiener Number

The Wiener number W was introduced in 1947 by Wiener as the *path number*.[76] The path number was defined as the number of bonds between all pairs of atoms in an acyclic molecule. Wiener also introduced the *polarity number p* of an acyclic structure, which is equal to the number of pairs of atoms separated by three bonds. By using a linear combination of the path number and the polarity number, Wiener was able to obtain a fair prediction of alkane boiling points.[76] In subsequent studies,[78,79] Wiener has extended the application of the path number and the polarity number to other physical parameters of alkanes such as heats of formation, heats of vaporization, molar volumes, and molar refractions. In view of the pioneering contribution of Wiener in recognizing the significance of the number of paths in a molecular skeleton, the term *Wiener number*[23,43,54] or *Wiener index*[19,42-44] has been adopted for the sum of the distances in all chemical structures. One should note here that the one-to-one correspondence between the (graph-theoretical) distance and the path number holds only for acyclic systems. Since the graph-theoretical distance is defined as the smallest number of bonds, i.e., the shortest path between the pair of atoms in a structure (see Chapter 4, Section II), the distance and the smallest path number between atoms coincide in cyclic structures.

In the original work by Wiener, the Wiener number was not, strictly speaking, formulated in graph-theoretical terms. It was defined within the framework of chemical graph theory later in 1971 by Hosoya,[1] who pointed out that the Wiener number can be obtained from the distance matrix of a graph. The Wiener number W = W(G) of a graph G is defined as the half-sum of the elements of the distance matrix **D**:

$$W = \sum_{k=1}^{N} \sum_{\ell=1}^{k} (\mathbf{D})_{k\ell} = (1/2) \sum_{k=1}^{N} \sum_{\ell=1}^{N} (\mathbf{D})_{k\ell} \qquad (43)$$

where $(\mathbf{D})_{k\ell}$ represents off-diagonal elements of **D**. The polarity number $p = p(G)$ of a graph G is given by

$$p = (1/2) \sum_{i} (p_3)_i \qquad (44)$$

where p_3 is the number of the paths of length 3 or the number of off-diagonal elements of **D** with the distance 3. Originally described as a polarity number,[76] p is more properly related to steric aspects of a structure.[80] As an example, the computation of the Wiener number and the polarity number for 2-methyl-4-propylhexane (the corresponding tree of which is shown in Figure 10) is given below.

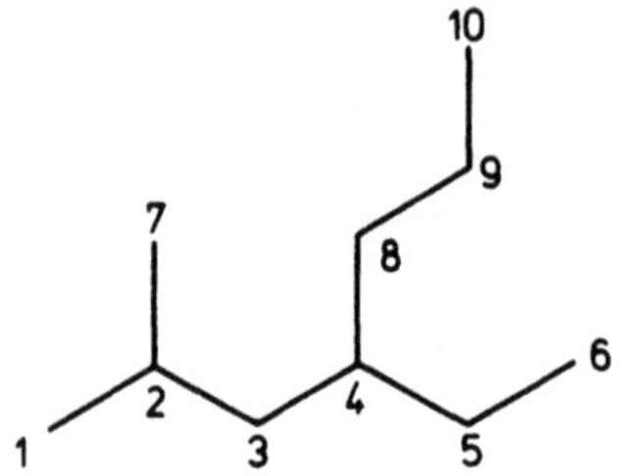

$$D(T) = \begin{bmatrix}
0 & 1 & 2 & 3 & 4 & 5 & 2 & 4 & 5 & 6 \\
1 & 0 & 1 & 2 & 3 & 4 & 1 & 3 & 4 & 5 \\
2 & 1 & 0 & 1 & 2 & 3 & 2 & 2 & 3 & 4 \\
3 & 2 & 1 & 0 & 1 & 2 & 3 & 1 & 2 & 3 \\
4 & 3 & 2 & 1 & 0 & 1 & 4 & 2 & 3 & 4 \\
5 & 4 & 3 & 2 & 1 & 0 & 5 & 3 & 4 & 5 \\
2 & 1 & 2 & 3 & 4 & 5 & 0 & 4 & 5 & 6 \\
4 & 3 & 2 & 1 & 2 & 3 & 4 & 0 & 1 & 2 \\
5 & 4 & 3 & 2 & 3 & 4 & 5 & 1 & 0 & 1 \\
6 & 5 & 4 & 3 & 4 & 5 & 6 & 2 & 1 & 0
\end{bmatrix}$$

T

$$\sum_{k=1}^{10} \sum_{l=1}^{10} 268 \quad \longrightarrow \quad W(T) = 134$$

FIGURE 10.   A tree T depicting 2-methyl-4-propylhexane and the computation of its Wiener and polarity numbers.

$$p_3 = 18 \quad \longrightarrow \quad p(T) = 9$$

The Wiener number of a graph G with N vertices can also be defined by means of the corresponding adjacency matrix:[29]

$$W = (1/2) \sum_{i=1}^{N} \sum_{\ell=1}^{\ell_{max}} (A_\ell^2)_{ii} \, \ell \tag{45}$$

where the matrix A_ℓ is defined as

$$(A_\ell)_{ij} = \begin{cases} 1 & \text{if the vertex } v_j \text{ is the } \ell\text{th} \\ & \text{neighbor of the vertex } v_i \\ 0 & \text{otherwise} \end{cases} \tag{46}$$

ℓ_{max} in Equation 45 corresponds to the length of the longest path in G.

Similarly, the polarity number can also be defined by means of the A_3 matrix

$$p = (1/2) \sum_{i=1}^{N} (A_3^2)_{ii} \tag{47}$$

The Wiener number appears to be a convenient measure of the compactness of the molecule.[35,42,54] The smaller the Wiener number, the larger the compactness of the molecule. This can be illustrated for the family of heptanes (see Figure 11).

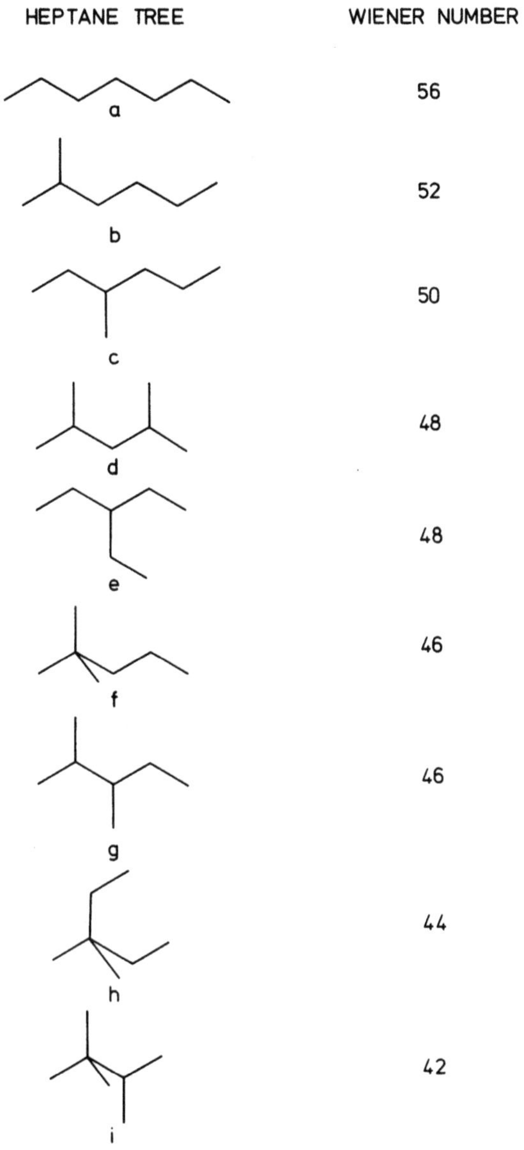

HEPTANE TREE   WIENER NUMBER

FIGURE 11.   Heptane trees and their Wiener numbers.

The least compact molecule is heptane (W[**a**] = 56) and the most compact molecule is 2,2,3-trimethylbutane (W[**i**] = 42). This result is also in agreement with the intuitive picture of these compounds. For two pairs of heptanes 2,4-dimethylpentane and 3-ethylpentane, and 2,2-dimethylpentane and 2,3-dimethylpentane, their Wiener numbers are identical (W[**d**] = W[**e**] = 48; W[**f**] = W[**g**] = 46). The meaning of this could be that the compactness of both of these pairs of C_7 alkanes is similar. In addition, it could signify that the degree of branching of these molecules is similar. Finally, it could be, for example, an accidental degeneracy of the Wiener numbers without any deeper meaning. Actually the pairs **d** and **e**, and **f** and **g** are the smallest molecular graphs with identical Wiener numbers. The Wiener number thus also appears to be a topological index of low degeneracy.[54]

For some classes of molecules the Wiener number can be computed by means of simple formulae.

(1) Linear alkanes with N atoms[54]

$$W = N(N^2 - 1)/6 \tag{48}$$

(2) Annulenes with N atoms[81]

$$W = \begin{cases} N^3/8 & N = \text{even} \\ N(N^2-1)/8 & N = \text{odd} \end{cases} \tag{49}$$

(3) Polyacenes with h number of hexagons[81,82]

$$W = (16h^3 + 36h^2 + 26h + 3)/3 \tag{50}$$

(4) Phenanthrenes and helicenes with h number of hexagons[81,82]

$$W = (8h^3 + 72h^2 - 26h + 27)/3 \tag{51}$$

The Wiener number is a rather popular index in the QSPR modeling.[3,4,10,12,13,35,37,42-44,54,76,78,79,81-93] A typical example of the QSPR model with the Wiener number is the relationship between boiling points (bp) of alkanes and their W numbers. A relationship of the form[94]

$$bp = a \, W^b + c \tag{52}$$

produced a very good agreement between $(bp)_{exp}$ and $(bp)_{calc}$. The statistical parameters (a, b, and c) were determined by performing a nonlinear least-squares recursion. The statistical characteristics of the above correlation are: $n = 150$ (all alkanes from C_1 to C_{10}), $a = 101.01$ (± 7.03), $b = 0.24627$ (± 0.000979), $c = -174.45$ (± 8.53), $r = 0.983$, $s = 8.9$, F = 2157. For purposes of comparison, the results for the correlation between the boiling points and the connectivity indices of alkanes are reported below. A relationship of the form of Equation 52 produced comparable statistical parameters: $n = 150$, $a = 117.72$ (± 6.26), $b = 0.6918$ (± 0.0227), $c = -176.61$ (± 7.21), $r = 0.988$, $s = 7.5$, F = 3116. If one uses the polynomial-type relationship between bp and χ than the statistical parameters can be improved.[94] Similarly, the relationship between bp and W improves if the polarity number is included in the correlation.[42]

2. The Platt Number

Platt[77,80] was also interested in devising a scheme for predicting physical parameters (molar volumes, boiling points, heats of formation, heats of vaporization) of alkanes. He introduced an index $F = F(G)$, which is equal to the total sum of edge-degrees in a graph G. The *edge-degree* of an edge e, $D(e)$, is the number of its adjacent edges. This index was named the *Platt number*.[10] The Platt number of a graph G is defined by

$$F = \sum_{i=1}^{M} D(e_i) \tag{53}$$

One may also regard the edge-degrees as edge-weights. Then, the Platt number may be considered as the sum of the weighted distances of the length one. The Platt number may also be defined by means of the adjacency matrix of G:[29]

$$F = \sum_{\{i,j\}} [(\mathbf{A}^2)_{ii} + (\mathbf{A}^2)_{ij} - 2] \tag{54}$$

Combining Equations 5 and 54, the Platt number may be reformulated in terms of the vertex-valencies,

$$F = \sum_{\{i,j\}} [D(i) + D(j) - 2] \tag{55}$$

The Platt number, given as Equation 55, then belongs to the preceding class of topological indices. Additionally, it reveals that the most of the topological indices may always be traced back to the adjacency matrix of a molecular graph.

3. The Gordon-Scantlebury Index

The Gordon-Scantlebury index,[95] $S = S(G)$, is defined as the number of distinct ways in which the chain fragment on three vertices, L_3, called *links*, may be removed from a graph G representing a carbon skeleton of a molecule:

$$S = \sum_{i} (L_3)_i \tag{56}$$

The Gordon-Scantlebury index thus represents the number of all paths of length two in G. Computation of the Gordon-Scantlebury index for a tree T is shown in Figure 12.

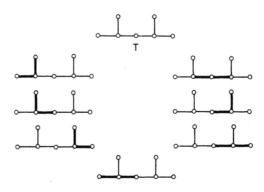

FIGURE 12.   The enumeration of the L_3 fragments in a tree T representing 2,4-dimethylpentane. The Gordon-Scantlebury index is the sum of all L_3 fragments, i.e., S(T) = 7.

The Gordon-Scantlebury index may also be defined by means of the adjacency matrix[29]

$$S = (1/2) \sum_{i=1}^{N} (\mathbf{A}^2)_{ii}(\mathbf{A}^2)_{ii} - M \qquad (57)$$

where M is the number of edges in G. Thus, the Gordon-Scantlebury index may also be expressed in terms of the vertex-valencies via Equation 5:

$$S = (1/2) \sum_{i} (D(i) \cdot D(i)) - M \qquad (58)$$

The Platt number and the Gordon-Scantlebury index are closely related quantities. It can be shown that the S index is equal to half of the value of F,

$$S = (1/2)F \qquad (59)$$

Similarly, it can be shown that S and M_1 indices are also related by

$$S = (1/2)M_1 - M \qquad (60)$$

4. The Balaban Index

Balaban[96-99] has proposed a topological index, denoted by J(G) and named the *Balaban index*,[100,101] which can be described as the average distance sum connectivity. The Balaban index J = J(G) of a graph G is defined as follows:[96]

$$J = \frac{M}{\mu + 1} \sum_{\{i,j\}} (d_i d_j)^{-1/2} \qquad (61)$$

where d_i ($i = 1,2, \ldots ,N$) is the distance sum, M the number of edges in G, and μ is the cyclomatic number of G. The *distance sum, d_i,* for a vertex i represents a sum of all entries in the corresponding row (or column) of the distance matrix $\mathbf{D}$,[96]

$$d_i = \sum_{j=1}^{N} (\mathbf{D})_{ij} \qquad (62)$$

The *cyclomatic number* (or the *circuit rank*), $\mu = \mu(G)$, of a polycyclic graph G is equal to the minimum number of edges necessary to be removed from G in order to convert it to the related acyclic graph,[102]

$$\mu = M - N + 1 \qquad (63)$$

Below, as an example, the computation of the Balaban index for a labeled bicyclic graph G depicting, for example, the π-system of benzocyclobutadiene (see Figure 13) is given.

In order to consider multigraphs (molecular structures with multiple bonds), Balaban[96] has suggested that the entries into the distance matrices of such graphs should be in terms of bond orders instead of graph-theoretical distances. Thus, the entry for a single connection (a single bond) should be 1, the entry for a double connection (a double bond) 1/2, etc. Barysz et al.[100] and Balaban et al.[99,103] have also extended the use of J to heterosystems by defining the distance matrix for weighted graphs.

The distance sum d_i was also separately used as a topological index, under the name the *distance sum index*.[104] This index appears to be a convenient measure of the *compactness* or *centrality* of a particular site in a molecule. The distance sum index may also be defined by means of the adjacency matrix of the corresponding graph:[29]

$$d_i = \sum_{\ell=1}^{\ell_{max}} (A_\ell^2)_{ii} \ell \qquad (64)$$

where the matrix $\mathbf{A}_\ell$ is defined by Equation 46. In this way the Wiener number can also be expressed in terms of the distance sums

$$W = (1/2) \sum_{i=1}^{N} d_i \qquad (65)$$

In the same way the Balaban index of G may be expressed via Equation 64 in terms of the adjacency matrix of G,[29]

$$J = \frac{M}{\mu + 1} \sum_{\{i,j\}} \left[\sum_{\ell=1}^{\ell_{max}} \sum_{\ell'=1}^{\ell'_{max}} (A_\ell^2)_{ii} \ell \, (A_{\ell'}^2)_{jj} \ell' \right]^{-1/2} \qquad (66)$$

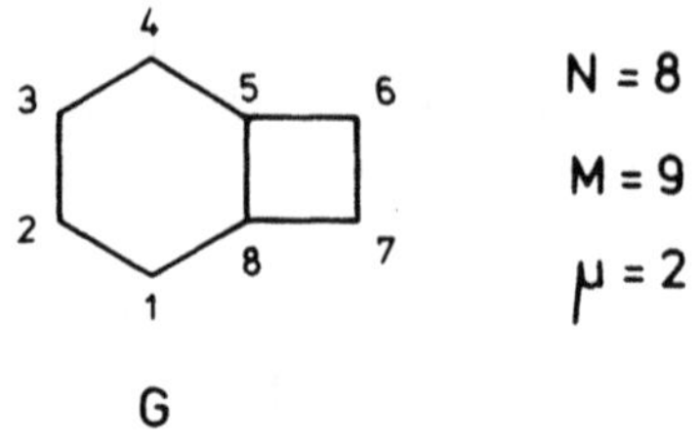

$N = 8$

$M = 9$

$\mu = 2$

G

FIGURE 13.   A labeled bicyclic graph G representing the π-system of benzocyclobutadiene and the computation of its Balaban index.

$$D = \begin{array}{c|cccccccc}
 & \textcircled{1} & \textcircled{2} & \textcircled{3} & \textcircled{4} & \textcircled{5} & \textcircled{6} & \textcircled{7} & \textcircled{8} \\
\hline
\textcircled{1} & 0 & 1 & 2 & 3 & 2 & 3 & 2 & 1 \\
\textcircled{2} & 1 & 0 & 1 & 2 & 3 & 4 & 3 & 2 \\
\textcircled{3} & 2 & 1 & 0 & 1 & 2 & 3 & 4 & 3 \\
\textcircled{4} & 3 & 2 & 1 & 0 & 1 & 2 & 3 & 2 \\
\textcircled{5} & 2 & 3 & 2 & 1 & 0 & 1 & 2 & 1 \\
\textcircled{6} & 3 & 4 & 3 & 2 & 1 & 0 & 1 & 2 \\
\textcircled{7} & 2 & 3 & 4 & 3 & 2 & 1 & 0 & 1 \\
\textcircled{8} & 1 & 2 & 3 & 2 & 1 & 2 & 1 & 0 \\
\end{array}
\begin{array}{l}
d_1 = 14 \\
d_2 = 16 \\
d_3 = 16 \\
d_4 = 14 \\
d_5 = 12 \\
d_6 = 16 \\
d_7 = 16 \\
d_8 = 12 \\
\end{array}$$

$J = 1.9215$

The Balaban index is a highly discriminating index, but is not unique.[105] For example, a pair of trees T_1 and T_2 (see Figure 14) possess the same value of the Balaban index $J = 3.7524$. The Balaban index has been used in several QSPR and QSAR studies.[20,41,96-98]

5. The Information-Theoretic Indices

Bonchev[19] and Bonchev and Trinajstić[54] have applied information theory to the problem of characterizing molecular structures.[84,85,106-111] Information theory provides a simple quantitative measure called the *information content* of a system. The information content $I = I(S)$ of a system S with N elements is defined by[112]

$$I = N \log_2 N - \sum_{i=1}^{n} N_i \log_2 N_i \qquad (67)$$

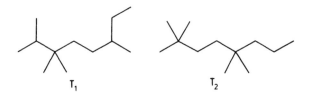

FIGURE 14.   A pair of trees T_1 and T_2, corresponding to 6-ethyl-2,3,3-trimethylheptane and 2,2,5,5-tetramethyloctane, which possess the same Balaban index.

where n is the number of different sets of elements, N_i is the number of elements in the ith set of elements, and the summation is over all sets of elements. The logarithm is taken at basis 2 for measuring the information content in *bits*.

Another information-theoretic measure is the *mean information content,* $\bar{I} = \bar{I}(S)$, of one element of the system, defined by means of the total information content[113]

$$\bar{I} = I/N = -\sum_{i=1}^{N} P_i \log_2 P_i \tag{68}$$

where $P_i = N_i/N$. This formula is known as the *Shannon equation.*[112]

The application of information theory to different systems or structures is based on the possibility of constructing a finite probability scheme for every system. For example, the distance matrix **D** of a given graph G has a convenient structure for the application of the information theory. The elements of the distance matrix $(\mathbf{D})_{ij}$ can be considered as the elements of a finite probability scheme associated with G. A certain distance of a value i $(1 \leq i \leq N-1)$ appears $2a_i$ times in the distance matrix. All elements of **D** may be partitioned into $n+1$ groups. The $n+1$ group contains only diagonal elements of **D** which are, of course, all equal to zero. Since the distance matrix is a symmetric matrix, the upper-triangle part of it preserves all the information about the system. The total number of elements in the upper triangle of **D** is $N(N-1)/2$.

Analysis of the distance matrix in terms of information theory produced the following topological indices:[19,54,114,115]

1. The information-theoretic indices, $I_D^E = I_D^E(G)$ and $\bar{I}_D^E = \bar{I}_D^E(G)$, containing the total and mean information on distances in a molecular graph G with N vertices:

$$I_D^E = \frac{N(N-1)}{N} \log_2 \frac{N(N-1)}{N} - \sum_{i=1}^{n} a_i - \log_2 a_i \tag{69}$$

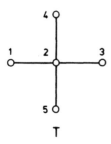

$$D = \begin{bmatrix} 0 & 1 & 2 & 2 & 2 \\ 1 & 0 & 1 & 1 & 1 \\ 2 & 1 & 0 & 2 & 2 \\ 2 & 1 & 2 & 0 & 2 \\ 2 & 1 & 2 & 2 & 0 \end{bmatrix}$$

FIGURE 15. A tree T depicting 2,2-dimethylpropane and its information-theoretic indices.

$$N = 5; \; N(N-1)/2 = 10; \; i = 1,2; \; a_1 = 4; \; a_2 = 6; \; W = 16$$

$$\bar{I}_D^E = 10 \log_2 10 - \{4 \log_2 4 + 6 \log_2 6\} = 9.71 \text{ bits}$$

$$\bar{I}_D^E = 9.71/10 = 0.971 \text{ bits}$$

$$I_D^W = 16 \log_2 16 - \{4 \cdot 1 \log_2 1 + 6 \cdot 2 \log_2 2\} = 52 \text{ bits}$$

$$I_D^W = 52/16 = 3.25 \text{ bits}$$

$$\bar{I}_D^E = \frac{2I_D^E}{N(N-1)} = -\sum_{i=1}^{n} \frac{2a_i}{N(N-1)} \log_2 \frac{2a_i}{N(N-1)} \tag{70}$$

2. The information-theoretic indices, $I_D^W = I_D^W(G)$ and $\bar{I}_D^W = \bar{I}_D^W(G)$, containing total and mean information on the distribution of the distances in a molecular graph G

$$I_D^W = W \log_2 W - \sum_{i=1}^{n} a_i \, i \log_2 i \tag{71}$$

$$\bar{I}_D^W = I_D^W/W = -\sum_{i=1}^{n} a_i(i/W)\log_2(i/W) \tag{72}$$

These indices in fact contain the information on the partitioning of the Wiener number into groups of distances of the same length.

The computation of these indices for a tree T (shown in Figure 15) is exemplified below. Only the upper triangle of the corresponding distance matrix is considered.

There are also a number of other information-theoretic indices available in the literature[19,111,115-117] and they have shown great potential for use in QSAR modeling.[16,118,119]

FIGURE 16. A tree T depicting 3-methylpentane.

6. The Bertz Index

Bertz[120-124] has proposed an elegant information-theoretic index, named the *Bertz index,*[19] for computing the complexity of molecules and reaction schemes. The Bertz index, denoted by $C(n)$, is given in terms of the graph invariant n,[120]

$$C(n) = 2n \log_2 n - \sum n_i \log_2 n_i \tag{73}$$

This formula is based on the reasonable assumption that the complexity of a molecular graph should increase with the size, branching, vertex-weights, edge-weights, etc.

Although n can represent any graph invariant, the choice of the specific invariant is governed by the problem at hand.

Bertz[120] has shown that the number of pairs of the adjacent edges in a molecular graph G is a very good choice for n, because it reflects both the *size* and the *symmetry* (in terms of equivalent pairs of the adjacent edges) of G. For example, the Bertz (complexity) index for a tree T depicting 3-methylpentane (see Figure 16) is given by

$$C(n) = 2 \cdot 5 \log_2 5 - 2 \cdot 2 \log_2 2 = 19.22 \tag{74}$$

The complexity index of Bertz has not so far been used in the structure-property or structure-activity correlations.

7. The Centric Index

In order to have a topological index which reflects the *shape* of alkanes, Balaban[125] has introduced the *centric index*. The maximum possible distance r_i between the vertex v_i and any other vertex v_j in a graph G is given by

$$r_i = \max_{v_j \in G} d(v_i, v_j) \tag{75}$$

Then

$$r = \min_{v_i \in G} r_i \tag{76}$$

is called the *radius* of the graph and each vertex v_i such that $r_i = r$ is called a *central vertex*. The *center* of G is the set of all central vertices. A classical result of the formal graph theory is that any tree has a center consisting of either one vertex or two adjacent vertices.[2]

The normalized centric index $C = C(T)$ of a tree is given by[125]

$$C = (1/2)[\beta(T) - 2N + U] \tag{77}$$

where

$$\beta(T) = \sum_i \delta_i^2 \tag{78}$$

$$U = [1 - (-1)^N]/2 = \begin{cases} 0 & N = \text{even} \\ 1 & N = \text{odd} \end{cases} \tag{79}$$

$\beta(T)$ represents a pruning sequence. Under the pruning of a tree, it is understood a stepwise process of removing all vertices of valency one (end-points) from T together with their incident edges until the center of T is found. The number of vertices, δ_j, removed at each step constitute the pruning sequence of a tree.

The centric index of a tree T may also be defined as

$$C = 1/2[\beta(T) - \beta(\text{chain})] \tag{80}$$

where β(chain) is the pruning sequence for a chain (*n*-alkane) with the same number of vertices as a corresponding branched tree (isomeric branched alkane). In Figure 17 the computation of the centric index for 2,4-dimethylpentane is shown.

The above procedure for determining the center of a tree is not applicable to polycyclic graphs. For polycyclic graphs, no simple procedure for determining the graph center could be found. However, Bonchev et al.[126,127] have succeeded in generalizing the concept of the graph center and have developed a procedure for determining the center of any acyclic or cyclic graph. Their procedure is based on the distance matrix of a graph. They have also referred to the center of a polycyclic graph G as the *polycenter* of G.

The polycenter of G is a vertex (or a set of vertices) which can be identified by the following procedure:

1. The radius r of a graph G should be minimum for a vertex v_i to be the center of G. This vertex can be identified by inspection of the distance matrix G. In the row of the matrix corresponding to this vertex the value of the maximum distance will be the smallest in comparison with all other vertices of G.

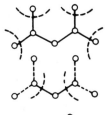

N = 7

1. Pruning process

Number of endpoints
removed at each pruning step

4

2

1

2. Center of a tree T

3. Pruning sequence

(4,2,1)

4. Centric index

$B(T) = 4^2 + 2^2 + 1^2 = 21$

$U = [1 - (-1)^7]/2 = 1$

$C(T) = (1/2)(21 - 13) = 4$

5. Calculation of the centric index by means of equation (80)

    * $B(T) = 21$

    * 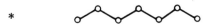 Chain with N = 7

    * Pruning process gives the following pruning sequence: (2,2,2,1)

    * $B(chain) = 13$

    * $C(T) = (1/2)(21 - 13) = 4$

FIGURE 17. The computation of the centric index for a tree T depicting 2,4-dimethylpentane.

2. If there are two or more vertices with the same value of the radius, the polycenter can be identified by means of the *distance sum index, d_i*. The smallest value of d_i points to a vertex which has total shortest distances to any other vertex, which appears to be a reasonable generalization of the concept of the graph center.

3. If both of the above criteria did not identify the graph polycenter, the *distance code* may be used for this purpose. The distance code represents the frequency of appearance of distances in a given row of the distance matrix. Among two or more vertices which have the same distance sum index, it is acceptable to discriminate on the basis of the frequency of distances and to take the vertex with the minimum number of larger distances as more central.

4. Application of the criteria (1 to 3) may still result in a situation that several nonequivalent vertices qualify as the graph center. Such vertices are referred to as the *pseudocenter*. The vertices in the pseudocenter can be used to construct a subgraph G' of G. Then criteria 1 to 3 are repeated on G'.

The determination of the graph polycenter by the above procedure is shown in Figure 18.

8. The Weighted ID Number

The weighted identification number was introduced by Szymanski et al.[58] and was denoted as the WID number. The incentive for the development of the WID number was the ID number introduced by Randić[55] and its successful use in QSAR studies.[59-64] However, isomeric trees were found with the same ID number in both formulations, i.e., as the connectivity ID number[56] and the prime ID number.[66] This fact stimulated Szymanski et al.[58] to search for an ID number with much higher selectivity than the two previous versions. In addition, there is another problem with the connectivity ID number and the prime ID number, that is the difficulty of enumerating the number of paths. This was then an additional constraint on the novel ID number.

The *weighted ID number* is defined in the following way.[14,58] The starting points are the distance matrix **D** of a graph G and the distance sums in **D** (see Equation 62). The weights of edges w_{ij} in G are defined as

$$w_{ij} = \begin{cases} (d_i d_j)^{-1/2} & \text{if vertices } v_i \text{ and } v_j \text{ are adjacent} \\ 0 & \text{otherwise} \end{cases} \qquad (81)$$

They represent the elements of the matrix of edge-weights, $\mathbf{W} = \mathbf{W}(G)$. Let $w = (v_{i_1}, v_{i_2}, \ldots, v_{i_{\ell+1}})$ be a walk of length ℓ. The weight of this walk is defined by

$$\prod_{j=1}^{\ell} (d_{i_j} d_{i_{j+1}})^{-1/2} = \prod_{j=1}^{\ell} w_{i_j i_{j+1}} \qquad (82)$$

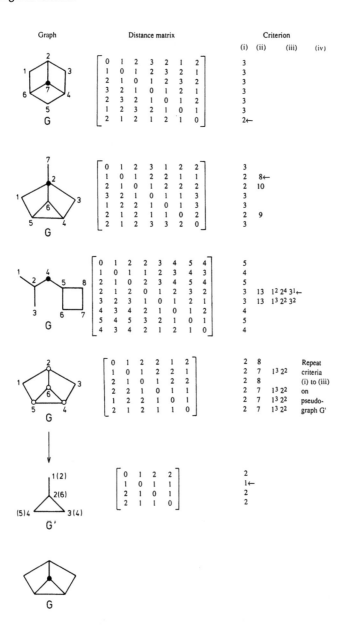

FIGURE 18. The identification of polycenters of polycyclic graphs by means of the procedure by Bonchev et al.[126,127] Black dots denote polycenters, whereas circles denote pseudocenters.

The weight of all walks of length ℓ between vertices v_{i_1} and $v_{i_{\ell+1}}$ is given by

$$\sum_i \prod_{j=1}^{\ell} w_{i_j i_{j+1}} \tag{83}$$

The entry w^ℓ of $\mathbf{W}^\ell$ is the sum of the all weighted walks of length ℓ. The WID number of G is then defined by means of an auxiliary ID number:

$$ID(G)^* = \sum_{i=1}^{N} \sum_{j=1}^{N} w_{ij}^* \tag{84}$$

where

$$(w_{ij}^*)_{1 \le i,j \le N} = \mathbf{W}^* \tag{85}$$

and

$$\mathbf{W}^* = \sum_{\ell=0}^{N-1} \mathbf{W}^\ell = \mathbf{I} + \sum_{\ell=1}^{N-1} \mathbf{W}^\ell \tag{86}$$

It can be seen that if $\mathbf{W} = \mathbf{0}$ (the zero matrix), then $\mathbf{W}^* = \mathbf{I}$ (the unit matrix) and

$$\sum_{i=1}^{N} \sum_{j=1}^{N} w_{ij}^* = N \tag{87}$$

In the case of $G = K_N$ (K_N = the complete graph)

$$ID^*(K_N) = N^2 \tag{88}$$

Therefore, the limits of the ID* number are

$$N \le ID^*(G) \le N^2 \tag{89}$$

for each graph G with N vertices. In other words, the ID* number is defined for all graphs from the null graph to the complete graph.

The weighted ID number is then defined as

$$WID = N - (1/N) + (1/N)^2 ID^*(G) \tag{90}$$

The limits of the WID number are

$$N \le WID < N + 1 \tag{91}$$

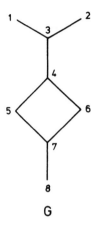

FIGURE 19.   A graph G depicting 2-isopropyl-1-methylcyclobutane.

The computation of the WID number, therefore, proceeds via the following scheme:

$$\mathbf{D} \to d_i \to \mathbf{W} \to \mathbf{W}^* \to \mathbf{ID}^* \to \mathbf{WID} \tag{92}$$

As an example, the computation of the WID number for a graph G (see Figure 19) representing 2-isopropyl-1-methylcyclobutane is shown in Table 5.

The WID number was tested for degeneracy on several classes of graphs.[58] In the field of 618,050 alkane trees (all alkane trees with up to 20 vertices), not a single degenerate pair was found. The WID number was also tested on the field of 1,346,024 trees (all trees with up to 20 vertices). Here was also the same outcome, no degenerate pairs were found. Then the WID was tested on the field of 38,472 (all strictly) planar polyhexes without holes, with up to 10 hexagons. Again, no degenerate pair was detected. Finally, the WID number was tested in the field of all systematically generated connected graphs with up to a certain number of vertices. In this case two pairs of graphs with six vertices were found with the same WID number. These graphs are given in Figure 20. Therefore, the WID number appears indeed to be an extremely discriminative index. Probably it is the most discriminative index to date among the known topological indices. The computation of the WID number is also simpler than the computation of either the connectivity ID number or the prime ID number, because it is much easier to enumerate (weighted) walks than (weighted) paths. The WID number is so far rather sparsely used in QSPR and QSAR studies.[14,128,129]

9. The Schultz Index

Schultz[130] has introduced a topological index, named the *Schultz index*,[131] for characterizing alkanes by an integer. This index, denoted by MTI (an

TABLE 5
Computation of the WID Number for a Graph (See Figure 17)
Depicting 1-Methyl-Isopropyl Cyclobutane

(1) The distance matrix and the distance sums of G

	①	②	③	④	⑤	⑥	⑦	⑧	
①	0	2	1	2	3	3	4	5	20
②	2	0	1	2	3	3	4	5	20
③	1	1	0	1	2	2	3	4	14
④	2	2	1	0	1	1	2	3	12
⑤	3	3	2	1	0	2	1	2	14
⑥	3	3	2	1	2	0	1	2	14
⑦	4	4	3	2	1	1	0	1	16
⑧	5	5	4	3	2	2	1	0	22

$D(G) =$ (rows ① through ⑧ as above)

(2) The matrix of edge-weights

$W(G) =$

0	0	0.05976	0	0	0	0	0
0	0	0.05976	0	0	0	0	0
0.05976	0	0.05976	0	0.07715	0	0	0
0	0	0.07715	0	0.07715	0.07715	0	0
0	0	0	0.07715	0	0	0.06681	0
0	0	0	0.07715	0	0	0.06681	0
0	0	0	0	0.06681	0.06681	00.0533	0
0	0	0	0	0	0	0.05330	0

(3) The matrix of weighted walks

$$W*(G) = \begin{bmatrix}
1.00362 & .00362 & .06056 & .00473 & .00037 & .00037 & .00005 & .000003 \\
.00362 & 1.00362 & .06056 & .00473 & .00037 & .00037 & .00005 & .000003 \\
.06056 & .06056 & 1.01334 & .07913 & .00615 & .00615 & .00082 & .00004 \\
.00473 & .00473 & .07913 & 1.01832 & .07927 & .07927 & .01061 & .00057 \\
.00037 & .00037 & .00615 & .07927 & 1.01068 & .01068 & .06843 & .00365 \\
.00037 & .00037 & .00615 & .07927 & .01068 & 1.01068 & .06843 & .00365 \\
.00005 & .00005 & .00082 & .01061 & .06843 & .06843 & 1.01201 & .05394 \\
.000003 & .000003 & .00004 & .00057 & .00365 & .00365 & .05394 & 1.00287
\end{bmatrix}$$

(4) The ID* number
     ID*(G) = 9.2883

(5) The WID number
     WID(G) = 8 − (1/8) + (1/64) 9.2883 = 8.0201

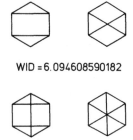

WID = 6.094608590182

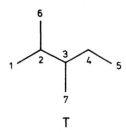

WID = 6.123192717321

FIGURE 20.   Two smallest pairs of graphs with the same WID numbers.

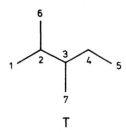

FIGURE 21.   A labeled tree T representing 2,3-dimethylpentane.

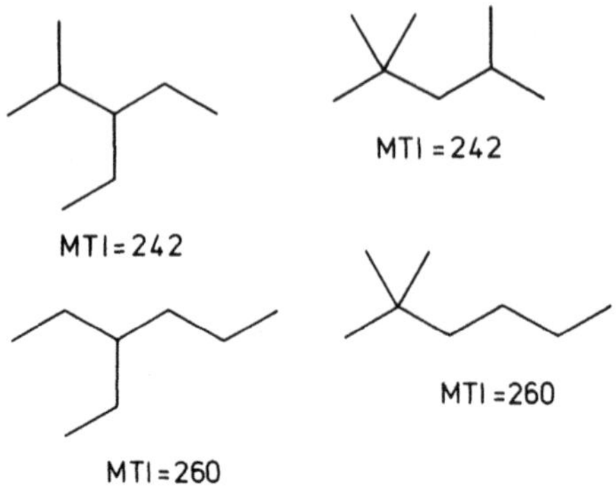

FIGURE 22. Two pairs of octane trees with identical Schultz indices.

acronym for *molecular topological index*), is based on the adjacency (N × N) matrix, **A**, the distance (N × N) matrix, **D**, and the valency (1 × N) matrix, **v**, of an alkane tree. The sum of elements $e_i(i = 1,2, \ldots N)$ of the row matrix **v**[**A** + **D**] gives the Schultz index

$$MTI = \sum_{i=1}^{N} e_i \qquad (93)$$

The computation of the MTI index for a tree T (see Figure 21), depicting 2,3-dimethylpentane, is given in Table 6.

The Schultz index has been shown to be a useful topological index for QSPR modeling,[130] but it is not a particularly discriminative index. Already in the octane family (18 members) appear two pairs of trees (corresponding to 3-ethyl-2-methylpentane and 2,2,4-trimethylpentane and to 3-ethylhexane and 2,2-dimethylhexane) with the same MTI values.[132] These alkane trees are given in Figure 22.

One might be tempted to redefine the MTI. The comparison between matrices **A** + **D** and **D** indicates that the distance matrix **D** is a richer mathematical structure (possesses higher information content)[54] than the matrix **A** + **D** because in the latter matrix there is not any distinction made between the distances of length 1 and 2. Therefore, one might redefine the MTI to be a sum of elements $e_i(i = 1,2, \ldots ,N)$ of the row matrix **v** **D** and denote it by (MTI)′. For example, this index for a 2,3-dimethylpentane tree is 144.

The (MTI)′ was hoped to be more discriminative than the MTI because the distance matrix **D** has higher information content than the sum matrix **A** + **D**, but this conjecture was not supported by numerical analysis. Already in the heptane family two pairs of nonisomorphic trees corresponding to 2,3-dimethylpentane and 2,2-dimethylpentane, [(MTI)′ = 142, MTI = 168 and 170], and to 3-ethylpentane and 2,4-dimethylpentane, [(MTI)′ = 150, MTI = 174 and 176] are found with the same (MTI)′ numbers (but possessing different MTI values).

TABLE 6
Computation of the Schultz Index for a Tree T (See Figure 19)
Depicting 2,3-Dimethylpentane

(1) The adjacency matrix of T

$$A(T) = \begin{bmatrix} 0 & 1 & 0 & 0 & 0 & 0 & 0 \\ 1 & 0 & 1 & 0 & 0 & 1 & 0 \\ 0 & 1 & 0 & 1 & 0 & 0 & 1 \\ 0 & 0 & 1 & 0 & 1 & 0 & 0 \\ 0 & 0 & 0 & 1 & 0 & 0 & 0 \\ 0 & 1 & 0 & 0 & 0 & 0 & 0 \\ 0 & 0 & 1 & 0 & 0 & 0 & 0 \end{bmatrix}$$

(2) The distance matrix of T

$$D(T) = \begin{bmatrix} 0 & 1 & 2 & 3 & 4 & 2 & 3 \\ 1 & 0 & 1 & 2 & 3 & 1 & 2 \\ 2 & 1 & 0 & 1 & 2 & 2 & 1 \\ 3 & 2 & 1 & 0 & 1 & 3 & 2 \\ 4 & 3 & 2 & 1 & 0 & 4 & 3 \\ 2 & 1 & 2 & 3 & 4 & 0 & 3 \\ 3 & 2 & 1 & 2 & 3 & 3 & 0 \end{bmatrix}$$

(3) The valency matrix of T
$v(T) = [1\ 3\ 3\ 2\ 1\ 1\ 1]$

(4) The $v[A+D]$ matrix of T
$v[A+D] = [27\ 19\ 18\ 22\ 30\ 27\ 25]$

(5) The Schultz index of T
$$MTI(T) = \sum_{i=1}^{7} e_i = 168$$

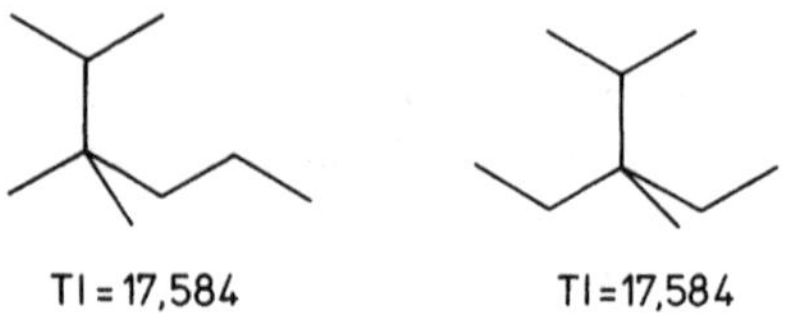

TI = 17,584   TI = 17,584

FIGURE 23.   A pair of nonisomorphic nonane trees with the same TI value.

10. The Determinant of the Adjacency-plus-Distance Matrix as a Topological Index

Schultz and co-workers[21] have introduced the *determinant of the adjacency-plus-distance matrix* as a topological index, TI, for numerical characterization of alkanes,

$$TI = \det|A + D| \tag{94}$$

This index is also a low-discriminatory topological index. The smallest pair of alkane trees with the same TI was found in the nonane family.[133] These nonisomorphic nonane trees (corresponding to 2-isopropyl-2-methylpentane and 3-isopropyl-3-methylpentane) with the identical TI values are shown in Figure 23. The TI index has also shown a considerable potential for use in QSPR studies.[21]

II. THE THREE-DIMENSIONAL WIENER NUMBER

The topological indices described in the preceding section are all obtained from the two-dimensional (2-D) diagrams (graphs) serving as models for molecules.[134,135] Molecules, however, are three-dimensional (3-D) objects. Hence, in addition to their topological and combinatorial contents, their 3-D character is of profound importance.[136] There have already appeared in the literature quite a few attempts to capture the 3-D character of a molecular structure by a single number.[94,137-146] These numbers are formally related to topological indices and are named *topographic indices*.[139] A good example to illustrate topographic indices is the *3-D Wiener number*.[94,141]

The 3-D Wiener number is based on the 3-D (geometric, topographic) distance matrix, much the same as the 2-D Wiener number is based on 2-D (graph-theoretical, topological) distance matrix. The 3-D *distance matrix, 3-*$D = 3\text{-}D(G)$, of a molecular structure G is a real symmetric $N \times N$ matrix whose matrix elements $(3\text{-}D)_{ij}$ represent the shortest Cartesian distances (in some arbitrary units of length) between atoms i and j in G. The 3-D Wiener number $3\text{-}W = 3\text{-}W(G)$ of G is then given by

$$3\text{-}W = (1/2) \sum_{i} \sum_{j} (3\text{-}D)_{ij} \tag{95}$$

The half-sum of the elements of the 3-D distance matrix is named the 3-D Wiener number because of its formal similarity to the 2-D Wiener number, see Equation 43, which may be rewritten as below to resemble Equation 95 even more.

$$2\text{-}W = (1/2) \sum_i \sum_j (2\text{-}\mathbf{D})_{ij} \tag{96}$$

The 3-D distance matrix can be constructed from the known geometry of a molecule. However, for many compounds their geometries are unknown. Therefore, the geometry of a molecule must be approximated in some way.[139,141,147] One way is by molecular mechanics (MM) computations.[148]

The input data for the MM computations may be selected, as will be shown in the case of alkanes, in the following way.[94,149] One first intuitively selects the most stable conformation of a particular alkane. This conformation is usually constructed using the standard bond lengths and bond angles. In the case of smaller and simpler alkanes, it is possible to consider all conformations. In the case of larger and more complex alkanes, the number of possible conformations increases enormously.[150] However, all conformations need not be considered. It is sufficient to take into account only those conformations which are present in the equilibrium mixture in reasonable amounts, probably greater than a few percents. A simple estimate shows that if one gauche-butane interaction contributes to the enthalpy ≈ 3.4 kJ/mol, then it is not necessary to consider conformations with more than three or four gauche-butane interactions relative to the minimum energy conformation. The minimum energy conformation of an alkane is taken to be the most extended conformation with the minimum number of gauche-butane interactions, excluding the "forbidden" pentane interactions. In Figure 24, structural arrangements for the gauche-butane (g) interactions and pentane (g^+g^-) interactions are shown.

The next step is to refine the input geometries by means of MM computations. In this way, the optimum geometry is obtained for each conformer considered. Every conformation thus refined may be used for the construction of the 3-D distance matrix. The 3-D distance matrix may be set up for the skeleton of a molecule or for the whole molecule including the hydrogen atoms. Once the 3-D distance matrix is set up, the 3-D Wiener number is immediately known via Equation 95. Schematically, the computational procedure for generating the 3-W numbers of alkanes may be presented as[149]

Intuitive conformation $\rightarrow$ Optimum $\rightarrow$ 3-D distance $\rightarrow$ 3-W(G)
of an alkane G   conformation of G   matrix of G

In Figure 25 as an illustrative example, the 3-D distance matrix of 2,2-dimethylpropane and the corresponding 3-D Wiener number are given.

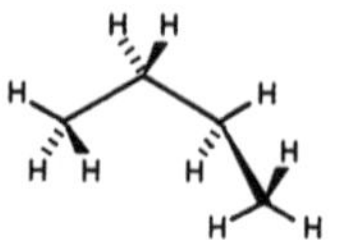

FIGURE 24.   The structural arrangements for the gauche-butane (g) interactions and pentane (g^+g^-) interactions.

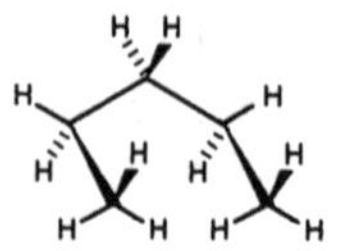

The 3-D Wiener number has several pleasing properties. For example, it appears as a very discriminative index. Thus far, there is not a single report of nonisomorphic graphs with identical 3-W numbers. It also possesses different values for different conformations of a given molecule. For example, in Figure 26 are given the anti-anti conformation (*aa*), the anti-gauche conformation (*ag*), and the gauche-gauche conformation (g^+g^+) of *n*-pentane, their 3-D distance matrices, and 3-D Wiener numbers.

As can be seen from Figure 26 the 3-D Wiener number decreases with the change in shape of each conformation. The most extended conformation (*aa*-pentane) has the largest 3-W value, while the most compact conformation (g^+g^+-pentane) has the smallest 3-W value. Obviously, the 3-D Wiener number decreases with increasing spheroidicity of a structure. This parallels the observation for the 2-D Wiener number which decreases with increasing branching of a structure. Therefore, it appears that the 3-D Wiener number may be a convenient index to model the shape of the molecule. Thus, it is expected to be most useful in the correlations with shape-dependent physical and biological properties. For example, the boiling points (bp) of alkanes can be quite accurately predicted by[94]

$$(bp)_{calc} = 395 \ (\pm 68)(3\text{-}W)^{0.0986(\pm 0.0103)} - 682(\pm 79) \qquad (97)$$

with the following statistical characteristics: $n = 54$, $r = 0.998$, $s = 5.8$, and $F = 5402$.

The 2-W (calculated only for carbon skeletons of alkanes) and 3-W (calculated for CH skeletons of alkanes) numbers are closely intercorrelated indices.

A linear relationship,

$$3\text{-}W = p(2\text{-}W) + q \qquad (98)$$

possesses the following statistical characteristics: $n = 150$, $p = 0.0849$ (± 0.0006), $q = -13.0049$ (± 0.9406), $r = 0.997$, $s = 3.6605$, and $F = 22995.1$.

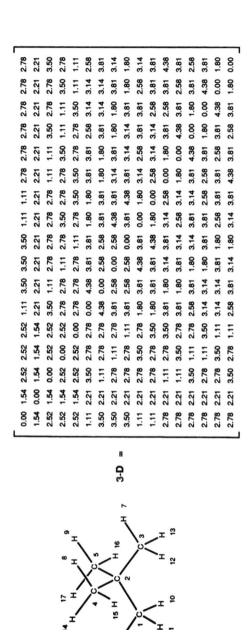

$$3\text{-}W = 377.74$$

FIGURE 25. The 3-D distance matrix and the 3-D Wiener number of 2,2-dimethylpropane.

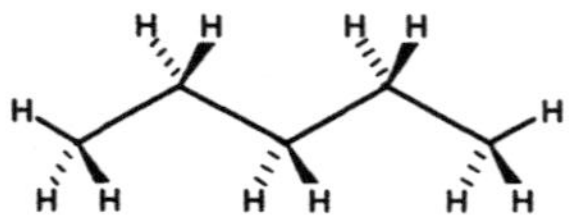

$$D(aa) =$$

0.000	1.534	2.544	3.914	5.087	1.115	1.114	1.114	2.174	2.174	2.777	2.777	4.213	4.212	6.036	5.249	5.249
1.534	0.000	1.537	2.548	3.914	2.197	2.198	2.198	1.116	1.116	2.178	2.178	2.787	2.787	4.739	4.244	4.244
2.544	1.537	0.000	1.537	2.544	3.517	2.816	2.816	2.181	2.181	1.116	1.116	2.181	2.181	3.517	2.816	2.816
3.914	2.548	1.537	0.000	1.534	4.739	4.244	4.243	2.787	2.787	2.178	2.178	1.116	1.116	2.197	2.198	2.198
5.087	3.914	2.544	1.534	0.000	6.036	5.249	5.249	4.212	4.213	2.777	2.777	2.174	2.174	1.115	1.114	1.114
1.115	2.197	3.517	4.739	6.036	0.000	1.800	1.800	2.519	2.519	3.802	3.801	4.904	4.904	6.918	6.247	6.246
1.114	2.198	2.816	4.244	5.249	1.800	0.000	1.803	3.113	3.113	3.151	3.151	4.437	4.788	6.247	5.172	5.477
1.114	2.198	2.816	4.243	5.249	1.800	1.803	0.000	3.113	3.113	2.539	2.585	4.788	4.436	6.246	5.477	5.171
2.174	1.116	2.181	2.787	4.212	2.519	3.113	3.113	0.000	1.798	3.103	3.103	2.559	3.128	4.904	4.436	4.788
2.174	1.116	2.181	2.787	4.213	2.519	3.113	3.113	1.798	0.000	2.529	3.103	3.127	2.558	4.904	4.787	4.436
2.777	2.178	1.116	2.178	2.777	3.802	3.151	2.539	3.103	2.529	0.000	1.798	2.529	3.103	3.801	2.586	3.151
2.777	2.178	1.116	2.178	2.777	3.801	3.151	2.585	3.103	3.103	1.798	0.000	3.103	2.529	3.801	3.151	2.586
4.213	2.787	2.181	1.116	2.174	4.904	4.437	4.788	2.559	3.127	2.529	3.103	0.000	1.798	2.519	2.539	3.113
4.212	2.787	2.181	1.116	2.174	4.904	4.788	4.436	3.128	2.558	3.103	2.529	1.798	0.000	2.519	3.113	2.539
6.036	4.739	3.517	2.197	1.115	6.918	6.247	6.246	4.904	4.904	3.801	3.801	2.519	2.519	0.000	1.800	1.800
5.249	4.244	2.816	2.198	1.114	6.247	5.172	5.477	4.436	4.787	2.586	3.151	2.539	3.113	1.800	0.000	1.803
5.249	4.244	2.816	2.198	1.114	6.246	5.477	5.171	4.788	4.436	3.151	2.586	3.113	2.539	1.800	1.803	0.000

$$3\text{-}W(aa) = 426.97$$

FIGURE 26A   The 3-D distance matrices and the 3-D Wiener numbers of *n*-pentane conformations.

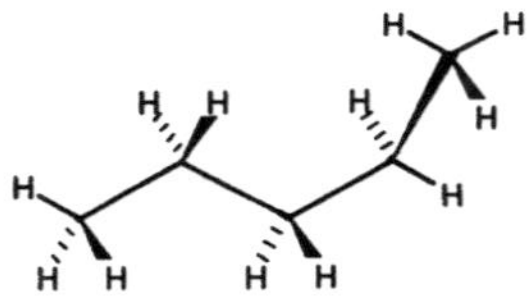

$$D\,(ag) = \begin{bmatrix}
0.000 & 1.535 & 2.543 & 3.929 & 4.583 & 1.115 & 1.114 & 1.114 & 2.175 & 2.167 & 2.736 & 2.801 & 4.707 & 4.211 & 5.650 & 4.723 & 4.350 \\
1.535 & 0.000 & 1.537 & 2.574 & 3.162 & 2.197 & 2.199 & 2.198 & 1.116 & 1.115 & 2.168 & 2.181 & 3.518 & 2.808 & 4.187 & 3.512 & 2.912 \\
2.543 & 1.537 & 0.000 & 1.538 & 2.570 & 3.516 & 2.819 & 2.832 & 2.181 & 2.187 & 1.117 & 1.116 & 2.171 & 2.184 & 3.533 & 2.813 & 2.901 \\
3.929 & 2.574 & 1.538 & 0.000 & 1.534 & 4.762 & 4.234 & 4.271 & 2.788 & 2.862 & 2.168 & 2.180 & 1.117 & 1.116 & 2.194 & 2.198 & 2.205 \\
4.583 & 3.162 & 2.570 & 1.534 & 0.000 & 5.251 & 5.185 & 4.774 & 3.490 & 2.864 & 3.512 & 2.800 & 2.164 & 2.179 & 1.115 & 1.114 & 1.114 \\
1.115 & 2.197 & 3.516 & 4.762 & 5.251 & 0.000 & 1.799 & 1.800 & 2.522 & 2.508 & 3.769 & 3.819 & 5.645 & 4.912 & 6.299 & 5.443 & 4.803 \\
1.114 & 2.199 & 2.819 & 4.234 & 5.185 & 1.799 & 0.000 & 1.803 & 2.539 & 3.108 & 3.108 & 3.190 & 4.858 & 4.401 & 6.197 & 5.394 & 5.094 \\
1.114 & 2.198 & 2.832 & 4.271 & 4.774 & 1.800 & 1.803 & 0.000 & 3.113 & 2.533 & 3.093 & 2.609 & 4.930 & 4.797 & 5.882 & 4.649 & 4.639 \\
2.175 & 1.116 & 2.181 & 2.788 & 3.490 & 2.522 & 2.539 & 3.113 & 0.000 & 1.799 & 2.542 & 3.105 & 3.776 & 2.558 & 4.349 & 4.122 & 3.126 \\
2.167 & 1.115 & 2.187 & 2.862 & 2.864 & 2.508 & 3.108 & 2.533 & 1.799 & 0.000 & 3.099 & 2.520 & 3.886 & 3.205 & 3.904 & 3.104 & 2.328 \\
2.736 & 2.168 & 1.117 & 2.168 & 3.512 & 3.769 & 3.108 & 3.093 & 2.542 & 3.099 & 0.000 & 1.793 & 2.438 & 2.527 & 4.344 & 3.799 & 3.921 \\
2.801 & 2.181 & 1.116 & 2.180 & 2.800 & 3.819 & 3.190 & 2.609 & 3.105 & 2.520 & 1.793 & 0.000 & 2.525 & 3.106 & 3.800 & 2.581 & 3.243 \\
4.707 & 3.518 & 2.171 & 1.117 & 2.164 & 5.645 & 4.858 & 4.930 & 3.776 & 3.886 & 2.438 & 2.525 & 0.000 & 1.792 & 2.476 & 2.555 & 3.108 \\
4.211 & 2.808 & 2.184 & 1.116 & 2.179 & 4.912 & 4.401 & 4.797 & 2.558 & 3.205 & 2.527 & 3.106 & 1.792 & 0.000 & 2.544 & 2.476 & 2.528 \\
5.650 & 4.187 & 3.533 & 2.194 & 1.115 & 6.299 & 6.197 & 5.882 & 4.349 & 3.904 & 4.344 & 3.800 & 2.476 & 2.544 & 0.000 & 1.799 & 1.792 \\
4.723 & 3.512 & 2.813 & 2.198 & 1.114 & 5.443 & 5.394 & 4.649 & 4.122 & 3.104 & 3.799 & 2.581 & 2.555 & 2.476 & 1.799 & 0.000 & 1.804 \\
4.350 & 2.912 & 2.901 & 2.205 & 1.114 & 4.803 & 5.094 & 4.639 & 3.126 & 2.328 & 3.921 & 3.243 & 3.108 & 2.528 & 1.792 & 1.804 & 0.000
\end{bmatrix}$$

$$3\text{-}W_{CH}\,(ag) = 413.20$$

FIGURE 26B

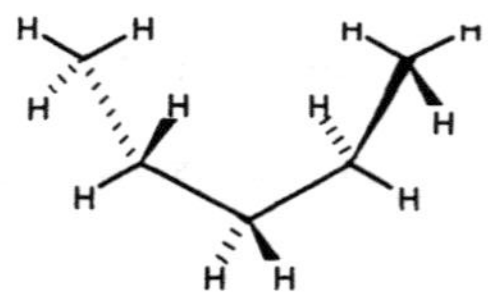

$$D(g^+g^+) =$$

0.000	1.534	2.570	3.140	3.774	1.115	1.114	1.114	2.179	2.165	2.830	3.517	4.139	2.852	4.611	4.421	3.321
1.534	0.000	1.538	2.601	3.140	2.195	2.204	2.197	1.115	1.117	2.169	2.171	3.538	2.900	4.184	3.438	2.891
2.570	1.538	0.000	1.538	2.570	3.533	2.904	2.808	2.190	2.169	1.117	1.117	2.170	2.190	3.533	2.808	2.904
3.140	2.601	1.538	0.000	1.534	4.184	2.892	3.438	2.900	3.538	2.169	1.117	1.115	1.115	2.194	2.197	2.204
3.774	3.140	2.570	1.534	0.000	4.611	3.321	4.421	2.852	4.139	2.830	3.517	2.165	2.179	1.115	1.114	1.114
1.115	2.195	3.533	4.184	4.611	0.000	1.792	1.800	2.543	2.475	3.812	4.350	5.217	3.915	5.456	5.192	3.964
1.114	2.204	2.904	2.892	3.321	1.792	0.000	1.805	2.529	3.107	3.304	3.907	3.930	2.358	3.964	4.171	2.775
1.114	2.197	2.808	3.438	4.421	1.800	1.805	0.000	3.115	2.558	2.612	3.819	4.254	2.997	5.191	5.102	4.171
2.179	1.115	2.190	2.900	2.852	2.543	2.529	3.115	0.000	1.785	3.101	2.505	3.332	3.894	3.915	2.997	2.358
2.165	1.117	2.169	3.538	4.139	2.475	3.107	2.558	1.785	0.000	2.474	1.788	2.469	2.504	4.350	4.254	3.929
2.830	2.169	1.117	2.169	2.830	3.812	3.304	2.612	3.101	2.474	0.000	1.788	2.469	3.101	3.813	3.819	3.906
3.517	2.171	1.117	1.117	3.517	4.350	3.907	3.819	2.505	1.788	1.788	0.000	2.474	3.101	3.813	2.612	3.304
4.139	3.538	2.170	1.115	2.165	5.217	3.930	4.254	3.332	2.469	2.469	2.474	0.000	1.785	2.475	2.558	3.108
2.852	2.900	2.190	1.115	2.179	3.915	2.358	2.997	3.894	2.504	3.101	3.101	1.785	0.000	2.543	3.115	2.529
4.611	4.184	3.533	2.194	1.115	5.456	3.964	5.191	3.915	4.350	3.813	3.813	2.475	2.543	0.000	1.800	1.792
4.421	3.438	2.808	2.197	1.114	5.192	4.171	5.102	2.997	4.254	3.819	2.612	2.558	3.115	1.800	0.000	1.805
3.321	2.891	2.904	2.204	1.114	3.964	2.775	4.171	2.358	3.929	3.906	3.304	3.108	2.529	1.792	1.805	0.000

$$3\text{-}W(g^+g^+) = 394.67$$

FIGURE 26C

There are only a few reports at this point in the literature on the use of the 3-D Wiener number in QSPR and QSAR modeling.[94,131,141,149,151-153] However, one expects that the use of the 3-W number as well as other topographic indices will considerably increase in the near future, especially in conjunction with computer graphics modeling.[154-157]

REFERENCES

1. **Hosoya, H.,** *Bull. Chem. Soc. Jpn.,* 44, 2332, 1971, 2nd printing.
2. **Harary, F.,** *Graph Theory,* Addison-Wesley, Reading, MA, 1971.
3. **Rouvray, D. H.,** *Congr. Numerantium,* 49, 161, 1985.
4. **Rouvray, D. H.,** *J. Mol. Struct. (Theochem.),* 185, 187, 1989.
5. **Trinajstić, N., Randić, M., and Klein, D. J.,** *Acta Pharm. Jugosl.,* 36, 267, 1986.
6. **Kier, L. B. and Hall, L. H.,** *Molecular Connectivity in Chemistry and Drug Research,* Academic Press, New York, 1976.
7. **Rouvray, D. H. and Balaban, A. T.,** in *Applications of Graph Theory,* Wilson, R. J. and Beineke, L. W., Eds., Academic Press, London, 1977, 177.
8. **Nizhnii, S. V. and Epshtein, N. A.,** *Russ. Chem. Rev.,* 47, 383, 1978.
9. **Balaban, A. T., Chiriac, A., Motoc, I., and Simon, Z.,** *Steric Fit in Quantitative Structure-Activity Relations,* Springer-Verlag, Berlin, 1980; **Balaban, A. T.,** in *Graph Theory and Topology in Chemistry,* King, R. B. and Rouvray, D. M., Eds., Elsevier, Amsterdam, 1987, 126.
10. **Sabljić, A. and Trinajstić, N.,** *Acta Pharm. Jugosl.,* 31, 189, 1981.
11. **Kier, L. B. and Hall, L. H.,** *Molecular Connectivity in Structure-Activity Analysis,* John Wiley & Sons, New York, 1986.
12. **Trinajstić, N., Sabljić, A., and Nikolić, S., Eds.,** *QSAR in Drug Research,* Special Subject Issue of *Acta Pharm. Jugosl.,* 36, 79–280, 1986; 37, 1–86, 1987.
13. **Stankevich, M. I., Stankevich, I. V., and Zefirov, N. S.,** *Russ. Chem. Rev.,* 57, 191, 1988.
14. **Trinajstić, N., Nikolić, S., and Carter, S.,** *Kem. Ind. (Zagreb),* 38, 469, 1989.
15. **Sabljić, A.,** *Environ. Health Perspect.,* 83, 179, 1989.
16. **Basak, S. C., Niemi, G. J., and Veith, G. D.,** in *Computational Chemical Graph Theory,* Rouvray, D. H., Ed., Nova Science Publishers, New York, 1990, 235.
17. **Kopp, H.,** *Ann. Chem. Pharm.,* 50, 71, 1844.
18. **Rouvray, D. H.,** in *Chemical Graph Theory: Introduction and Fundamentals,* Bonchev, D. and Rouvray, D. H., Eds., Abacus Press/Gordon & Breach Science Publishers, New York, 1991, 1.
19. **Bonchev, D.,** *Information Theoretic Indices for Characterization of Chemical Structures,* Research Studies Press, Chichester, 1983.
20. **Balaban, A. T. and Feroiu, V.,** *Rep. Mol. Theory,* 1, 133, 1990.
21. **Schultz, H. P., Schultz, E. B., and Schultz, T. P.,** *J. Chem. Inf. Comput. Sci.,* 30, 27, 1990; **Schultz, H. P. and Schultz, T. P.,** *J. Chem. Inf. Comput. Sci.,* 31, 144, 1991.
22. **Hall, L. H.,** in *Computational Chemical Graph Theory,* Rouvray, D. H., Ed., Nova Science Publishers, New York, 1990, 201.
23. **Gutman, I. and Polansky, O. E.,** *Mathematical Concepts in Organic Chemistry,* Springer-Verlag, Berlin, 1986, chap. 11.

24. **Labanowski, J. K., Motoc, I., and Dammkoehler, R. A.,** *Comput. Chem.,* 15, 47, 1991.
25. **Motoc, I. and Balaban, A. T.,** *Math. Chem. (Mülheim/Ruhr),* 5, 197, 1979; *Rev. Roum. Chim.,* 26, 593, 1981.
26. **Motoc, I., Balaban, A. T., Mekenyan, O., and Bonchev, D.,** *Math. Chem. (Mülheim/ Ruhr),* 13, 369, 1982.
27. **Kovačević, K., Plavšić, D., Trinajstić, N., and Horvat, D.,** in *MATH/CHEM/COMP 1988,* Graovac, A., Ed., Elsevier, Amsterdam, 1989, 213.
28. **Roberts, F. S.,** *Discrete Mathematical Models,* Prentice-Hall, Englewood Cliffs, CA, 1975, 8.
29. **Barysz, M., Plavšić, D., and Trinajstić, N.,** *Math. Chem. (Mülheim/Ruhr),* 19, 89, 1986.
30. **Gutman, I. and Trinajstić, N.,** *Chem. Phys. Lett.,* 17, 535, 1972.
31. **Gutman, I., Ruščić, B., Trinajstić, N., and Wilcox, C. F., Jr.,** *J. Chem. Phys.,* 62, 3399, 1975.
32. **Randić, M.,** *J. Am. Chem. Soc.,* 97, 6609, 1975.
33. **Sabljić, A., Trinajstić, N., and Maysinger, D.,** *Acta Pharm. Jugosl.,* 31, 71, 1981.
34. **Sabljić, A. and Protić-Sabljić, M.,** *Mol. Pharmacol.,* 23, 213, 1983; **Sabljić, A.,** *J. Agric. Food Chem.,* 32, 243, 1984.
35. **Rouvray, D. H.,** *Sci. Am.,* 254, 40, 1986.
36. **Rouvray, D. H.,** *Acta Pharm. Jugosl.,* 36, 239, 1986.
37. **Seybold, P. G., May, M. A., and Gargas, M. L.,** *Acta Pharm. Jugosl.,* 36, 253, 1986.
38. **Sabljić, Z.,** *Z. Gesamte Hyg. Ihre Grenzgeb.,* 33, 493, 1987.
39. **Sabljić, A.,** *Environ. Sci. Technol.,* 21, 358, 1987.
40. **Sabljić, A.,** in *QSAR in Environmental Toxicology — II,* Kaiser, K. L. E., Ed., D. Reidel Publishing, Dordrecht, 1987, 309.
41. **Randić, M., Sabljić, A., Nikolić, S., and Trinajstić, N.,** *Int. J. Quantum Chem.: Quantum Biol. Symp.,* 15, 267, 1988; **Randić, M., Hansen, P. J., and Jurs, P. C.,** *J. Chem. Inf. Comput. Sci.,* 28, 60, 1988.
42. **Seybold, P. G., May, M., and Bagal, U. A.,** *J. Chem. Educ.,* 64, 575, 1987.
43. **Hansen, P. J. and Jurs, P. C.,** *J. Chem. Educ.,* 65, 574, 1988.
44. **Needham, D. E., Wei, I-C., and Seybold, P. G.,** *J. Am. Chem. Soc.,* 110, 4186, 1988.
45. **Kier, L. B. and Hall, L. H.,** *Rep. Mol. Theory,* 1, 121, 1990.
46. **Sabljić, A., Gusten, H., Schönherr, J., and Riederer, M.,** *Environ. Sci. Technol.,* 24, 1321, 1990.
47. **Sabljić, A.,** in *Practical Applications of Quantitative Structure-Activity Relationships (QSAR) in Environmental Chemistry and Toxicology,* Karcher, W. and Devillers, J., Eds., Kluwer, Dordrecht, 1990, 61.
48. **Kier, L. B., Hall, L. H., Murray, W. J., and Randić, M.,** *J. Pharm. Sci.,* 65, 1226, 1976.
49. **Kier, L. B. and Hall, L. H.,** *J. Pharm. Sci.,* 65, 1806, 1976.
50. **Singh, V. K., Tewari, V. P., Gupta, D. K., and Srivastava, A. K.,** *Tetrahedron,* 40, 2859, 1984.
51. **Murrell, J. N., Kettle, S. F. A., and Tedder, J. M.,** *Valence Theory,* John Wiley & Sons, London, 1965, 262.
52. **Randić, M., Jeričević, Ž., Sabljić, A., and Trinajstić, N.,** *Acta Phys. Polon.,* A74, 317, 1988.
53. **Gupta, S. P. and Singh, P.,** *Bull. Chem. Soc. Jpn.,* 52, 2745, 1979.
54. **Bonchev, D. and Trinajstić, N.,** *J. Chem. Phys.,* 67, 4517, 1977.
55. **Randić, M.,** *J. Chem. Inf. Comput. Sci.,* 24, 164, 1984.
56. **Szymanski, K., Müller, W. R., Knop, J. V., and Trinajstić, N.,** *J. Chem. Inf. Comput. Sci.,* 25, 413, 1985.
57. **Knop, J. V., Müller, W. R., Jeričević, Ž., and Trinajstić, N.,** *J. Chem. Inf. Comput. Sci.,* 21, 91, 1981.

58. **Szymanski, K., Müller, W. R., Knop, J. V., and Trinajstić, N.,** *Int. J. Quantum Chem.: Quantum Chem. Symp.,* 20, 173, 1986.
59. **Randić, M.,** *Int. J. Quantum Chem.: Quantum Biol. Symp.,* 11, 137, 1984.
60. **Randić, M.,** in *Molecular Basis of Cancer. Part A. Macromolecular Structure, Carcinogens and Oncogens,* Rein, R., Ed., Alan R. Liss, New York, 1985, 309.
61. **Carter, S., Trinajstić, N., and Nikolić, S.,** *Acta Pharm. Jugosl.,* 37, 37, 1987.
62. **Carter, S., Trinajstić, N., and Nikolić, S.,** *Med. Sci. Res.,* 16, 185, 1988.
63. **Carter, S., Nikolić, S., and Trinajstić, N.,** in *MATH/CHEM/COMP 1988,* Graovac, A., Ed., Elsevier, Amsterdam, 1989, 255.
64. **Trinajstić, N., Nikolić, S., and Carter, S.,** *Period. Biol.,* 92, 431, 1990.
65. **Randić, M.,** *J. Chem. Inf. Comput. Sci.,* 36, 134, 1986.
66. **Szymanski, K., Müller, W. R., Knop, J. V., and Trinajstić, N.,** *Croat. Chem. Acta,* 59, 719, 1986.
67. **Lang, S.,** *Algebra,* Addison-Wesley, Reading, MA, 1969.
68. **Cvetković, D. M. and Gutman, I.,** *Croat. Chem. Acta,* 49, 115, 1977.
69. **Lovász, L. and Pelikán, J.,** *Period. Math. Hung.,* 3, 175, 1973.
70. **Hosoya, H.,** in *Mathematics and Computational Concepts in Chemistry,* Trinajstić, N., Ed., Horwood, Chichester, 1986, 110.
71. **Mizutani, K., Kawasaki, K., and Hosoya, H.,** *Natl. Sci. Rep. Ochanomizu Univ. (Tokyo),* 22, 39, 1971; **Kawasaki, K., Mizutani, K., and Hosoya, H.,** *Natl. Sci. Rep. Ochanomizu Univ. (Tokyo),* 22, 181, 1971.
72. **Yamaguchi, T., Suzuki, M., and Hosoya, H.,** *Natl. Sci. Rep. Ochanomizu Univ. (Tokyo),* 26, 39, 1975.
73. **Hosoya, H. and Ohkami, N.,** *J. Comput. Chem.,* 4, 585, 1983.
74. **Hosoya, H., Kawasaki, K., and Mizutani, K.,** *Bull. Chem. Soc. Jpn.,* 45, 3415, 1972.
75. **Adler, N., Kovačić-Beck, L., and Trinajstić, N.,** in *MATH/CHEM/COMP 1988,* Graovac, A., Ed., Elsevier, Amsterdam, 1989, 225.
76. **Wiener, H.,** *J. Am. Chem. Soc.,* 69, 17, 1947; 69, 2636, 1947.
77. **Platt, J. R.,** *J. Chem. Phys.,* 15, 419, 1947.
78. **Wiener, H.,** *J. Chem. Phys.,* 15, 766, 1947.
79. **Wiener, H.,** *J. Phys. Chem.,* 52, 425, 1948; 52, 1082, 1948.
80. **Platt, J. R.,** *J. Phys. Chem.,* 56, 328, 1952.
81. **Polansky, O. E.,** in *MATH/CHEM/COMP 1988,* Graovac, A., Ed., Elsevier, Amsterdam, 1989, 167.
82. **Gutman, I., Kennedy, J. V., and Quintas, L. V.,** *Chem. Phys. Lett.,* 173, 403, 1990.
83. **Bonchev, D., Mekenyan, O., Protić, G., and Trinajstić, N.,** *J. Chromatogr.,* 176, 149, 1979.
84. **Bonchev, D., Mekenyan, O., and Trinajstić, N.,** *Int. J. Quantum Chem.,* 17, 845, 1980.
85. **Mekenyan, O., Bonchev, D., and Trinajstić, N.,** *Int. J. Quantum Chem.,* 19, 929, 1981.
86. **Rouvray, D. H.,** in *Chemical Applications of Topology and Graph Theory,* King, R. B., Ed., Elsevier, Amsterdam, 1983, 159.
87. **Adler, N., Babić, D., and Trinajstić, N.,** *Fresenius Z. Anal. Chem.,* 322, 426, 1985.
88. **Rouvray, D. H.,** in *Mathematics and Computational Concepts in Chemistry,* Trinajstić, N., Ed., Horwood, Chichester, 1986, 295.
89. **Bošnjak, N., Adler, N., Perić, M., and Trinajstić, N.,** in *Modelling of Structure and Properties of Molecules,* Maksić, Z. B., Ed., Horwood, Chichester, 1987, 103.
90. **Hanson, M. P. and Rouvray, D. H.,** in *Graph Theory and Topology in Chemistry,* King, R. G. and Rouvray, D. H., Eds., Elsevier, Amsterdam, 1987, 201.
91. **Bonchev, D., Mekenyan, O., and Polansky, O. E.,** in *Graph Theory and Topology in Chemistry,* King, R. B. and Rouvray, D. H., Eds., Elsevier, Amsterdam, 1987, 209.
92. **Adler, N. and Kovačić-Beck, L.,** in *Graph Theory and Topology in Chemistry,* King, R. B. and Rouvray, D. H., Eds., Elsevier, Amsterdam, 1987, 194.

93. **Lukovits, I.,** *J. Chem. Soc. Perkin Trans. II*, 84, 1667, 1988; *Quant. Struct.-Act. Relat.*, 9, 227, 1990; *Rep. Mol. Theory*, 1, 127, 1990.
94. **Mihalić, Z. and Trinajstić, N.,** *J. Mol. Struct. (Theochem)*, 232, 65, 1991.
95. **Gordon, M. and Scantlebury, G. R.,** *Trans. Faraday Soc.*, 60, 604, 1964.
96. **Balaban, A. T.,** *Chem. Phys. Lett.*, 89, 399, 1982; *Pure Appl. Chem.*, 55, 199, 1983; *Math. Chem. (Mülheim/Ruhr)*, 21, 115, 1986.
97. **Balaban, A. T., Motoc, I., Bonchev, D., and Mekenyan, O.,** *Topics Curr. Chem.*, 114, 21, 1983.
98. **Balaban, A. T., Niculescu-Duvaz, I., and Simon, Z.,** *Acta Pharm. Jugosl.*, 37, 7, 1987.
99. **Balaban, A. T. and Ivanciuc, O.,** in *MATH/CHEM/COMP 1988*, Graovac, A., Ed., Elsevier, Amsterdam, 1989, 193.
100. **Barysz, M., Jashari, G., Lall, R. S., Srivastava, V. K., and Trinajstić, N.,** in *Chemical Applications of Topology and Graph Theory*, King, R. B., Ed., Elsevier, Amsterdam, 1983, 222.
101. **Mekenyan, O., Bonchev, D., Sabljić, A., and Trinajstić, N.,** *Acta Pharm. Jugosl.*, 37, 75, 1987.
102. **Wilson, R. J.,** *Introduction to Graph Theory*, Oliver & Boyd, Edinburgh, 1972, 46.
103. **Balaban, A. T. and Filip, P.,** *Math. Chem. (Mülheim/Ruhr)*, 16, 163, 1984.
104. **Seybold, P. G.,** *Int. J. Quantum Chem.: Quantum Biol. Symp.*, 10, 95, 1983; 10, 103, 1983.
105. **Balaban, A. T. and Quintas, L. V.,** *Math. Chem. (Mülheim/Ruhr)*, 14, 213, 1983.
106. **Bonchev, D. and Trinajstić, N.,** *Int. J. Quantum Chem.: Quantum Chem. Symp.*, 12, 293, 1978.
107. **Bonchev, D., Mekenyan, O., Knop, J. V., and Trinajstić, N.,** *Croat. Chem. Acta*, 52, 361, 1979.
108. **Mekenyan, O., Bonchev, D., and Trinajstić, N.,** *Int. J. Quantum Chem.*, 18, 369, 1980.
109. **Bonchev, D., Mekenyan, D., and Trinajstić, N.,** *J. Comput. Chem.*, 2, 127, 1981.
110. **Bonchev, D. and Trinajstić, N.,** *Int. J. Quantum Chem.: Quantum Chem. Symp.*, 16, 463, 1982.
111. **Barysz, M., Trinajstić, N., and Knop, J. V.,** *Int. J. Quantum Chem.: Quantum Chem. Symp.*, 17, 441, 1983.
112. **Brillouin, L.,** *Science and Information Theory*, Academic Press, New York, 1956.
113. **Shannon, C. and Weaver, W.,** *Mathematical Theory of Communication*, University of Illinois Press, Urbana, IL, 1949.
114. **Bonchev, D., Kamenski, D., and Kamenska, V.,** *Bull. Math. Biol.*, 38, 119, 1976.
115. **Bonchev, D.,** *Math. Chem. (Mülheim/Ruhr)*, 7, 65, 1979.
116. **Sarkar, R., Roy, A. B., and Sarkar, P. K.,** *Math. Biosci.*, 39, 299, 1978.
117. **Raychaudhury, C., Ray, S. K., Ghosh, J. J., Roy, A. B., and Basak, S. C.,** *J. Comput. Chem.*, 5, 581, 1984.
118. **Basak, S. C., Monsrud, L. J., Rosen, V. R., Frane, C. M., and Magnusson, V. R.,** *Acta Pharm. Jugosl.*, 31, 81, 1986.
119. **Basak, S. C.,** *Med. Sci. Res.*, 15, 605, 1987.
120. **Bertz, S. H.,** *J. Am. Chem. Soc.*, 103, 3599, 1981.
121. **Bertz, S. H.,** *J. Chem. Soc. Chem. Commun.*, 818, 1981.
122. **Bertz, S. H.,** *J. Am. Chem. Soc.*, 104, 5801, 1982.
123. **Bertz, S. H.,** *Bull. Math. Biol.*, 45, 849, 1983.
124. **Bertz, S. H.,** in *Chemical Applications of Topology and Graph Theory*, King, R. B., Ed., Elsevier, Amsterdam, 1983, 206; *Discrete Appl. Math.*, 19, 65, 1988.
125. **Balaban, A. T.,** *Theoret. Chim. Acta*, 53, 355, 1979.
126. **Bonchev, D., Balaban, A. T., and Mekenyan, O.,** *J. Chem. Inf. Comput. Sci.*, 20, 106, 1980; **Bonchev, D., Mekenyan, O., and Balaban, A. T.,** *J. Chem. Inf. Comput. Sci.*, 29, 91, 1989.

127. **Bonchev, D., Balaban, A. T., and Randić, M.,** *Int. J. Quantum Chem.,* 19, 61, 1981; erratum *Int. J. Quantum Chem.,* 22, 441, 1982; **Bonchev, D.,** *J. Mol. Struct. (Theochem.),* 185, 155, 1989.

128. **Trinajstić, N., Nikolić, S., and Horvat, D.,** *Kem. Ind. (Zagreb),* 36, 493, 1987.

129. **Bogdanov, B., Nikolić, S., Sabljić, A., Trinajstić, N., and Carter, S.,** *Int. J. Quantum.: Quantum Biol. Symp.,* 14, 325, 1987; **Carter, S., Nikolić, S., and Trinajstić, N.,** *Int. J. Quantum Chem.: Quantum Biol. Symp.,* 16, 323, 1989.

130. **Schultz, H. P.,** *J. Chem. Inf. Comput. Sci.,* 29, 227, 1989.

131. **Mihalić, Z., Veljan, D., Amić, D., Nikolić, S., Plavšić, D., and Trinajstić, N.,** *J. Math. Chem.,* in press.

132. **Müller, W. R., Szymanski, K., Knop, J. V., and Trinajstić, N.,** *J. Chem. Inf. Comput. Sci.,* 30, 160, 1990.

133. **Knop, J. V., Müller, W. R., Szymanski, K., and Trinajstić, N.,** *J. Chem. Inf. Comput. Sci.,* 31, 83, 1991.

134. **Trindle, C.,** *Croat. Chem. Acta,* 57, 1231, 1984.

135. **Trinajstić, N.,** in *MATH/CHEM/COMP 1987,* Lacher, R. C., Ed., Elsevier, Amsterdam, 1988, 83.

136. **Turro, N. J.,** *Angew. Chem. Int. Ed. Engl.,* 25, 882, 1986.

137. **Mekenyan, O., Peitchev, D., Bonchev, D., Trinajstić, N., and Bangov, I.,** *Drug Res.,* 30, 176, 1986.

138. **Mekenyan, O., Peitchev, D., Bonchev, D., Trinajstić, N., and Dimitrova, J.,** *Drug Res.,* 36, 629, 1986.

139. **Randić, M.,** in *MATH/CHEM/COMP 1987,* Lacher, R. C., Ed., Elsevier, Amsterdam, 1988, 101.

140. **Randić, M.,** *Int. J. Quantum Chem.: Quantum Biol. Symp.,* 15, 201, 1989.

141. **Bogdanov, B., Nikolić, S., and Trinajstić, N.,** *J. Math. Chem.,* 3, 299, 1983; 5, 305, 1990.

142. **Jurs, P. C. and Edwards, P. A.,** in *Computational Chemical Graph Theory,* Rouvray, D. H., Ed., Nova Science Publishers, New York, 1990, 279.

143. **Delance, E., Doucet, J. P., and Dubois, J.-E.,** in *Modelling of Molecular Structure and Properties,* Rivail, J.-L., Ed., Elsevier, Amsterdam, 1990, 755.

144. **Randić, M.,** *J. Math. Chem.,* 4, 157, 1990.

145. **Randić, M.,** in *Concepts and Applications of Molecular Similarity,* Johnson, M. A. and Maggiora, G. M., Eds., John Wiley & Sons, New York, 1990, 77.

146. **Randić, M., Jerman-Blažič, B., and Trinajstić, N.,** *Comput. Chem.,* 14, 237, 1990.

147. **Balasubramanian, K.,** *Chem. Phys. Lett.,* 169, 224, 1990.

148. **Burkert, K. and Allinger, N. L.,** Molecular Mechanics, ACS, Washington, D.C., 1982.

149. **Nikolić, S., Trinajstić, N., Mihalić, M., and Carter, S.,** *Chem. Phys. Lett.,* 179, 21, 1991.

150. **Balaban, A. T.,** *Rev. Roum. Chim.,* 21, 1049, 1976.

151. **King, J. W., Kassel, R. J., and King, B. B.,** *Int. J. Quantum. Chem.: Quantum Biol. Symp.,* 17, 27, 1990.

152. **Bošnjak, N., Mihalić, Z., and Trinajstić, N.,** *J. Chromatogr.,* 540, 430, 1991.

153. **King, J. W. and Kassel, R. J.,** *Int. J. Quantum Chem.: Quantum Biol. Symp.,* in press.

154. **Motoc, I., Dammkoehler, R. A., and Marshall, G. R.,** in *Mathematics and Computational Concepts in Chemistry,* Trinajstić, N., Ed., Ed., Horwood, Chichester, 1986, 227.

155. **Hol, W. G. J.,** *Angew. Chem. Int. Ed. Engl.,* 25, 767, 1986.

156. **Riche, C. and Moy, L.,** in *Modelling of Molecular Structures and Properties,* Rivail, J.-L., Ed., Elsevier, Amsterdam, 1990, 281.

157. **Testa, B. and Mayer, J. A.,** *Acta Pharm. Jugosl.,* 40, 315, 1990.

Chapter 11

ISOMER ENUMERATION

The enumeration of isomeric structures is one of the oldest uses of graph theory in chemistry.[1-4] However, besides graph theory the basic mathematical tools necessary for isomer enumeration studies are combinatorial theory[5,6] and group theory.[7] *Isomers* are chemical compounds with an identical molecular formula (and molecular weight) which display at least some differing (physical and/or chemical) properties and which are stable for periods of time that are long in comparison with those during which measurements of their properties are made.[1] The term *isomerism* and the definition of isomers was introduced by Berzelius[8] in 1830. Compounds without isomers are called *unimers*.[9]

Traditionally, isomers have been classified as either structural isomers or stereoisomers.[10] *Structural isomers* or *constitutional isomers* differ in their structures, i.e., in the manner of bonding the atoms in the molecule.[11] *Stereoisomers* have identical structures but differ in configuration or conformation, i.e., in the spatial architecture of the molecule.[12,13] Both classes may be further divided into subclasses.[10] The emphasis in this chapter will be placed on the enumeration of structural isomers. They were first recognized in the last century by Butlerov.[14]

There are a number of methods available for the isomer enumeration. Here the enumeration methods developed by Cayley,[15-19] Henze and Blair,[20-24] Pólya,[25,26] and Knop et al.[27-30] will be presented.

I. THE CAYLEY GENERATING FUNCTIONS

Cayley[17] was first to attempt to enumerate the isomeric alkanes $C_N H_{2N+2}$ and alkyl radicals $C_N H_{2N+1}$. He represented the carbon skeletons of alkanes by *trees* in which the maximum vertex valency is four. Similarly, alkyl radicals were depicted by *rooted trees* in which again the maximum vertex valency is four. Trees (rooted trees) corresponding to alkanes (alkyl radicals) are sometimes called the *Cayley (rooted) trees*.[31] The pioneering contribution of Arthur Cayley (born: Richmond, Surrey, August 16, 1821 — died: Cambridge, January 26, 1895) to chemical graph theory was recently knowledgably reviewed by Rouvray.[2]

The graph-theoretical representation of isomeric pentanes $C_5 H_{10}$ by means of trees is given in Figure 1. The related pentyl radicals $C_5 H_9$, which are depicted by rooted trees, are shown in Figure 2.

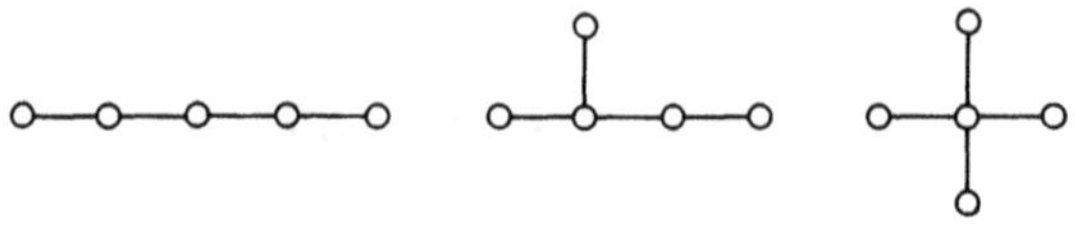

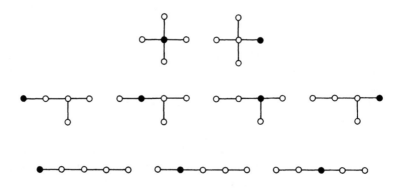

FIGURE 1.   Trees representing isomeric pentanes.

FIGURE 2.   Rooted trees representing pentyl radicals.

A. ENUMERATION OF TREES

The mathematical theory of trees was introduced in the last century.[15,16,32-37] Cayley[15] first used the term *tree* in 1857, although Kirchhoff[32] first utilized the concept in his fundamental work on electrical networks in 1847.

Cayley developed a generating function for enumeration of rooted trees,[15]

$$(1 - x)^{-A_0}(1 - x^2)^{-A_1}(1 - x^3)^{-A_2}(1 - x^4)^{-A_3} \ldots (1 - x^N)^{-A_{N-1}}$$

$$= A_0 + A_1 x + A_2 x^2 + A_3 x^3 + \ldots + A_N x^N \tag{1}$$

where x is a variable, N the number of vertices in the rooted tree, A_N the number or rooted trees for a given N, and $A_0 = 1$ by definition. The enumeration of rooted trees with up to 13 vertices by means of a generating function (Equation 1) is given in Table 1.

Ten years later, Jordan[34] discovered the existence of the *center* and *bicenter* of the tree. Every tree has a center or a bicenter, but not both. Thus, trees with a center are called *centric trees,* while those with a bicenter are called *bicentric trees.* Examples of the centric tree and the bicentric tree are shown in Figure 3.

Cayley made use of the work by Jordan and enumerated the centric and bicentric trees.[36] The sum of centric and bicentric trees produced the total number of isomeric trees for a given N. His results are given in Table 2.

Later (in 1881) Cayley[37] improved his method for enumerating trees by using the generating functions and already known A_N numbers for the enumeration of rooted trees. Thus, the trees T_N can be counted by means of the following formulae:

$$T_1 = 1 \tag{2}$$

$$T_2 = (1/2)\, A_1\, (A_1 + 1) \tag{3}$$

$$T_3 = \text{the coefficient at } x^2 \text{ in } (1 - x)^{-A_1} \tag{4}$$

$$T_4 = (1/2)A_2\, (A_2 + 1) + \text{the coefficient}$$
$$\text{at } x^3 \text{ in } (1 - x)^{-A_1} \tag{5}$$

$$T_5 = \text{the coefficient at } x^4 \text{ in}$$
$$(1 - x)^{-A_1} (1 - x^2)^{-A_2} \tag{6}$$

$$T_6 = (1/2)A_3(A_3 + 1) + \text{the coefficient}$$
$$\text{at } x^5 \text{ in } (1 - x)^{-A_1} (1 - x^2)^{-A_1} \tag{7}$$

$$T_7 = \text{the coefficient at } x^6 \text{ in}$$
$$(1 - x)^{-A_1} (1 - x^2)^{-A_2} (1 - x^3)^{-A_3} \tag{8}$$

$$T_8 = (1/2)A_4(A_4 + 1) + \text{the coefficient}$$
$$\text{at } x^7 \text{ in } (1 - x)^{-A_1} (1 - x^2)^{-A_2}(1 - x^3)^{-A_3} \tag{9}$$

TABLE 1
Enumeration of Rooted Trees with up to 13 Vertices by Means of the Cayley Generating Function

	x^{1}	x^{2}	x^{3}	x^{4}	x^{5}	x^{6}	x^{7}	x^{8}	x^{9}	x^{10}	x^{11}	x^{12}	Cayley generating function
$A_1 = 1$	(1)	–	–	–	–	–	–	–	–	–	–	–	$(1-x^2)^{-1}$
$A_2 = 1$	1	(2)	2	3	3	4	4	5	5	6	6	7	$(1-x^3)^{-2}$
$A_3 = 1$	1	2	(4)	5	7	11	13	17	23	27	33	42	$(1-x^4)^{-4}$
$A_4 = 1$	1	2	4	(9)	11	19	29	47	61	91	125	180	$(1-x^5)^{-9}$
$A_5 = 1$	1	2	4	9	(20)	28	47	83	142	235	341	531	$(1-x^6)^{-20}$
$A_6 = 1$	1	2	4	9	20	(48)	67	123	222	415	741	1301	$(1-x^7)^{-48}$
$A_7 = 1$	1	2	4	9	20	48	(115)	171	318	607	1173	2261	$(1-x^8)^{-115}$
$A_8 = 1$	1	2	4	9	20	48	115	(286)	433	837	1633	3296	$(1-x^9)^{-286}$
$A_9 = 1$	1	2	4	9	20	48	115	286	(719)	1123	2205	4440	$(1-x^{10})^{-719}$
$A_{10} = 1$	1	2	4	9	20	48	115	286	719	(1842)	2924	5878	$(1-x^{11})^{-1842}$
$A_{11} = 1$	1	2	4	9	20	48	115	286	719	1842	(4766)	7720	$(1-x^{12})^{-4766}$
$A_{12} = 1$	1	2	4	9	20	48	115	286	719	1842	4766	(12486)	$(1-x^{13})^{-12486}$

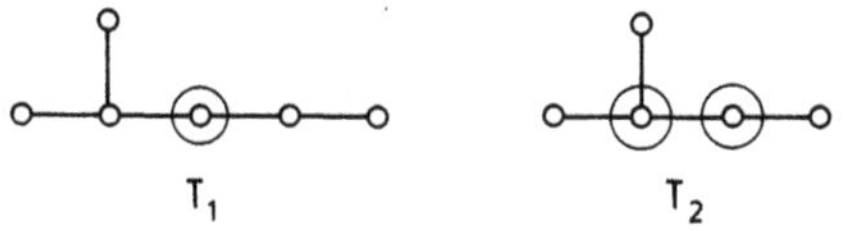

FIGURE 3.   Examples of the centric tree (T_1) and the bicentric tree (T_2).
Centers of these trees are denoted by larger circles.

TABLE 2
Enumeration of Trees with up to 13 Vertices by Counting the Centric and Bicentric Trees

Number of vertices	1	2	3	4	5	6	7	8	9	10	11	12	13
Number of centric trees	1	0	1	1	2	3	7	12	27	55	127	284	682
Number of bicentric trees	0	1	0	1	1	3	4	11	20	51	108	267	618
Number of trees		1	1	1	2	3	6	11	23	47	106	235	551	1301

After Cayley, A., *Rep. Br. Assoc. Adv. Sci.*, 95, 257, 1875.

T_9 = the coefficient at x^8 in

$$(1 - x)^{-A_1} (1 - x^2)^{-A_2}$$

$$(1 - x^3)^{-A_3} (1 - x^4)^{-A_4} \tag{10}$$

and so on. The use of the Cayley counting formulae is exemplified for the case of trees with 7 vertices in Table 3.

Application of the Cayley counting formulae is rather tedious and errors are easily made. Actually Cayley made several numerical errors in his work. Cayley was very close to deriving a general formula for the number of trees in terms of the number of rooted trees, entirely by means of generating functions, but did not find it. However, half a century later Otter[38] derived an elegant formula for counting trees in terms of rooted trees,

$$T(x) = A(x) - (1/2)[A^2(x) - A(x^2)] \tag{11}$$

where $A(x)$ is the counting series for the rooted trees. The enumeration of trees by means of the Otter counting formula is carried out as follows:

$$A(x) = x + x^2 + 2x^3 + 4x^4 + 9x^5 + 20x^6 + 48x^7 + \ldots \tag{12}$$

$$A^2(x) = x^2 + 2x^3 + 5x^4 + 12x^5 + 30x^6 + 74x^7 + \ldots \tag{13}$$

TABLE 3
Enumeration of Trees with 7 Vertices Using the Cayley Counting Formulae

$(1-x)^{-A_1} = 1 + A_1 x + (1/2) A_1 (A_1+1) x^2 + (1/6) A_1 (A_1+1) (A_1+2) x^3$
$\qquad + (1/24) A_1 (A_1+1) (A_1+2) (A_1+3) x^4$
$\qquad + (1/120) A_1 (A_1+1) (A_1+2) (A_1+3) (A_1+4) x^5$
$\qquad + (1/720) A_1 (A_1+1) (A_1+2) (A_1+3) (A_1+4) (A_1+5) x^6 + ...$

$(1-x^2)^{-A_2} = 1 + A_2 x^2 + (1/2) A_2 (A_2+1) x^4 + (1/6) A_2 (A_2+1) (A_2+2) x^4 + ...$

$(1-x^3)^{-A_3} = 1 + A_3 x^3 + (1/2) A_3 (A_3+1) x^6 + ...$

$T_7 = (1/720) A_1 (A_1+1) (A_1+2) (A_1+3) (A_1+4) (A_1+5) + (1/6) A_2 (A_2+1) (A_2+2)$
$\qquad + (1/2) A_3 (A_3+1) + A_1 A_2 A_3 + (1/4) A_1 A_2 (A_1+1) (A_2+1) + (1/6) A_1 A_3 (A_1+1) (A_1+2)$
$\qquad + (1/24) A_1 A_2 (A_1+1) (A_1+2) (A_1+3) = 11$

$$A(x^2) = x^2 + x^4 + 2x^6 + 4x^8 + ... \qquad (14)$$

$$(1/2)[A^2(x) - A(x^2)] = x^3 + 2x^4 + 6x^5 + 14x^6 + 37x^7 + ... \quad (15)$$

$$T(x) = x + x^2 + x^3 + 2x^4 + 3x^5 + 6x^6 + 11x^7 + ... \qquad (16)$$

The coefficients in Equation 16 represent the counts of trees with a given number of vertices.

B. ENUMERATION OF ALKANES

Cayley[16] was the first to realize the potentiality of the mathematical theory of trees for the enumeration of acyclic hydrocarbons. He enumerated alkanes[17] and alkyl radicals[18] with up to 13 carbon atoms, but the number of isomers obtained for C_{12} and C_{13} alkanes (357 and 799) and C_{13} alkyl radicals (7638) were incorrect. In addition, he stated that no compact formula could be found for the isomer enumeration of alkanes. However, in spite of its shortcomings, the work of Cayley had a considerable impact on chemists of his time. Almost immediately after the Cayley paper on enumeration of alkanes,[17] a work by Schiff[39] appeared in which he correctly counted distinct alkanes, alkenes, and alkyl radicals with up to 10 carbon atoms. Schiff also attempted to calculate the number of dodecanes $C_{12}H_{26}$, obtaining the same erroneous value as Cayley. The errors in computing the number of C_{12} and C_{13} alkanes were first corrected (C_{12}: 355 and C_{13}: 802) by Herrmann[40] five years later. It is interesting to note that Losanitsch[41] 18 years later was arguing that the number of C_{12} alkanes is neither 357 nor 355, but 354. There was quite a discussion between Herrmann[42,44] and Losanitsch[43] about whose number is correct. Herrmann, of course, produced a correct value (355).[27,45] The correct value for the $C_{13}H_{22}$ radicals is 7639.[27,45] However, none of the above authors, and several others,[46-48] were able to produce a reliable method for enumeration

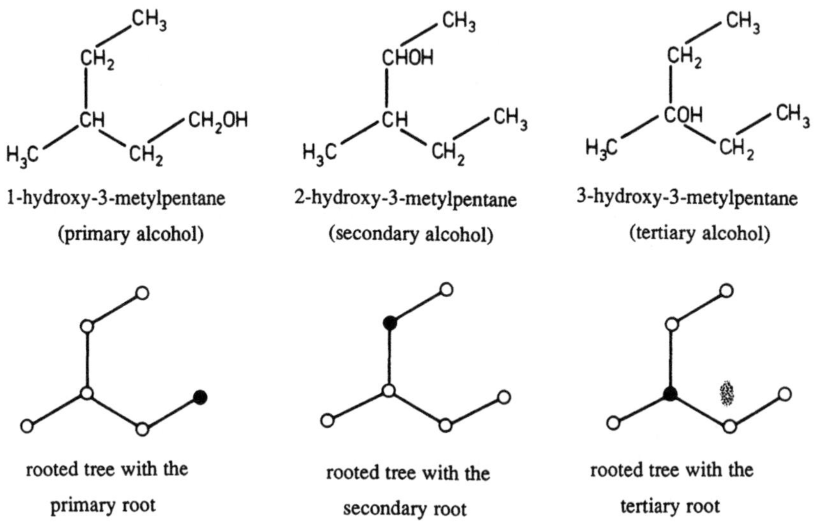

1-hydroxy-3-metylpentane
(primary alcohol)

2-hydroxy-3-metylpentane
(secondary alcohol)

3-hydroxy-3-metylpentane
(tertiary alcohol)

rooted tree with the
primary root

rooted tree with the
secondary root

rooted tree with the
tertiary root

FIGURE 4.   Example of primary, secondary, and tertiary alcohols and the corresponding rooted trees.

of alkanes with large N. Their methods have been unwieldy and often led to incorrect results. The first significant advance after Cayley came in 1931 when Henze and Blair,[20-24] at the University of Texas at Austin, developed recursion formulae for enumeration of alkanes and related acyclic structures.

II. THE HENZE-BLAIR APPROACH

The Henze-Blair approach[20-24] is based on the following suppositions: (1) the Cayley conclusion that no formula could be derived for isomer counts for members of homologous series; (2) free rotation about single CC bonds does not bring forth new isomeric structures; and (3) if the isomer enumeration for a given member of an homologous series having N atoms could be established, an appropriate recursion formula would produce the isomer count for the next member having $N+1$ atoms.

The essence of the Henze-Blair approach[20] will be illustrated on the enumeration of alcohols $C_NH_{2N+1}OH$. Alcohols may conveniently be represented by the rooted trees. The rooted trees can be classified in those with a *primary root* (a rooted vertex with valency one), in those with a *secondary root* (a rooted vertex with valency two), in those with a *tertiary root* (a rooted vertex with valency three), and in those with a *quarternary root* (a rooted vertex with valency four). In the case of alcohols, the corresponding rooted trees with the primary (secondary, tertiary) root represent the primary (secondary, tertiary) alcohols. Examples of primary, secondary, and tertiary alcohols and the corresponding rooted trees are shown in Figure 4.

The number of rooted trees with the primary root, secondary root, and tertiary root containing N vertices are denoted by p_N, s_N, and t_N, respectively. The total number of isomeric rooted trees (e.g., alcohols) T_N^* is then given by

$$T_N^* = p_N + s_N + t_N \tag{17}$$

The replacement of an $-OH$ group in any alcohol by a $-CH_2OH$ results always in the formation of a primary alcohol. Thus, the number of primary alcohols p_N is given by,

$$p_N = T_{N-1}^* \tag{18}$$

The secondary alcohol may be imagined to consist of two alkyl radicals R_i and R_j connected to the $>CHOH$ group. The number of carbon atoms (i and j) in R_i and R_j is obviously $N-1$, i.e., $i+j = N-1$. This may be utilized for deriving equations for s_N:

$$s_N = \begin{cases} T_1^* T_{N-2}^* + T_2^* T_{N-3}^* + \ldots \qquad N = \text{even} \\[4pt] \quad + T_{(N-2)/2}^* T_{N/2}^* \\[8pt] T_1^* T_{N-2}^* + T_2^* T_{N-3}^* + \ldots \qquad N = \text{odd} \\[4pt] \quad + T_{(N-3)/2}^* T_{(N+1)/2}^* \\[8pt] \quad + (1/2) T_{(N-1)/2}^* [1 + T_{(N-1)/2}^*] \end{cases} \tag{19}$$

In the case of tertiary alcohols formation of a molecule may be imagined to arise from joining three alkyl radicals R_i, R_j, and R_k to the $>COH$ group. The number of carbon atoms (i, j, and k) in R_i, R_j, and R_k is again $N-1$, i.e., $i+j+k = N-1$. The equations for t_N are then given by

$$t_N = \begin{cases} \Sigma\, T_i^* \, T_j^* \, T_k^*; & R_i \neq R_j \neq R_k;\ i + j + k = N - 1 \\[6pt] (1/2)\Sigma T_i^* \, (1 + T_i^*)\, T_j^*; & R_i = R_j \neq R_k;\ 2i + k = N - 1 \\[6pt] (1/6) T_i^* \, (1 + T_i^*)(2 + T_i^*); & R_i = R_j = R_k;\ 3i = N - 1 \end{cases} \tag{20}$$

Enumeration of isomeric alcohols $C_N H_{2N+1} OH$ with up to $N = 6$, using the above formulae, is given in Table 4.

Henze and Blair calculated correctly the number of primary, secondary, and tertiary alcohols with up to 20 carbon atoms. Then they extended their approach to alkanes.[21] They have separated alkanes $C_N H_{2N+2}$ into classes

TABLE 4
Enumeration of Isomeric Alcohols $C_NH_{2N+1}OH$ with up to N = 6 Using the Henze-Blair Approach

(1) $N = 1$
$p_1 = T_0^*, T_0^* = 1$ by definition
$s_1 = 0$
$t_1 = 0$
$T_1^* = p_1 + s_1 + t_1 = 1$

(2) $N = 2$
$p_2 = T_1^* = 1$
$s_2 = 0$
$t_2 = 0$
$T_2^* = p_2 + s_2 + t_2 = 1$

(3) $N = 3$
$p_3 = T_2^* = 1$
$s_3 = (1/2) T_1^* (1 + T_1^*) = 1$
$t_3 = 0$
$T_3^* = p_3 + s_3 + t_3 = 2$

(4) $N = 4$
$p_4 = T_3^* = 2$
$s_4 = T_1^* T_2^* (2 + T_1^*) = 1$
$t_4 = (1/6) T_1^* (1 + T_1^*) = 1$
$T_4^* = p_4 + s_4 + t_4 = 4$

(5) $N = 5$
$p_5 = T_4^* = 4$
$s_5 = T_1^* T_3^* + (1/2)T_2^*(1 + T_2^*) = 3$
$t_5 = (1/2)T_1^*(1 + T_1^*)T_2^* = 1$
$T_5^* = p_5 + s_5 + t_5 = 8$

(6) $N = 6$
$p_6 = T_5^* = 8$
$s_6 = T_1^* T_4^* + T_2^* T_3^* = 6$
$t_6 = (1/2) T_1^*(1 + T_1^*) T_3^* + (1/2) T_2^* (1 + T_2^*) T_1^* = 3$
$T_6^* = p_6 + s_6 + t_6 = 17$

according to whether the number of carbon atoms N in them is *even* or *odd*. Each class is further divided into two groups. Alkanes with N = even are partitioned into *Group A'* containing those alkanes which can be divided into parts (alkyl radicals) with N/2 atoms each and *Group B'* containing those alkanes which cannot be divided into two equal parts. Alkanes with N = odd are similarly partitioned into *Group A"* containing those alkanes which can be divided into two parts (alkyl radicals), one with (N + 1)/2 carbon atoms and the other with (N − 1)/2 carbon atoms, respectively, and *Group B"* containing those odd-membered alkanes which cannot be divided in this way.

The alkane isomers in *Group A'* can be enumerated using the following formula:

$$(1/2) \, T_{N/2}^* \, (1 + T_{N/2}^*) \qquad (21)$$

where the previously determined values for the rooted trees are utilized.

The alkane isomers in *Group A″* can be enumerated by means of the formula:

$$(1/2)\ T^*_{(N-1)/2}\ (1\ +\ 2T^*_{(N+1)/2}\ +\ T^*_{(N-1)/2}) \tag{22}$$

The alkane isomers in *Group B′* and *Group B″* are of two types: (1) those in which *three* branches are attached to the specified carbon atom, and (2) those in which *four* branches are connected to the specified carbon atom. The total number of isomers in *Group B′* and in *Group B″* is obtained by summing up the numbers of isomers of both types.

Type (1) consists of these *three* possible cases:

Case 1 All three branches are of different length:

$$\sum T^*_i T^*_j T^*_k \tag{23}$$

where $i + j + k = N - 1$ and $i > j > k$. For N = even, $i \le$ (N/2) − 1 and for N = odd, $i \le (N-3)/2$

Case 2 Two branches are of the same length and different from the third:

$$(1/2) \sum T^*_i T^*_j (1\ +\ T^*_i) \tag{24}$$

where $2i + j = N - 1$. For N = even, $i,j \le$ (N/2) − 1 and for N = odd, $i,j \le (N-3)/2$

Case 3 All three branches are of the same size:

$$(1/6) \sum T^*_i (1\ +\ T^*_i)(2\ +\ T^*_i) \tag{25}$$

where $3i = N - 1$

Type (2) consists of *five* possible cases:

Case 1 All four branches are of different length:

$$\sum T^*_h T^*_i T^*_j T^*_k \tag{26}$$

where $h + j + k = N - 1$ and $h > i > j > k$. For N = even, $h \le$ (N/2) − 1 and for N = odd, $h = (N-3)/2$

Case 2 Two branches are of the same length while each of the other two is of different length:

$$(1/2) \sum T^*_i T^*_j T^*_k (1\ +\ T^*_k) \tag{27}$$

TABLE 5
Enumeration of Isomeric Dodecanes $C_{12}H_{26}$ Using the Henze-Blair Formulae

(1)   *Group A'* $= (1/2)\, T_6^*(1 + T_6^*) = 153$

(2)   *Group B'*
   Type (1):   Case 1 $= T_3^* T_4^* T_4^* = 32$

   Case 2 $= (1/2)\, T_3^* T_4^*(1 + T_4^*) + (1/2)\, T_4^* T_3^*(1 + T_3^*) + (1/2)\, T_4^* T_3^*(1 + T_3^*)$
   $= 80$

   Type (2):   Case 1 $= T_3^* T_3^* T_4^* = 16$

   Case 2 $= (1/2)\, T_4^* T_3^* T_3^*(1 + T_3^*) + (1/2)\, T_3^* T_4^* T_3^*(1 + T_3^*)$
   $\quad\quad + (1/2)\, T_3^* T_4^* T_3^*(1 + T_3^*) + (1/2)\, T_3^* T_3^* T_4^*(1 + T_4^*) = 62$

   Case 3 $= (1/6)\, T_3^* T_3^*(1 + T_3^*)(2 + T_3^*) + (1/6)\, T_3^* T_3^*(1 + T_3^*)(2 + T_3^*) = 12$

(3)   The total number of isomeric dodecanes
   $T_{12} = 355$

where $2i + j = k = N - 1$ and $i > j > k$

Case 3   Three branches are of the same length and different from the fourth:

$$(1/6) \sum T_i^* T_j^*(1 + T_i^*)(2 + T_i^*) \tag{28}$$

where $3i + j = N - 1$ and $i > j$

Case 4   All four branches are of the same length:

$$(1/24) \sum T_i^*(1 + T_i^*)(2 + T_i^*)(3 + T_i^*) \tag{29}$$

where $4i = N - 1$

Case 5   Four branches are partitioned into two sets containing two branches each. The individual members of each set are of the same length and different from the members of the other set:

$$(1/4) \sum T_i^* T_j^*(1 + T_i^*)(1 + T_j^*) \tag{30}$$

where $2i + 2j = N - 1$ and $i > j$

The total number of structural isomers of alkanes with N carbon atoms is equal to the sum of the number of isomers in *Group A' (Group A")* and the number of isomers obtained for each of the cases in *Group B' (Group B")*. Enumeration of isomeric dodecanes $C_{12}H_{26}$, using the Henze-Blair formulae, is shown in Table 5.

Henze and Blair[21] have calculated the number of alkane isomers with up to $N = 20$, but have also listed the number of structural isomers for $C_{25}H_{52}$, $C_{30}H_{62}$, and $C_{40}H_{82}$ alkanes. A year later, Perry[49] augmented their work by producing the numbers of isomeric alcohols with the carbon content from N

= 21 to N = 30 and the numbers of isomeric alkanes from $C_{21}H_{44}$ to $C_{39}H_{70}$ and all alkane isomers with the common formula $C_{60}H_{122}$. Perry has also detected a numerical error in the work by Henze and Blair on the enumeration of alkanes.[21] They have reported erroneously the number of the $C_{19}H_{40}$ isomers to be 147,284 instead of 148,284 (which may have been a typographical error).

Although the application of the Henze-Blair approach involves tedious calculations, it has been used considerably for the enumeration of structural isomers of various compounds (saturated hydrocarbons, unsaturated hydrocarbons, alkynes, alcohols, etc.)[22-24,50-54] and was even extended to count the stereoisomeric alcohols.[55] Actually 40 years after the introduction, the computer program was written[56] based on the Henze-Blair approach and the computations have been carried out for isomeric alkanes with up to 50 carbon atoms.

III. THE PÓLYA ENUMERATION METHOD

Pólya in 1937 proposed the most powerful enumeration method that is available to chemists.[1,2,25,26,57-65] Although the Pólya method rests on some results by Redfield,[66-73] and Lunn and Senior,[74] it gave the first recipe for the systematic derivation of counting series by making an integrated use of symmetry properties of molecules, generating functions, and weighting factors.

The Pólya enumeration method counts isomers by taking into account the symmetry operations, namely the proper rotation axes, of the structure. The symmetry is expressed as the *cycle index* Z(H),

$$Z(H) = (1/h) \sum' n_{i_1 i_2 \ldots i_p} f_1^{i_1} f_2^{i_2} \ldots f_p^{i_p} \tag{31}$$

where Z(H) is the cycle index for a permutation group H of order h, p is the number of vertices permuted, the f_p are variables, while $n_{i_1 i_2} \ldots i_p$ represents the number of permutation of H which consists of i_1 cycles of order one, i_2 cycles of order two, etc. A cycle of order n is one involving the interchange of n vertices after performance of the symmetry operation. The prime over the summation in Equation 31 indicates that the condition

$$\sum_{i=1}^{P} h_i = P \tag{32}$$

always holds, where P is the total number of permutations. The cycle indices for all the symmetry operations are added and the total is divided by the number of operations for a given molecule. As an example the derivation of the cycle index for each symmetry operation which may be performed on the benzene ring is shown in Table 6.

TABLE 6
The Derivation of the Cycle Index for Each Symmetry Operation Which May be Executed on the Benzene Ring (Which Belongs to the Point Group Symmetry D_{6h})

Symmetry operation	Vertex interchange	Permutation grouping	Cycle index
E	$\begin{pmatrix} 1\ 2\ 3\ 4\ 5\ 6 \\ 1\ 2\ 3\ 4\ 5\ 6 \end{pmatrix}$	$\binom{1}{1}\binom{2}{2}\binom{3}{3}\binom{4}{4}\binom{5}{5}\binom{6}{6}$	f_1^6
C_6^+	$\begin{pmatrix} 1\ 2\ 3\ 4\ 5\ 6 \\ 6\ 1\ 2\ 3\ 4\ 5 \end{pmatrix}$	$\begin{pmatrix} 1\ 2\ 3\ 4\ 5\ 6 \\ 6\ 1\ 2\ 3\ 4\ 5 \end{pmatrix}$	f_6^1
C_6^-	$\begin{pmatrix} 1\ 2\ 3\ 4\ 5\ 6 \\ 2\ 3\ 4\ 5\ 6\ 1 \end{pmatrix}$	$\begin{pmatrix} 1\ 2\ 3\ 4\ 5\ 6 \\ 2\ 3\ 4\ 5\ 6\ 1 \end{pmatrix}$	f_6^1
C_3^+	$\begin{pmatrix} 1\ 2\ 3\ 4\ 5\ 6 \\ 5\ 6\ 1\ 2\ 3\ 4 \end{pmatrix}$	$\begin{pmatrix} 1\ 3\ 5 \\ 5\ 1\ 3 \end{pmatrix}\begin{pmatrix} 2\ 4\ 6 \\ 6\ 2\ 4 \end{pmatrix}$	f_3^2
C_3^-	$\begin{pmatrix} 1\ 2\ 3\ 4\ 5\ 6 \\ 3\ 4\ 5\ 6\ 1\ 2 \end{pmatrix}$	$\begin{pmatrix} 1\ 3\ 5 \\ 3\ 5\ 1 \end{pmatrix}\begin{pmatrix} 2\ 4\ 6 \\ 4\ 6\ 2 \end{pmatrix}$	f_3^2
C_2	$\begin{pmatrix} 1\ 2\ 3\ 4\ 5\ 6 \\ 4\ 5\ 6\ 1\ 2\ 3 \end{pmatrix}$	$\begin{pmatrix} 1\ 4 \\ 4\ 1 \end{pmatrix}\begin{pmatrix} 2\ 5 \\ 5\ 2 \end{pmatrix}\begin{pmatrix} 3\ 6 \\ 6\ 3 \end{pmatrix}$	f_2^3
$\sigma_v^{(1)}$	$\begin{pmatrix} 1\ 2\ 3\ 4\ 5\ 6 \\ 1\ 6\ 5\ 4\ 3\ 2 \end{pmatrix}$	$\binom{1}{1}\binom{4}{4}\begin{pmatrix} 2\ 6 \\ 6\ 2 \end{pmatrix}\begin{pmatrix} 3\ 5 \\ 5\ 3 \end{pmatrix}$	$f_1^2 f_2^2$
$\sigma_v^{(2)}$	$\begin{pmatrix} 1\ 2\ 3\ 4\ 5\ 6 \\ 5\ 4\ 3\ 2\ 1\ 6 \end{pmatrix}$	$\binom{3}{3}\binom{6}{6}\begin{pmatrix} 1\ 5 \\ 5\ 1 \end{pmatrix}\begin{pmatrix} 2\ 4 \\ 4\ 2 \end{pmatrix}$	$f_1^2 f_2^2$
$\sigma_v^{(3)}$	$\begin{pmatrix} 1\ 2\ 3\ 4\ 5\ 6 \\ 3\ 2\ 1\ 6\ 5\ 4 \end{pmatrix}$	$\binom{2}{2}\binom{5}{5}\begin{pmatrix} 1\ 3 \\ 3\ 1 \end{pmatrix}\begin{pmatrix} 4\ 6 \\ 6\ 4 \end{pmatrix}$	$f_1^2 f_2^2$
$\sigma_v^{(4)}$	$\begin{pmatrix} 1\ 2\ 3\ 4\ 5\ 6 \\ 6\ 5\ 4\ 3\ 2\ 1 \end{pmatrix}$	$\begin{pmatrix} 1\ 6 \\ 6\ 1 \end{pmatrix}\begin{pmatrix} 2\ 5 \\ 5\ 2 \end{pmatrix}\begin{pmatrix} 3\ 4 \\ 4\ 3 \end{pmatrix}$	f_2^3
$\sigma_v^{(5)}$	$\begin{pmatrix} 1\ 2\ 3\ 4\ 5\ 6 \\ 2\ 1\ 6\ 5\ 4\ 3 \end{pmatrix}$	$\begin{pmatrix} 1\ 2 \\ 2\ 1 \end{pmatrix}\begin{pmatrix} 3\ 6 \\ 6\ 3 \end{pmatrix}\begin{pmatrix} 4\ 5 \\ 5\ 4 \end{pmatrix}$	f_2^3
$\sigma_v^{(6)}$	$\begin{pmatrix} 1\ 2\ 3\ 4\ 5\ 6 \\ 4\ 3\ 2\ 1\ 6\ 5 \end{pmatrix}$	$\begin{pmatrix} 1\ 4 \\ 4\ 1 \end{pmatrix}\begin{pmatrix} 2\ 3 \\ 3\ 2 \end{pmatrix}\begin{pmatrix} 5\ 6 \\ 6\ 5 \end{pmatrix}$	f_2^3

The cycle index for benzene is given by

$$Z = (1/12)(f_1^6 + 4f_2^3 + 2f_3^2 + 2f_6^1 + 3f_1^2f_2^2) \tag{33}$$

In order to obtain the number of structural isomers of benzene when k hydrogen atoms are substituted by monovalent atoms and groups R for each of the component cycle indices f_k, a substitution of the type

$$f_k^\ell = (1 + x^k)^\ell \tag{34}$$

is made into Equation 33. The reason for this substitution is the creation of the power series in x in which the coefficient at x^k in the expansion gives directly the number of isomers which can be formed when k hydrogen atoms in the benzene molecule are substituted by monovalent atoms or groups. The substitution of Equation 34 into Equation 33 yields the counting polynomial

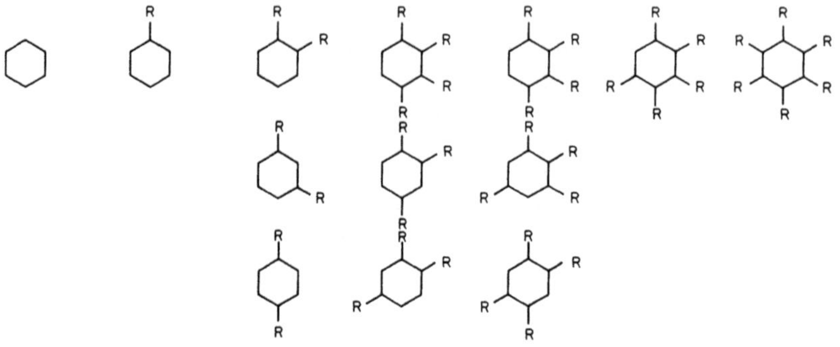

1   x $3x^2$ $3x^3$ $3x^4$ x^5 x^6

FIGURE 5. All structural isomers formed when the hydrogen atoms in benzene are successively substituted with the monovalent atoms or groups R.

TABLE 7
Cycle Indices for Several Benzenoid Hydrocarbons

Benzenoid hydrocarbon	Symmetry group	Cycle index
Naphthalene	D_{2h}	$(1/4)\ (f_1^8 + 3\ f_2^4)$
Anthracene	D_{2h}	$(1/4)\ (f_1^{10} + f_1^2 f_2^4 + 2\ f_2^5)$
Phenanthrene	C_{2v}	$(1/2)\ (f_1^{10} + f_2^5)$
Tetracene	D_{2h}	$(1/4)\ (f_1^{12} + 3\ f_2^6)$
Triphenylene	D_{3h}	$(1/6)\ (f_1^{12} + 3\ f_2^6 + 2\ f_3^4)$

$$Z = (1/12)[(1 + x)^6 + 4(1 + x^2)^3 + 2(1 + x^3)^2$$
$$+ 2(1 + x^6) + 3(1 + x)^2(1 + x^2)^2]$$
$$= 1 + x + 3x^2 + 3x^3 + 3x^4 + x^5 + x^6 \qquad (35)$$

The exponent of each term gives the number of monovalent atoms or groups R substituted into the benzene nucleus while the coefficients give the corresponding number of isomers. For example, since the coefficient at x^3 in Equation 35 is *three*, it means that the trisubstitution of benzene by the univalent atom or group R will produce *three* structural isomers $C_6H_3R_3$. All possible structural isomers formed when the hydrogen atoms in benzene are successively substituted by monovalent atoms or groups R counted by Equation 35 are presented in Figure 5.

The cycle indexes and the corresponding counting polynomials for several benzenoid hydrocarbons are given in Tables 7 and 8, respectively.

TABLE 8
The Counting Polynomials for Several Benzenoid Hydrocarbons

Benzenoid hydrocarbon	Maximal number of substituents	Counting polynomial
Naphthalene	8	$1 + 2x + 10x^2 + 14x^3 + 22x^4 + 10x^6 + 2x^7 + x^8$
Anthracene	10	$1 + 3x + 15x^2 + 32x^3 + 60x^4 + 66x^5 + 60x^6 + 32x^7 + 15x^8 + 3x^9 + x^{10}$
Phenanthrene	10	$1 + 5x + 25x^2 + 60x^3 + 110x^4 + 126x^5 + 110x^6 + 60x^7 + 25x^8 + 5x^9 + x^{10}$
Tetracene	12	$1 + 3x + 21x^2 + 55x^3 + 135x^4 + 198x^5 + 236x^6 + 198x^7 + 135x^8 + 55x^9 + 21x^{10} + 3x^{11} + x^{12}$
Triphenylene	12	$1 + 2x + 14x^2 + 38x^3 + 90x^4 + 132x^5 + 166x^6 + 132x^7 + 90x^8 + 38x^9 + 14x^{10} + 2x^{11} + x^{12}$

TABLE 9
Derivation of the Cycle Index for Each Symmetry Operation Which May Be Performed on the Cyclobutadiene Ring (Which Belongs to the Point Group Symmetry D_{4h})

Symmetry operation	Vertex interchange	Permutation grouping	Cycle index
E	$\begin{pmatrix} 1\ 2\ 3\ 4 \\ 1\ 2\ 3\ 4 \end{pmatrix}$	$\begin{pmatrix} 1 \\ 1 \end{pmatrix}\begin{pmatrix} 2 \\ 2 \end{pmatrix}\begin{pmatrix} 3 \\ 3 \end{pmatrix}\begin{pmatrix} 4 \\ 4 \end{pmatrix}$	f_1^4
C_4^+	$\begin{pmatrix} 1\ 2\ 3\ 4 \\ 4\ 1\ 2\ 3 \end{pmatrix}$	$\begin{pmatrix} 1\ 2\ 3\ 4 \\ 4\ 1\ 2\ 3 \end{pmatrix}$	f_4^1
C_4^-	$\begin{pmatrix} 1\ 2\ 3\ 4 \\ 2\ 3\ 4\ 1 \end{pmatrix}$	$\begin{pmatrix} 1\ 2\ 3\ 4 \\ 2\ 3\ 4\ 1 \end{pmatrix}$	f_4^1
C_2	$\begin{pmatrix} 1\ 2\ 3\ 4 \\ 3\ 4\ 1\ 2 \end{pmatrix}$	$\begin{pmatrix} 1\ 3 \\ 3\ 1 \end{pmatrix}\begin{pmatrix} 2\ 4 \\ 4\ 2 \end{pmatrix}$	f_2^2
$\sigma_v^{(1)}$	$\begin{pmatrix} 1\ 2\ 3\ 4 \\ 1\ 4\ 3\ 2 \end{pmatrix}$	$\begin{pmatrix} 1 \\ 1 \end{pmatrix}\begin{pmatrix} 3 \\ 3 \end{pmatrix}\begin{pmatrix} 2\ 4 \\ 4\ 2 \end{pmatrix}$	$f_1^2 f_2^1$
$\sigma_v^{(2)}$	$\begin{pmatrix} 1\ 2\ 3\ 4 \\ 3\ 2\ 1\ 4 \end{pmatrix}$	$\begin{pmatrix} 2 \\ 2 \end{pmatrix}\begin{pmatrix} 4 \\ 4 \end{pmatrix}\begin{pmatrix} 1\ 3 \\ 3\ 1 \end{pmatrix}$	$f_1^2 f_2^1$
$\sigma_v^{(3)}$	$\begin{pmatrix} 1\ 2\ 3\ 4 \\ 2\ 1\ 4\ 3 \end{pmatrix}$	$\begin{pmatrix} 1\ 2 \\ 2\ 1 \end{pmatrix}\begin{pmatrix} 3\ 4 \\ 4\ 3 \end{pmatrix}$	f_2^2
$\sigma_v^{(4)}$	$\begin{pmatrix} 1\ 2\ 3\ 4 \\ 4\ 3\ 2\ 1 \end{pmatrix}$	$\begin{pmatrix} 1\ 4 \\ 4\ 1 \end{pmatrix}\begin{pmatrix} 2\ 3 \\ 3\ 2 \end{pmatrix}$	f_2^2

The question how many structural isomers can be obtained if the substitution is accomplished with two different monovalent atoms or groups R_1 and R_2 will be answered for the case of cyclobutadiene. The derivation of the cycle index for cyclobutadiene is given in Table 9.

The cycle index for cyclobutadiene is given by

$$Z = (1/8)(f_1^4 + 3f_2^2 + 2f_4^1 + 2f_1^2 f_2^1) \tag{36}$$

The substitution of the expression

$$f_k^\ell = (1 + x^k + y^k)^\ell \tag{37}$$

into Equation 36 leads to the counting polynomial when k hydrogen atoms of cyclobutadiene are exchanged with the monovalent atoms or groups R_1 and k hydrogen atoms of cyclobutadiene are simultaneously exchanged with the different monovalent atoms or groups R_2. This counting polynomial is given by

$$\begin{aligned} Z = (1/8)[&(1 + x + y)^4 + 3(1 + x^2 + y^2)^2 \\ &+ 2(1 + x^4 + y^4) + 2(1 + x + y)^2(1 + x^2 + y^2)] \\ = &1 + x^4 + x^3 y + 2x^2 y^2 + xy^3 + y^4 \end{aligned} \tag{38}$$

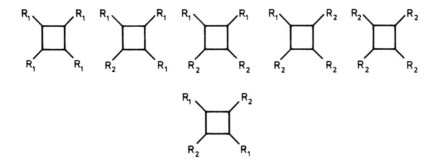

FIGURE 6. All structural isomers formed when the hydrogen atoms in cyclobutadiene are successively substituted with two different monovalent atoms or groups R_1 and R_2.

Diagrams of all structural isomers formed when the hydrogen atoms in cyclobutadiene are substituted with two different monovalent atoms or groups R_1 and R_2 are shown in Figure 6.

The Pólya enumeration method was generalized and extended by several people[1,4,26,75-85] who, thus, have extended the power and the spectrum of uses of the method.

Since its introduction the Pólya method has been, and is, widely used for enumeration of various isomeric, inorganic and organic, structures.[1,4] Thus, for example, it was employed for enumeration of isotopic isomers,[4,86,87] cyclic molecules,[87,88] benzenoid hydrocarbons,[89] porphyrins,[90] chiral and achiral alkanes,[91] ferrocenes,[92] clusters,[93] various inorganic structures,[94] etc. The Pólya enumeration method is certainly the most powerful counting technique available, although in recent years the computer-oriented methods are used in many instances because of the advantages that the high-speed computers have brought to this area of research:[27-30,95-100] high accuracy of numerical work and the possibility to display the generated structures.

IV. THE ENUMERATION METHOD BASED ON THE N-TUPLE CODE

One of the most powerful computer-oriented approaches is the enumeration approach based on the *N-tuple code* which has been developed by Knop et al.[27-30,101-103] The N-tuple code represents a set of non-negative integers smaller than N and was developed for trees.[27] Each number in the N-tuple represents the valency of a vertex in the tree or in its subtrees. The N-tuple code can be derived for a given tree by means of the following convention: one first has to identify the vertices of the highest valency and select among them one that will result in a code that yields lexicographically the largest number.

The first entry in the N-tuple is the valency of the first vertex considered, i.e., the vertex with the highest valency. After this first vertex or the *starting*

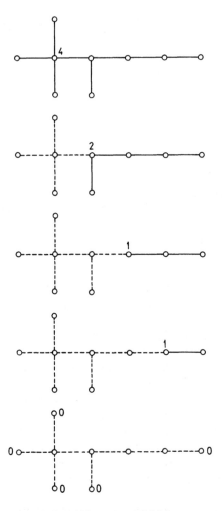

N-TUPLE CODE : 4 2 1 1 0 0 0 0 0

FIGURE 7.   The derivation of the N-tuple code for a tree T depicting 2,2,3-trimethylhexane.

vertex is located, that vertex and incident edges are removed and the generated subtrees examined. In the case of rooted trees, a root-vertex is the starting vertex at which the N-tuple begins. Thus, a given (rooted) tree with $N > 1$ vertices and M edges incident to the starting vertex (the root-vertex) give rise to M (rooted) subtrees. Then their codes are derived and combined in such a way that the result corresponds to the maximum possible N-tuple. The construction of the N-tuple code for a tree depicting the carbon skeleton of 2,2,3-trimethylhexane is shown in Figure 7.

The N-tuple code given in this figure is lexicographically the highest possible N-tuple for a tree representing 2,2,3-trimethylhexane. All other N-tuples that can be associated with this tree, such as {1 3 2 1 1 0 0 0 0}, {3 3 0 0 0 1 1 0 0}, and {1 2 3 0 0 0 1 0 0} are lexicographically smaller. It is important to point out that two nonisomorphic (rooted) trees cannot have the same N-tuple.[28,30]

The N-tuple codes were used for the enumeration and generation of acyclic (molecular) graphs and a computer program was prepared for this purpose.[27,30] In Table 10 the number of isomeric trees and rooted trees with up to 50 vertices are given.

In Table 11 the number of isomeric alkanes and alcohols with up to 50 carbon atoms are given.

The N-tuple codes possess a number of interesting properties. Two of them will be mentioned here. One of these properties is the following: N-tuple codes order isomeric structures in accordance with their mode of branching. In Figure 8 this is exemplified by giving the ordering of trees corresponding to the carbon skeletons of heptanes, C_7H_{16}, according to their N-tuple codes. First is given the most branched heptane tree (which therefore possess the lexicographically largest N-tuple code) and last is shown the unbranched tree (which possesses the lexicographically smallest N-tuple code) corresponding to *n*-heptane. All other heptane trees appear between these extremes. Other theoretical measures of branching[104-108] support the ordering given in Figure 8, although there are some differences in ordering trees 2 to 8. For example, the connectivity index[104] orders the heptane trees as follows: 1, 2, 3, 5, 4, 6, 7, 8, 9.

Another very useful property of N-tuple codes is that they induce a unique labeling of vertices (or atoms) in a tree (or acyclic system).[109] Since every digit in an N-tuple code belongs to a single vertex, the sequential appearance of the digits dictates a sequential labeling of vertices. In Figure 9 a labeling of a tree corresponding to 2,2,3-trimethylhexane is given. Labels at the terminal vertices (7, 8, and 9) may be exchanged because of symmetry. It is worth noting that this labeling scheme represents a single choice among many possibilities. Because of that the above scheme favorably compares with other proposals for the canonical labeling of atoms in acyclic molecules.[110-112]

TABLE 10
Number of Trees and Rooted Trees with up to 50 Vertices

N	No. of trees	No. of rooted trees
1	1	1
2	1	1
3	1	2
4	2	4
5	3	9
6	6	20
7	11	48
8	23	115
9	47	286
10	106	719
11	235	1842
12	551	4766
13	1301	12486
14	3159	32973
15	7741	87811
16	19320	235381
17	48629	634847
18	123867	1721159
19	317955	4688676
20	823065	12826228
21	2144505	35221832
22	5623756	97055181
23	14828074	268282855
24	39299897	743724984
25	104636890	2067174645
26	279793450	5759636510
27	751065460	16083734329
28	2023443032	45007066269
29	5469566585	126186554308
30	14830871802	354426847597
31	40330829030	997171512998
32	109972410221	2809934352700
33	300628862480	7929819784355
34	823779631721	22409533673568
35	2262366343746	63411730258053
36	6226306037178	179655930440464
37	17169677490714	509588049810620
38	47436313524262	1447023384581029
39	131290543779126	4113254119923150
40	363990257783343	11703780079612453
41	1010748076717151	33333125878283632
42	2810986483493475	95020085893954917
43	7828986221515605	271097737169671824
44	21835027912963086	774088023431472074
45	60978390985918906	2212039245722726118
46	170508699155987862	6325843306177425928
47	477355090753926460	18103111141539779470
48	1337946100045842285	51842285219378800562
49	3754194185716399992	148558992149369434381
50	10545233702911509534	425976989835141038353

TABLE 11
Number of Isomeric Alkanes and Alcohols with up to 50 Carbon Atoms

N	No. of alkanes	No. of primary alcohols	No. of secondary alcohols	No. of tertiary alcohols	Total no. of alcohols
1	1	1	0	0	1
2	1	1	0	0	1
3	1	1	1	0	2
4	2	2	1	1	4
5	3	3	3	1	8
6	5	4	6	3	17
7	9	8	15	7	39
8	18	17	33	17	89
9	35	39	82	40	211
10	75	89	194	102	507
11	159	211	482	249	1238
12	355	507	1188	631	3057
13	802	1238	2988	1594	7639
14	1858	3057	7528	4074	19241
15	4347	7639	19181	10443	48865
16	10359	19241	49060	26981	124906
17	24894	48865	126369	69923	321198
18	60523	124906	326863	182158	830219
19	148284	321198	849650	476141	2156010
20	366319	830219	2216862	1249237	5622109
21	910726	2156010	5806256	3287448	14715813
22	2278658	5622109	15256265	8677074	38649152
23	5731580	14715813	40210657	22962118	101821927
24	14490245	38649152	106273050	60915508	269010485
25	36797588	101821927	281593237	161962845	712566567
26	93839412	269010485	747890675	431536102	1891993344

TABLE 11 (continued)
Number of Isomeric Alkanes and Alcohols with up to 50 Carbon Atoms

N	No. of alkanes	No. of primary alcohols	No. of secondary alcohols	No. of tertiary alcohols	Total no. of alcohols
27	240215803	1891993344	1990689459	1152022025	5034704828
28	617105614	5034704828	5309397294	3081015684	13425117806
29	1590507121	13425117806	14187485959	8253947104	35866550869
30	4111846763	35866550869	37977600390	22147214029	9591365288
31	10660307791	9591365288	101827024251	59514474967	257332864506
32	27711253769	257332864506	273442837014	160152652585	690928354105
33	72214088660	690928354105	735356029184	431536968270	1857821351559
34	188626236139	1857821351559	1980245349791	1164238905803	5002305607153
35	493782952902	5002305607153	5339453162253	3144681306263	13486440075669
36	1295297588128	13486440075669	14414507646239	8503434708370	36404382430278
37	3404490780161	36404382430278	38958262395690	23018134344315	98380779170283
38	8964747474595	98380779170283	105407071465709	62370701364485	26615855200477
39	23647478933969	26615855200477	285486823673472	169162601157498	720807976831447
40	62481801147341	720807976831447	773973501324306	459220572506066	1954002050661819
41	165351455535782	1954002050661819	2100240521050067	1247708120305177	5301950692017063
42	438242894769226	5301950692017063	5704211125099465	3392829794022689	14398991611139217
43	1163169707886427	14398991611139217	15505573111151541	9233204029174994	39137768751465752
44	3091461011836856	39137768751465752	42182179142471817	25146006764593896	106465954658531465
45	8227162372221203	106465954658531465	114842744354613849	68532690093294099	289841389106439413
46	21921834086683418	289841389106439413	312893758322611809	186906970120044539	789642117549095761
47	58481806621987010	789642117549095761	853091050332332960	510081778090226835	2152814945971655556
48	156192366647590639	2152814945971655556	2327481497654824913	1392929364734851485	5873225808361331954
49	417612400765382272	5873225808361331954	6354149922621658161	3806119516574048959	16033495247557039074
50	1117743651746953270	16033495247557039074	17357923632762137245	10406136061618401441	43797554941937577760

HEPTANE TREE   N-TUPLE CODE

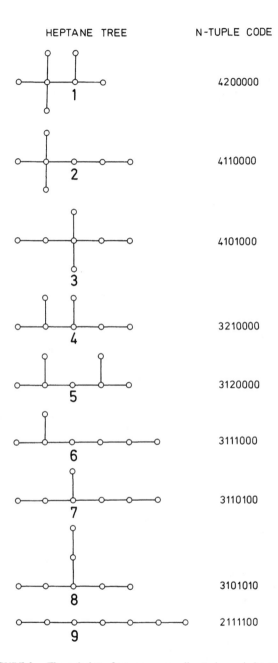

4200000

4110000

4101000

3210000

3120000

3111000

3110100

3101010

2111100

FIGURE 8.   The ordering of trees corresponding to isomeric heptanes.

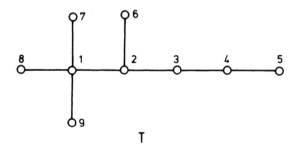

FIGURE 9.   A unique labeling of vertices in a tree T depicting 2,2,3-trimethylhexane.

REFERENCES

1. **Rouvray, D. H.,** *Chem. Soc. Rev.,* 3, 355, 1974; *Endeavour,* 34, 28, 1975.
2. **Rouvray, D. H.,** *J. Mol. Struct. (Theochem),* 185, 1, 1989; in *Chemical Graph Theory: Introduction and Fundamentals,* Abacus Press/Gordon & Breach Science Publishers, Bonchev, D. and Rouvray, D. H., Eds., New York, 1991, 1.
3. **Biggs, N. L., Lloyd, E. K., and Wilson, R. J.,** *Graph Theory 1736–1936,* Clarendon, Oxford, 1976.
4. **Balaban, A. T.,** in *Chemical Graph Theory: Introduction and Fundamentals,* Bonchev. D. and Rouvray, D. H., Eds., Abacus Press/Gordon & Breach, New York, 1991, 177.
5. **Lin, C. L.,** *Introduction to Combinatorial Mathematics,* McGraw-Hill, New York, 1968.
6. **Cohen, D. I. A.,** *Basic Techniques of Combinatorial Theory,* John Wiley & Sons, New York, 1978.
7. **Hamermesh, M.,** *Group Theory and Its Application to Physical Problems,* Addison-Wesley, Reading, MA, 1967.
8. **Berzelius, J. J.,** *Ann. Phys. Chem.,* 19 (der ganzen Folge 95), 305, 1830.
9. **Senior, J. K.,** *J. Chem. Phys.,* 19, 865, 1951.
10. **Slanina, Z.,** *Contemporary Theory of Chemical Isomerism,* Academia, Praha, 1986.
11. **Pimentel, G. C. and Spartley, R. C.,** *Understanding Chemistry,* Holden-Day, San Francisco, 1971, 733.
12. **Mislow, K.,** *Introduction to Stereochemistry,* W. A. Benjamin, New York, 1966.
13. **Brocas, J., Gielen, M., and Willem, R.,** *The Permutational Approach to Dynamic Stereochemistry,* McGraw-Hill, New York, 1983.
14. **Butlerov, A. M.,** *Z. Chem.,* 4, 549, 1861; 5, 298, 1862.
15. **Cayley, A.,** *Philos. Mag.,* 13, 172, 1857.
16. **Cayley, A.,** *Philos. Mag.,* 47, 444, 1874.
17. **Cayley, A.,** *Ber. Dtsch. Chem. Ges.,* 8, 1056, 1875.
18. **Cayley, A.,** *Philos. Mag.,* Ser. 5, 3, 34, 1877.
19. **Cayley, A.,** *Q. J. Math.,* 23, 376, 1889.
20. **Henze, H. R. and Blair, C. M.,** *J. Am. Chem. Soc.,* 53, 3042, 1931.
21. **Henze, H. R. and Blair, C. M.,** *J. Am. Chem. Soc.,* 53, 3077, 1931.
22. **Blair, C. M. and Henze, H. R.,** *J. Am. Chem. Soc.,* 54, 1098, 1932.
23. **Blair, C. M. and Henze, H. R.,** *J. Am. Chem. Soc.,* 54, 1538, 1932.
24. **Henze, H. R. and Blair, C. M.,** *J. Am. Chem. Soc.,* 55, 680, 1933.
25. **Pólya, G.,** *Acta Math.,* 68, 145, 1937.

26. **Pólya, G. and Read, R. C.**, *Combinatorial Enumeration of Groups, Graphs and Chemical Compounds*, Springer-Verlag, New York, 1987.
27. **Knop, J. V., Müller, W. R., Jeričević, Ž., and Trinajstić, N.**, *J. Chem. Inf. Comput. Sci.*, 21, 94, 1981.
28. **Trinajstić, N., Jeričević, Ž., Knop, J. V., Müller, W. R., and Szymanski, K.**, *Pure Appl. Chem.*, 55, 379, 1983.
29. **Knop, J. V., Müller, W. R., Szymanski, K., and Trinajstić, N.**, *Computer Generation of Certain Classes of Molecules*, SKTH, Zagreb, 1985.
30. **Trinajstić, N., Nikolić, S., Knop, J. V., Müller, W. R., and Szymanski, K.**, *Computational Chemical Graph Theory: Characterization, Enumeration and Generation of Chemical Structures by Computer Methods*, Simon & Schuster, New York, 1991.
31. **Domb, C.**, in *Phase Transitions and Critical Phenomena*, Vol. 3, Domb, C. and Green, M. S., Eds., Academic Press, London, 1974, 1.
32. **Kirchhoff, G.**, *Ann. Phys. Chem.*, 72, 497, 1847.
33. **Cayley, A.**, *Philos. Mag.*, 18, 374, 1859.
34. **Jordan, C.**, *J. Reine Angew. Math.*, 70, 185, 1869.
35. **Sylverster, J. J.**, *Proc. R. Inst. G. Br.*, 7, 179, 1873–1875.
36. **Cayley, A.**, *Rep. Br. Assoc. Adv. Sci.*, 45, 257, 1875.
37. **Cayley, A.**, *Am. J. Math.*, 4, 266, 1881.
38. **Otter, R.**, *Ann. Math.*, 49, 583, 1948.
39. **Schiff, H.**, *Ber. Dtsch. Chem. Ges.*, 8, 1542, 1875.
40. **Herrmann, F.**, *Ber. Dtsch. Chem. Ges.*, 13, 792, 1880.
41. **Losanitsch, S. M.**, *Ber. Dtsch. Chem. Ges.*, 30, 1917, 1898.
42. **Herrmann, F.**, *Ber. Dtsch. Chem. Ges.*, 30, 2423, 1898.
43. **Losanitsch, S. M.**, *Ber. Dtsch. Chem. Ges.*, 30, 3059, 1898.
44. **Herrmann, F.**, *Ber. Dtsch. Chem. Ges.*, 31, 91, 1898.
45. **Read, R. C.**, in *Chemical Applications of Graph Theory*, Balaban, A. T., Ed., Academic Press, London, 1976, 25.
46. **Tiemann, A.**, *Ber. Dtsch. Chem. Ges.*, 26, 1595, 1893.
47. **Delannoy, M.**, *Bull. Chem. Soc. Fr.*, 239, 1894.
48. **Goldberg, A.**, *Chem.-Ztg.*, 22, 395, 1898.
49. **Perry, D.**, *J. Am. Chem. Soc.*, 54, 2918, 1932.
50. **Coffman, D. D., Blair, C. M., and Henze, H. R.**, *J. Am. Chem. Soc.*, 55, 252, 1933.
51. **Coffman, D. D.**, *J. Am. Chem. Soc.*, 55, 695, 1933.
52. **Henze, H. R. and Blair, C. M.**, *J. Am. Chem. Soc.*, 56, 157, 1934.
53. **Rancke-Madsen, E.**, *Acta Chem. Scand.*, 4, 1450, 1950.
54. **Kornilov, M. Y.**, *Zh. Struct. Khim*, 8, 373, 1967.
55. **Allen, E. S. and Diehl, H.**, *Iowa State Coll. J. Sci.*, 16, 161, 1942.
56. **Davis, C. C., Cross, K., and Ebel, M.**, *J. Chem. Educ.*, 48, 675, 1971.
57. **Polansky, D. E.**, *Math. Chem. (Mülheim/Ruhr)*, 1, 11, 1975.
58. **Balaban, A. T.**, *Math. Chem. (Mülheim/Ruhr)*, 1, 33, 1875.
59. **Harary, F., Palmer, E. M., Robinson, R. W., and Read, R. C.**, in *Chemical Applications of Graph Theory*, Balaban, A. T., Ed., Academic Press, London, 1976, 11.
60. **Read, R. C.**, in *Chemical Applications of Graph Theory*, Balaban, A. T., Ed., Academic Press, London, 1976, 25.
61. **Rouvray, D. H. and Balaban, A. T.**, in *Applications of Graph Theory*, Wilson, R. J. and Beineke, L. W., Eds., Academic Press, London, 1979, 177.
62. **Slanina, Z.**, in *Advances in Quantum Chemistry*, Vol. 13, Lowdin, P.-O., Ed., Academic Press, London, 1981, 89.
63. **Dimitriev, I. S.**, *Molecules Without Chemical Bonds*, Mir, Moscow, 1981, 138.
64. **Kasum, D. and Trinajstić, N.**, *Kem. Ind. (Zagreb)*, 34, 245, 1985.
65. **Read, R. C.**, *Math. Mag.*, 60, 275, 1987.
66. **Redfield, J. H.**, *Am. J. Math.*, 49, 433, 1927.
67. **Davidson, R. A.**, *J. Am. Chem. Soc.*, 103, 312, 1981.

68. **Harary, F. and Robinson, R. W.,** *J. Graph Theory,* 8, 191, 1984.
69. **Redfield, J. H.,** *J. Graph Theory,* 8, 204, 1984.
70. **Hall, J. I., Palmer, E. M., and Robinson, R. W.,** *J. Graph Theory,* 8, 225, 1984.
71. **Lloyd, E. K.,** in *Graph Theory and Topology in Chemistry,* King, R. B. and Rouvray, D. H., Eds., Elsevier, Amsterdam, 1987, 537.
72. **Lloyd, E. K.,** *Discrete Appl. Math.,* 19, 289, 1988.
73. **Lloyd, E. K.,** in *MATH/CHEM/COMP 1988,* Graovac, A., Ed., Elsevier, Amsterdam, 1989, 85.
74. **Lunn, A. C. and Senior, J. K.,** *J. Phys. Chem.,* 33, 1027, 1929.
75. **Riordan, J.,** *J. Soc. Indian Appl. Math.,* 5, 225, 1957.
76. **de Bruijn, N. G.,** *Indagationes Math.,* 21, 59, 1959.
77. **Kennedy, B. A., McQuarrie, C. H., and Brubaker, C. H.,** *Inorg. Chem.,* 3, 265, 1964.
78. **Harary, F. and Palmer, E. M.,** *Proc. Natl. Acad. Sci. U.S.A.,* 54, 680, 1965.
79. **Harary, F. and Palmer, E. M.,** *J. Comb. Theory,* 1, 157, 1966.
80. **Williamson, S. G.,** *J. Comb. Theory,* 8, 162, 1970.
81. **McDaniel, D. H.,** *Inorg. Chem.,* 11, 2678, 1972.
82. **Palmer, E. D.,** in *New Directions in the Theory of Graphs,* Harary, F., Ed., Academic Press, New York, 1973, 187.
83. **Harary, F. and Palmer, E. D.,** *Graphical Enumeration,* Academic Press, New York, 1973, chap. 2.
84. **White, D. E.,** *Discrete Math.,* 13, 277, 1975.
85. **White, D. E. and Williamson, S. G.,** *Proc. Am. Math. Soc.,* 55, 233, 1976.
86. **Balaban, A. T.,** *J. Labelled Comp.,* 6, 211, 1970.
87. **Balaban, A. T.,** in *Chemical Applications of Graph Theory,* Balaban, A. T., Ed., Academic Press, London, 1976, 63.
88. **Balaban, A. T. and Harary, F.,** *Rev. Roum. Chim.,* 12, 1511, 1967.
89. **Rouvray, D. H.,** *J. S. Afr. Chem. Inst.,* 26, 141, 1973; 27, 20, 1974.
90. **Balaban, A. T.,** *Rev. Roum. Chim.,* 20, 227, 1975.
91. **Robinson, R. W., Harary, F., and Balaban, A. T.,** *Tetrahedron,* 32, 355, 1976.
92. **Rinehart, K. and Motz, K. L.,** *Chem. Ind.,* 1150, 1957.
93. **King, R. B.,** *J. Am. Chem. Soc.,* 94, 95, 1972.
94. **Krivoshei, I. V.,** *Zh. Strukt. Khim.,* 4, 757, 1963; 6, 322, 1965; 7, 430, 1966; 8, 321, 1967.
95. **Lederberg, J., Sutherland, G. L., Buchanan, B. G., Feigenbaum, E. A., Robertson, A. V., Duffield, A. M., and Djerassi, C.,** *J. Am. Chem. Soc.,* 91, 2973, 1969.
96. **Masinter, L. M., Sridharan, N. S., Lederberg, J., and Smith, D. H.,** *J. Am. Chem. Soc.,* 96, 7702, 1974.
97. **Davis, R. E. and Freyd, P. J.,** *J. Chem. Educ.,* 66, 278, 1989.
98. **Müller, W. R., Szymanski, K., Knop, J. V., Nikolić, S., and Trinajstić, N.,** *J. Comput. Chem.,* 11, 223, 1990.
99. **Hendrickson, J. B. and Parks, C. A.,** *J. Chem. Inf. Comput. Sci.,* 31, 107, 1991; **Knop, J. V., Müller, W. R., Szymanski, K., and Trinajstić, N.,** *J. Chem. Inf. Comput. Sci.,* 30, 159, 1990.
100. **Cyvin, B. N., Brunvoll, J., and Cyvin, S. J.,** *Top. Curr. Chem.,* in press.
101. **Trinajstić, N., Klein, D. J., and Randić, M.,** *Int. J. Quantum Chem.: Quantum Chem. Symp.,* 20, 699, 1986.
102. **Knop, J. V., Müller, W. R., Szymanski, K., Nikolić, S., and Trinajstić, N.,** in *Computational Chemical Graph Theory,* Rouvray, D. H., Ed., Nova Science Publisher, New York, 1990, 9.
103. **Trinajstić, N.,** *Rep. Mol. Theory,* 1, 185, 1990.
104. **Randić, M.,** *J. Am. Chem. Soc.,* 97, 6609, 1975.
105. **Gutman, I., Ruščić, B., Trinajstić, N., and Wilcox, C. F., Jr.,** *J. Chem. Phys.,* 62, 3399, 1975.

106. **Bonchev, D. and Trinajstić, N.,** *J. Chem. Phys.,* 67, 4517, 1977.
107. **Bertz, S.,** *Discrete Appl. Math.,* 19, 65, 1988.
108. **Barysz, M., Knop, J. V., Pejaković, S., and Trinajstić, N.,** *Pol. J. Chem.,* 59, 405, 1985.
109. **Randić, M.,** *J. Chem. Inf. Comput. Sci.,* 26, 136, 1986.
110. **Randić, M.,** *J. Chem. Phys.,* 60, 3920, 1974.
111. **Hendrickson, J. B. and Toczko, A. G.,** *J. Chem. Inf. Comput. Sci.,* 23, 171, 1983.
112. **Kvasnička, V. and Pospichal, J.,** *J. Chem. Inf. Comput. Sci.,* 30, 99, 1990.

INDEX

A

Q

Q_n parameters, for conjugated-circuit model, 206

Quantitative structure-activity relationships (QSAR), topological indices used in, 225, 226, 232, 235, 237, 248, 250, 254, 257, 269

Quantitative structure-property relationships (QSPR),topological indices used in, 225, 226, 232, 237, 240, 244, 248, 257, 260, 262, 269

Quantum-chemical graphs, as weighted graphs, 36

Quaternary root, 281

Quinodimethanes, isomeric, TRE values for, 156

π-Quinoline
acyclic polynomial for, 138, 140
vertex- and edge-weighted graphs and subgraphs for, 138

R

Radialenes, total π-electron energy for, 119

Radicals, conjugated
DREs of, 130
topological resonance energies of, 150–152

Radius, of a graph, 252

Randić aromaticity postulate, 215–216

Randić connectivity index, 226–232

Randić fragmentation method, 177

Random walks, 63

Reaction graphs, as weighted graphs, 36

Rectangular model, for benzenoid hydrocarbons, 173
use in enumeration of Kekulé structures, 175

Reference polynomial, 134, 143

Reference structure, in topological resonance energy theory, 133

Regular graphs, 13, 14

Resonance energy
of benzenoid hydrocarbons, 161
of π-electrons, 203–205
topological, 125
use to predict aromaticity in conjugated molecules, 125

Resonance integrals, 87

Resonance theory, 97, 125, 161
correspondence to Hückel molecular orbital (HMO) theory, 187

Ring current intensity, 148

Ring currents, of [4m + 2] π-electron annulenes, relationship with Hückel resonance energies, 145–147

Rings, 13

R_n parameters, for conjugated-circuit model, 205

Root, 13

Rooted graphs, see Vertex-weighted graphs

Rooted tree(s), 13, 14, 36
alcohol depiction by, 281–282
alkyl radical depiction as, 275
enumeration by N-tuple code, 293, 294
enumeration of, 278

Root-vertex, 13

Ruedenberg charge density-bond order matrix, 96

S

Sachs formula
adaptation in topological resonance energy theory, 133
application to Möius [N]cycle, 75
extension to Möbius systems, 66–67
extension to weighted graphs, 67–70
pairing theorem introduction by, 99
results obtained by, 70, 71
truncated, 137

Sachs graphs, 64, 65
acyclic, 133
for Hückel and Möbius annulenes, 110–112
unweighted and weighted, 67, 68

Sachs method
application to simple graphs, 65–66
for computing the characteristic polynomial, 64–72

Schlegel graphs, 16

Schultz index, as topological index, 257, 260–262

Secondary alcohols, derivation of number of, 282

Secondary root, for alcohols, 281

Second-order connectivity index, 228–229

Secular determinant, 88

Secular polynomial, 61

Self-avoiding directed peak-to-valley paths, in path counting method, 178

Milton Keynes UK
Ingram Content Group UK Ltd.
UKHW021627071024
449327UK00020BA/1223